Compliments
Glenn S...

THE LIBRARY
OF
THE UNIVERSITY
OF CALIFORNIA

GIFT OF

PROF. GLENN T. SEABORG

Compliments of
THE WELCH FOUNDATION
Houston, Texas

THE ELEMENTS
BEYOND URANIUM

THE ELEMENTS BEYOND URANIUM

GLENN T. SEABORG
Nuclear Science Division
Lawrence Berkeley Laboratory
Berkeley, California

WALTER D. LOVELAND
Department of Chemistry
Oregon State University
Corvallis, Oregon

A Wiley-Interscience Publication
JOHN WILEY & SONS, INC.
New York · Chichester · Brisbane · Toronto · Singapore

In recognition of the importance of preserving what has
been written, it is a policy of John Wiley & Sons, Inc. to
have books of enduring value published in the United
States printed on acid-free paper, and we exert our best
efforts to that end.

Copyright © 1990 by John Wiley & Sons, Inc.

All rights reserved. Published simultaneously in Canada.

Reproduction or translation of any part of this work
beyond that permitted by Section 107 or 108 of the
1976 United States Copyright Act without the permission
of the copyright owner is unlawful. Requests for
permission or further information should be addressed to
the Permissions Department, John Wiley & Sons, Inc.

Library of Congress Cataloging in Publication Data:

Seaborg, Glenn Theodore, 1912–
   The elements beyond uranium / Glenn T. Seaborg and Walter D.
   Loveland
   p. cm.
   Includes bibliographical references.
   ISBN 0-471-89062-6
   1. Transuranium elements.   I. Loveland, Walter D.   II. Title.
QD172.T7S35   1990
546′ . 44—dc20   90-12643
                                                                     CIP

Printed in the United States of America

10 9 8 7 6 5 4 3 2 1

*To
Albert Ghiorso
friend, colleague,
intrepid investigator of
the transuranium elements*

PREFACE

This is the 12th of the volumes on the transuranium elements for which one of the authors (G.T.S.) has served as either author, co-author, editor or co-editor. It is unique because it is the only one among these that covers all aspects, albeit in a concise manner, of these elements–discovery, chemical properties, nuclear properties, nuclear synthesis reactions, experimental techniques, presence in nature, superheavy elements, practical applications, and predictions for the future. With two exceptions, these previous volumes bore specialized themes–chemical properties, nuclear properties, compilation of research papers. Although the volumes published in 1958 and 1963 were somewhat broad in scope, the intervening decades have seen great changes in science generally, and very large increases in information, understanding and new concepts of the transuranium elements in particular. This, in itself, would be ample justification for an up-to-date survey of this field. In addition, we must recognize the important role these elements play in chemistry, in physics, in science generally, and, therefore, in the affairs of man.

The first of these previous volumes, published in 1949, the two-volume *The Transuranium Elements: Research Papers*, with co-editors Joseph J. Katz and Winston M. Manning, consisted of 162 papers written by participants in the wartime Plutonium Project. Next, in 1954, came *The Actinide Elements*, with co-editor Katz, a survey volume covering the chemical and nuclear properties as they were known by that date. *The Chemistry of the Actinide Elements*, with co-author Katz, came in 1957; then, the summary volumes *The Transuranium Elements* in 1958 and *Man-Made Transuranium Elements*, in 1963; the two-volume *The Nuclear Properties of the Heavy Elements*, with co-authors Earl K. Hyde and Isadore Perlman, in 1964; *Transuranium Elements: Products of Modern Alchemy* (a Benchmark book), a compilation of 122 original and key publications, in 1978; and the two-volume second edition of *The Chemistry of the Actinide Elements*, with co-editors Katz and Lester R. Morss, in 1986.

This book which bears the unique title *The Elements Beyond Uranium* to avoid confusion with the titles of previous volumes, is being published on the fiftieth anniversary of the discovery of the transuranium elements. Our goal in writing this volume is to provide a primer for the transuranium elements. Our aim is to convey the essence of the ideas and the blend of theory and experiment that characterizes the study of these elements. However, we have included some more advanced material for those who would like a somewhat deeper immersion in the subject. Our hope is that the reader can use this book for an introductory treatment of a subject of interest and can use the end-of-chapter general references as a guide to more advanced and more detailed articles. We also include a substantial, but not exhaustive, number of references to the original literature. The aforementioned Benchmark book can serve as a convenient source book for the key publications in this field. We also hope the practicing scientists who deal with the transuranium elements might use this volume as a quick refresher course for the rudiments of relatively unfamiliar aspects of the chemistry and physics of these synthetic elements and as an information booth for directions for more detailed inquiries.

We began the writing of this book over five years ago. In the ensuing years, several important developments have forced us to revise our ideas. We have no doubt that, in the not too distant future, some portions of the book will become out of date. Only time will tell if we have been clever enough to emphasize the truly fundamental aspects of the behavior of the transuranium elements.

We wish to thank P. Armbruster, J. Bloom, G.R. Choppin, A. Ghiorso, R. Hoff, D.C. Hoffman, E.K. Hulet, J.V. Kratz, J.M. Nitschke, T.H. Pigford, and L.P. Somerville for helpful discussions and comments concerning various chapters.

GLENN T. SEABORG
WALTER D. LOVELAND

Berkeley, California
Corvallis, Oregon
July 1990

CONTENTS

1 **Introduction**   1

    General References, 5

2 **Discovery (Synthesis) of New Elements**   7

    2.1 The Limits of the Periodic Table, 7
    2.2 Criteria for the Discovery (Synthesis and Identification) of New Chemical Elements, 8
    2.3 Synthesis of the First Man-Made Transuranium Element, Neptunium, 8
    2.4 Plutonium, 11
    2.5 Americum and Curium, 17
    2.6 Berkelium and Californium, 21
    2.7 Einsteinium and Fermium, 28
    2.8 Mendelevium, 38
    2.9 Nobelium–the First of the "Controversial" Elements, 46
    2.10 Lawrencium–the Last Actinide Element, 49
    2.11 Rutherfordium and Hahnium, 51
    2.12 Element 106, 54
    2.13 Element 107, 56
    2.14 Element 108, 57
    2.15 Element 109, 59
    2.16 Element 110, 60
    References, 61

3 Chemical Properties — 65

3.1 The Evolution of the Periodic Table, 65
3.2 Electronic Structures of the Gaseous Transuranium Atoms: f Electron Chemistry, 71
   3.2.1 Nonrelativistic Orbitals, 71
   3.2.2 Relativistic Orbitals, 75
   3.2.3 Electron Configurations, 79
3.3 Ionic Radii, 80
3.4 Optical and Magnetic Properties, 82
3.5 Oxidation States, 84
3.6 Hydrolysis, 89
3.7 Complex Ions, 90
3.8 Chemical Separations of the Transuranium Elements, 92
   3.8.1 The Bismuth Phosphate Process, 93
   3.8.2 Solvent Extraction, 94
   3.8.3 Ion Exchange, 99
   3.8.4 Fast Methods of Separation, 101
3.9 The Metallic State, 104
3.10 Solid Compounds, 108
3.11 Organometallic Compounds, 110
3.12 Chemistry of the Transactinides, 113
General References, 116
References, 117

4 Nuclear Structure and Radioactive Decay Properties — 121

4.1 Nuclear Shapes and Sizes, 121
4.2 Nuclear Masses, 123
4.3 Fission Barriers, 126
   4.3.1 Introductory Comments, 126
   4.3.2 The Fissionability of Nuclei, 127
   4.3.3 The Calculation of the Fission Barrier Height, 130
4.4 Spontaneous Fission, 139
   4.4.1 Ground State Spontaneous Fission Systematics, 139
   4.4.2 Spontaneously Fissioning Isomers, 146
4.5 Alpha-Particle Decay, 149
4.6 Heavy Particle Radioactivity, 153
4.7 Nuclear Structure, 154
   4.7.1 Intrinsic States, 154
   4.7.2 Collective States, 158
      4.7.2.1 Rotational Levels, 158
      4.7.2.2 Vibrational Levels, 161

4.7.3 Pairing Effects, 162
4.7.4 Energy Levels and Illustrations of Nuclear Structure Models, 164
4.7.5 Alpha-Decay Hindrance Factors, 167
4.8 Electromagnetic Transition Rates, 168
4.8.1 Transitions in Even-Even Nuclei, 170
4.8.2 Transitions in Odd-Mass Nuclei, 172
4.9 Beta-Decay Rates and Energetics, 173
4.10 Ground State and Low Energy Fission Properties of the Transuranium Nuclei, 176
4.10.1 Fission Product Mass Distribution, 176
4.10.2 Total Kinetic Energy Release in Fission, 184
4.10.3 Charge Distributions in Low Energy and Spontaneous Fission, 189
4.10.4 The Excitation Energies of the Fission Fragments, 191
References, 194

5 Experimental Techniques 199

5.1 Availability of Materials, 199
5.2 Chemical Manipulations, 201
5.3 Physical Techniques, 203
5.3.1 Transuranium Targetry, 203
5.3.2 Identification of Transuranium Reaction Products, 204
5.3.2.1 General Considerations, 204
5.3.2.2 Chemical Methods, 204
5.3.2.3 The Helium Jet, Drums, Tapes, and Wheels, 205
5.3.2.4 Magnetic Spectrometers, Velocity Filters, 208
5.3.2.5 Time-of-Flight (TOF), Decay-in-Flight (DIF), and Blocking Techniques, 212
5.4 Health and Safety Aspects of Transuranium Element Use, 213
References, 214

6 Nuclear Synthetic Techniques 217

6.1 General Considerations, 217
6.2 Compound Nucleus Formation, 219
6.2.1 Neutrons, 219
6.2.1.1 Qualitative Description, 219
6.2.1.2 Production Calculations, 224

　　　　　　6.2.2　Light Charged Particles, 226
　　　　　　　　　6.2.2.1　Energetics, 226
　　　　　　　　　6.2.2.2　Cross Sections, 228
　　　　　　6.2.3　Heavy Ions, 234
　　　　　　　　　6.2.3.1　Energetics, 234
　　　　　　　　　6.2.3.2　Cross Sections, 235
　　　　6.3　Direct Reactions, 243
　　　　　　6.3.1　General Considerations, Light Charged Particles, 243
　　　　　　6.3.2　Heavy Ions, 245
　　　　6.4　Deep Inelastic Transfer, 250
　　　　General References, 263
　　　　References, 264

7　Superheavy Elements　　　　　　　　　　　　　269

7.1　Introduction, 269
7.2　Properties of the Superheavy Elements, 271
7.3　Laboratory Synthesis of Superheavy Elements, 276
7.4　What's Next? 285
　　　References, 288

8　Presence in Nature　　　　　　　　　　　　　　291

　　　8.1　Natural Abundances, 291
　　　8.2　Nucleosynthesis of the Transuranium Elements, 292
　　　8.3　Sources and Distribution of Man-Made Transuranium Elements in the Environment, 296
　　　8.4　Transport and Fate in the Environment, 297
　　　References, 298

9　Practical Applications　　　　　　　　　　　　　301

　　　9.1　Nuclear Power, 301
　　　9.2　Nuclear Weapons, 309
　　　9.3　Radionuclide Power Sources, 315
　　　9.4　Industrial Applications of the Transuranium Nuclides, 316
　　　9.5　Production of the Transuranium Elements, 319
　　　References, 319

10　Reflections　　　　　　　　　　　　　　　　　321

　　　References, 327

APPENDIX I	Tables of Radioactive Decay Properties of the Transuranium Nuclei	329
	Table A1. Radioactive Decay Properties of Neptunium Isotopes, 330	
	Table A2. Radioactive Decay Properties of Plutonium Isotopes, 332	
	Table A3. Radioactive Decay Properties of Americium Isotopes, 334	
	Table A4. Radioactive Decay Properties of Curium Isotopes, 336	
	Table A5. Radioactive Decay Properties of Berkelium Isotopes, 337	
	Table A6. Radioactive Decay Properties of Californium Isotopes, 338	
	Table A7. Radioactive Decay Properties of Einsteinium Isotopes, 340	
	Table A8. Radioactive Decay Properties of Fermium Isotopes, 342	
	Table A9. Radioactive Decay Properties of Mendelevium Isotopes, 343	
	Table A10. Radioactive Decay Properties of Nobelium Isotopes, 344	
	Table A11. Radioactive Decay Properties of Lawrencium Isotopes, 344	
	Table A12. Radioactive Decay Properties of Rutherfordium Isotopes, 345	
	Table A13. Radioactive Decay Properties of Hahnium-109 Isotopes, 346	

Name Index 347

Subject Index 355

**THE ELEMENTS
BEYOND URANIUM**

1

INTRODUCTION

The story of the elements beyond uranium, the transuranium elements, is one of the most dramatic in the history of science. These elements hold a unique place amongst the chemical elements. They are man-made, the realization of the alchemist's dream of transmutation. The explanation of their unusual and complex chemical properties challenges the modern chemist. The special nuclear properties of these elements give them special importance in the affairs of man.

The endeavor to produce these elements beyond uranium, a quest born of scientific curiosity, was destined to be the trigger for a series of events which, within a decade, were to rock the world, forever altering it and to impinge upon the consciousness of every literate human being. These events were the discoveries that led to the exploitation of nuclear energy, particularly for use in weapons of mass destruction. Many scientific discoveries have had a profound effect upon man but the discovery of one of these transuranium elements seems unparalleled in its dramatic impact: the announcement of the existence of plutonium was the nuclear bomb dropped over Nagasaki.

The story of these elements, their discovery, their chemical and nuclear properties, and their role in our society is a fascinating one. As we write these words in 1989, there are seventeen known transuranium elements, representing approximately an 18% expansion in the building blocks of nature. A list of these elements by name, their chemical symbols, and atomic numbers is shown in Table 1.1. The names of the first eleven

TABLE 1.1 The Transuranium Elements

Atomic Number	Element	Symbol
93	Neptunium	Np
94	Plutonium	Pu
95	Americium	Am
96	Curium	Cm
97	Berkelium	Bk
98	Californium	Cf
99	Einsteinium	Es
100	Fermium	Fm
101	Mendelevium	Md
102	Nobelium	No
103	Lawrencium	Lr
104	Rutherfordium	Rf
105	Hahnium	Ha
106	106	
107	107	
108	108	
109	109	

elements (neptunium through lawrencium) are well established in the scientific community. Considerable controversy surrounds the names of elements 104 and 105 which is related to the question of who discovered these elements. (The privilege of naming a chemical element is traditionally given to its discoverers.) The names for elements 104 and 105 (rutherfordium and hafnium) shown in Table 1.1 are the names suggested by scientists at the University of California at Berkeley who claim to have made the first definitive identification of these elements. Competing claims for priority in the discovery of these elements have been made by scientists at the Joint Institute of Nuclear Research at Dubna who have suggested the names of kurchatovium (Ku) and nielsbohrium (Ns) for elements 104 and 105. A potentially similar situation exists with discovery of elements 106–109 and the investigators have refrained from naming these elements. In this book we shall refer to elements that have not been given names in the traditional manner by their discoverers, as well as undiscovered elements, by simply using the atomic number as the chemical symbol. Thus the isotope of element 106 containing 263 nucleons will be $^{263}106$, while the hexacarbonyl of element 106 would be designated $106(CO)_6$

The transuranium elements complete the actinide series of the chemical elements and (potentially) the 6d-7p transition series (see periodic table overleaf). They represent one of the few modifications of the periodic table as postulated by Mendeleev in 1872. Their chemical behavior is complex because their outer electron configurations involve s, p, d, and f electrons with significant relativistic effects (see Chapter 3) due to their large atomic numbers.

The demography of the transuranium elements is chronicled in Figure 1.1. The first transuranium element, neptunium, was discovered in 1940

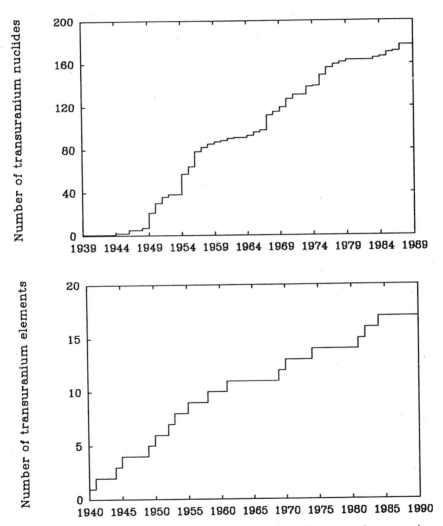

Figure 1.1   The growth of the number of transuranium elements and transuranium nuclei as a function of time.

Heavy Element Halflives

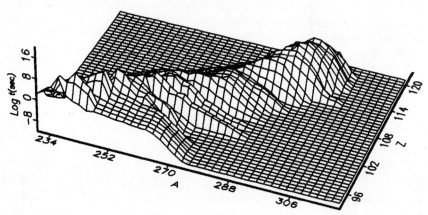

Figure 1.2 A chart showing the known and undiscovered transuranium nuclides. For the details of how this chart was prepared and a detailed discussion of the trends shown in it, see Chapter 4.

and the second element, plutonium, was discovered in 1941. The third and fourth members of this series, americium and curium, were discovered during World War II at the wartime Metallurgical Laboratory at the University of Chicago. In the 1950s there was a rapid increase in the number of these elements but as the atomic number of the new species rose, their lifetimes became shorter and they became progressively harder to produce. These new circumstances required great sophistication and cleverness from the discovery teams and the rate of discovery of new elements was much less in the 1960s and 1970s. The 1980s have seen a resurgence of activity in the study of these elements, culminating in the discovery of elements 107, 108, and 109. We now believe that barring the discovery of a new class of transuranium elements, the superheavy elements (see Chapter 7), scientists will be able only to make very few more transuranium elements. (The discovery of superheavy elements would dramatically alter our perception of the periodic table and its limits.)

All the transuranium elements are radioactive with over 160 known nuclides. The rate of discovery of new transuranium nuclides has paralleled the elemental discovery rate (Figure 1.1). However, because we have only identified a small fraction of the potential transuranium nuclides (Figure 1.2), this situation may change. (In Figure 1.2, we show the possible transuranium nuclides that are stable against instantaneous neutron and proton emission along with estimates of the lifetimes of unidentified nuclides). If we assume that it is possible to identify any nucleus with a half-life greater than 10^{-6} s, then the number of undiscovered members of the transuranium family is greater than 500, a truly staggering number. Realizing that we have *identified* only ~25% of the members of this family and we know so little about so many of these nuclides, it is apparent that the story of these transuranium elements is a continuing saga with a continual growth in the number of chapters.

GENERAL REFERENCES

While much has been written about the elements beyond uranium, five general references can be particularly recommended. They are:

Sea 63   G.T. Seaborg, The Man-Made Transuranium Elements (Prentice-Hall, Englewood Cliffs, 1963). An elementary introduction written for high school students.

Sea 78   G.T. Seaborg, Ed., *The Transuranium Elements—Products of Modern Alchemy* (Dowden, Hutchinson-Ross, Stroudsberg, 1978). An annotated collection of the most important, "benchmark" papers on transuranium elements.

KSM 86   J.J. Katz, G.T. Seaborg, and L. Morss, Eds., *The Chemistry of the Actinide Elements* (Chapman and Hall, London, 1986). An advanced, definitive account of the chemistry of the transuranium elements.

Ede 82   N.M. Edelstein, Ed., *Actinides in Perspective* (Pergamon, Oxford, 1982). An advanced account of the chemical and nuclear properties of these elements.

Gme   *Gmelin Handbuch der Anorganischen-Chemie* (Springer-Verlag, Berlin). This classic reference work has many authoritative articles about the transuranium elements. Of specific interest are the volumes on the transuranium elements (Transurane) and some of the volumes dealing with uranium.

2

DISCOVERY (SYNTHESIS) OF NEW ELEMENTS

2.1 THE LIMITS OF THE PERIODIC TABLE

The heaviest known element that occurs in macroscopic quantities in nature is uranium, element 92. This is the last element whose lifetime is comparable to the age of the earth. To understand how far we might extend the "natural" periodic table through laboratory experiments, consider the following argument due to Huizenga (Hui 79). The liquid drop model of the nucleus predicts that a nucleus will instantaneously fission when

$$E_c = 2E_s \qquad (2.1)$$

where E_c and E_s are the Coulomb and surface energy of the nucleus, respectively. The quantities E_c and E_s are given by

$$E_c = \frac{3}{5}\frac{(Ze)^2}{R} = k_c \frac{Z^2}{A^{1/3}} \qquad (2.2)$$

$$E_s = 4\pi R^2 \gamma = k_s A^{2/3} \qquad (2.3)$$

where γ is the nuclear surface tension ($\sim 1\,\text{MeV/fm}^2$), Z is the atomic number and R is the nuclear radius (which is proportional to $A^{1/3}$ where A is the mass number). The limiting value of the atomic number, Z_{LIMIT}, is then

$$Z^2_{\text{LIMIT}} = 2(k_s/k_c)A_{\text{LIMIT}} \tag{2.4}$$

If we remember that the neutron/proton ratio in heavy nuclei is ~1.5/1, then

$$Z_{\text{LIMIT}} = 5(k_s/k_c) \tag{2.5}$$

Thus the upper bound to the periodic table is proportional to the ratio of two fundamental constants related to the strength of the nuclear (surface) and electromagnetic forces. The ratio k_s/k_c is about 20–25 and thus we expect 100–125 chemical elements. Clearly only a moderate extension of the "natural" periodic table by man would be expected.

2.2 CRITERIA FOR THE DISCOVERY (SYNTHESIS AND IDENTIFICATION) OF NEW CHEMICAL ELEMENTS

Having established the approximate extent to which the natural periodic table can be extended, it is important to define the extent of experimental proof required to definitively establish the production and identification of a new element. An international group of scientists has proposed such criteria (Har 76). The basic criterion for the discovery of a new element is the experimentally verified proof that the atomic number of the new element is different from the atomic numbers of all previous elements. Establishment of the atomic number can be by chemical means, by identification of the characteristic X-rays in the decay of the new species, or by establishment of genetic decay relationships through α-particle decay chains in which the isotope of the new element is identified by the observation of previously known decay products. The use of systematic relationships about the nuclear reactions used to produce the new species can be used as supportive evidence for the discovery, but such relationships are not so generally well established as to lead to unambiguous identification of the atomic number of the new species. The claim to have discovered a new element should be published in a refereed journal with sufficient data to enable the reader to judge whether the evidence is consistent with these criteria. A name for the new element should not be proposed by the discoverers until the observation is confirmed.

2.3 SYNTHESIS OF THE FIRST MAN-MADE TRANSURANIUM ELEMENT, NEPTUNIUM

The first scientific attempts to prepare the elements beyond uranium were by Enrico Fermi, Emilio Segre and co-workers in Rome in 1934, shortly after the existence of the neutron was discovered. This group of investigators irradiated uranium with slow neutrons and found a number of radioactive products, which were thought to be new elements. However, chemical studies by O. Hahn and F. Strassman showed these species were isotopes of the known elements formed by the fission of uranium into two approximately equal parts. This discovery of nuclear fission in December of 1938 which led to the "nuclear age" was thus a by-product of man's quest for the transuranium elements.

With poetic justice, the actual discovery of the first transuranium element came as part of an experiment to study the nuclear fission process. E.M. McMillan, working at the University of California at Berkeley in the spring of 1939, was trying to measure the energies of the two recoiling fragments from the neutron-induced fission of uranium. He placed a thin layer of uranium oxide on one piece of paper. Next to this he stacked very thin sheets of cigarette paper to stop and collect the uranium fission fragments. In the course of his studies he found there was another radioactive product of the reaction—one that did not recoil enough to escape the uranium layer as did the fission products. He suspected that this product was formed by the capture of a neutron by the more abundant isotope of uranium, $^{238}_{92}U$. McMillan and P.H. Abelson (McM 40) (Figure 2.1), who joined him in this research, were able to show in 1940, by chemical means, this product is an isotope of element 93, $^{239}_{93}Np$, formed in the following sequence:

$$^{238}_{92}U + ^{1}_{0}n \rightarrow ^{239}_{92}U + \gamma \qquad (2.6)$$

$$^{239}_{92}U \xrightarrow[t_{1/2}=23.5\text{ min}]{\beta^-} ^{239}_{93}Np\,(t_{1/2} = 2.36\text{ days}) \qquad (2.7)$$

(Here the presently known half-lives are used.)

In 1940 it was not apparent what the electron configuration of neptunium would be; consequently the chemical properties of the new element could not be predicted. Uranium was known to have some similarity to tungsten, and at that time, it was thought that element 93 might resemble rhenium, the element below tungsten in the periodic table. There was the possibility, however, that neptunium might be a member of a new heavy element transition series. The experimental studies of McMillan and Abelson showed that neptunium resembles uranium, not rhenium, in its chemical properties. This was the first definitive evidence that an inner electron shell, the 5f shell, is filled in the transuranium region of elements. The consequence of such filling, as in the rare earth elements, is the outer

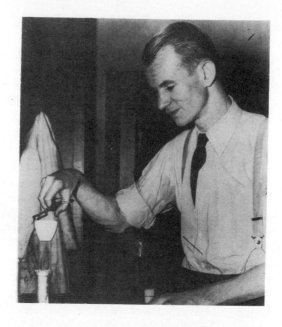

Figure 2.1 The discoverers of neptunium, Edwin M. McMillan and Philip H. Abelson.

electrons, which primarily determine chemical behavior, remain much the same; this leads to a series of chemically similar elements.

The first isolation of a macroscopic weighable quantity of neptunium did not occur until 1944 when L.B. Magnusson and T.J. La Chappelle (Mag 48), working at the wartime Metallurgical Laboratory, isolated long-lived $^{237}_{93}$Np from uranium irradiated in a nuclear reactor. The nuclear reaction used in the production of $^{237}_{93}$Np was the $^{238}_{92}$U$(n, 2n)$ $^{237}_{92}$U reaction followed by β^- decay of $^{237}_{92}$U to $^{237}_{93}$Np. About 10 μg of neptunium oxide were isolated.

Neptunium, the element beyond uranium, was named after the planet Neptune because this planet is beyond the planet Uranus for which uranium was named.

2.4 PLUTONIUM

Plutonium was the second transuranium element to be discovered. By bombarding uranium with charged particles, in particular, deuterons (^{2}H), utilizing the 60-inch cyclotron at the University of California at Berkeley, G.T. Seaborg, McMillan, J.W. Kennedy, and A.C. Wahl (Figure 2.2) succeeded in preparing a new isotope of neptunium, $^{238}_{93}$Np, which decayed by β^- emission to $^{238}_{94}$Pu, that is,

$$^{238}_{92}\text{U} + ^{2}_{1}\text{H} \rightarrow ^{238}_{93}\text{Np} + 2\,^{1}_{0}n \tag{2.8}$$

$$^{238}_{93}\text{Np} \xrightarrow[t_{1/2}=2.12\ \text{days}]{\beta^-} {}^{238}\text{Pu}(t_{1/2} = 87.7\ \text{years}) \tag{2.9}$$

(Here again the best presently known half-lives are used.)

The story of the discovery of plutonium is best told in the words of Seaborg (Sea 72).

> ... during the summer and fall of 1940, Ed McMillan started looking for the daughter product of the 2.3-day activity, which obviously would be the isotope of element 94 with mass number 239 (94-239). He had also bombarded uranium with deuterons. Before he could finish this project, he was called away to work on radar at M.I.T.
>
> During this time my interest in the transuranium elements continued. Since Ed McMillan and I lived only a few rooms apart in the Faculty Club, we saw each other quite often, and, as I recall, much of our conversation, whether in the laboratory, at meals, in the hallway, or even going in and out of the shower, had something to do with element 93 and the search for element 94.

Figure 2.2 Codiscoverers of plutonium, Joseph W. Kennedy (December 25, 1940), Arthur C. Wahl and Glenn T. Seaborg. Seaborg and Wahl are shown with the sample of ^{239}Pu in which fission was demonstrated in 1941 (the cigar box was that of G.N. Lewis).

When I learned that Ed McMillan had gone, I wrote to him asking whether it might not be a good idea if we carried on the work he had started, especially the deuteron bombardment of uranium. He readily assented.

Our first deuteron bombardment of uranium was conducted on Dec. 14, 1940. What we bombarded was a form of uranium oxide, U_3O_8, which was literally plastered onto a copper backing plate. From this bombarded material Art Wahl isolated a chemical fraction of element 93. The radioactivity of this fraction was measured and studied. We observed that it had different characteristics than the radiation from a sample of pure 93-239. The beta particles, which in this case were due to a mixture of 93-239 and the new isotope of element 93 with mass number 238 (93-238), had a somewhat higher energy than the radiation from pure 93-239 and there was more gamma radiation. But the composite half-life was about the same, namely, 2 days. However, the sample also differed in another very important way from a sample of pure 93-239. Into this sample there grew an alpha-particle-emitting radioactivity. A proportional counter was used to count the alpha particles to the exclusion of the beta particles. This work led us to the conclusion that we had a daughter of the new isotope 93-238—a daughter with a half-life of about 50 years and with the atomic number 94. This is much shorter lived than the now known half-life of 94-239, which is 24,400 years. The shorter half-life means a higher intensity of alpha-particle emission, which explains why it was so much easier to identify what proved to be the isotope of element 94 with the mass number 238 (94-238). (Later it was proved that the true half-life of what we had, that is, 94-238, is about 90 years.)

On Jan. 28, 1941, we sent a short note to Washington (Sea 46a) describing our initial studies on element 94; this note also served for later publication in The Physical Review under the names of Seaborg, McMillan, Kennedy, and Wahl. We did not consider, however, that we had sufficient proof at that time to say we had discovered a new element and felt that we had to have chemical proof to be positive. So, during the rest of January and into February, we attempted to identify this alpha activity chemically.

Our attempts proved unsuccessful for some time. We did not find it possible to oxidize the isotope responsible for this alpha radioactivity. Then I recall that we asked Professor Wendell Latimer, whose office was on the first floor of Gilman Hall, to suggest the strongest oxidizing agent he knew for use in aqueous solution. At his suggestion we used peroxydisulphate with argentic ion as a catalyst.

On the stormy night of Feb. 23, 1941, working in room 307, Gilman Hall, in an experiment that ran well into the next morning, Art Wahl performed the oxidation which gave us proof that what we had made was chemically different from all other known elements. Now another note went to Washington, later published in The Physical Review (Sea 46b) under the authorship of Seaborg, Wahl and Kennedy, dated March 7, 1941, indicating that it was extremely probable that this alpha activity was due to an isotope of element 94.

The chemical properties of elements 93 and 94 were studied by the tracer method at the University of California for the next year and a half.

These first two transuranium elements were referred to simply as "element 93" and "element 94" or by code names, until the spring of 1942, at which time the first detailed reports on this work were written. The early work, even in those days, was carried on under a self-imposed cover of secrecy, in view of the potential military application of element 94. Throughout 1941, element 94 was referred to by the code name of "copper," which was satisfactory until it was necessary to introduce the element copper into some of the experiments. This posed the problem of distinguishing between the two. For a while, plutonium was referred to as "copper" and the real copper as "honest-to-God copper."

This offered more and more difficulties as time went on, and element 94 was finally christened "plutonium" with the chemical symbol "Pu" in March of 1942 (Sea 48). Plutonium was named after the planet Pluto, following the pattern used in naming neptunium. Pluto is the second and last known planet beyond Uranus. These appellations were suggested rather than the more logical "plutium" and "Pl".

Because of its ability to undergo fission and thereby serve as a source of nuclear energy similar to $^{235}_{92}U$, the plutonium isotope with the mass number 239 is the one of major importance. The search for this isotope, as a decay product of $^{239}_{93}Np$, was being conducted by J.W. Kennedy, E. Segre, A.C. Wahl, and G.T. Seaborg (Figures 2.2 and 2.3) simultaneously with experiments leading to the discovery of plutonium. The isotope $^{239}_{94}Pu$ was identified and its possibilities as a nuclear energy source were established during the spring of 1941. $^{239}_{94}Pu$ was produced by the decay of $^{239}_{93}Np$, which in turn was produced from $^{238}_{92}U$ by neutrons, using the reaction

$$^{238}_{92}U + ^{1}_{0}n \rightarrow ^{239}_{92}U + \gamma \qquad (2.10)$$

$$^{239}_{92}U \xrightarrow[t_{1/2}=23.5\,\text{min}]{\beta^-} ^{239}_{93}Np \xrightarrow[t_{1/2}=2.35\,\text{days}]{\beta^-} ^{239}_{94}Pu(t_{1/2} = 24{,}110\,\text{years}) \qquad (2.11)$$

A sample of uranyl nitrate weighing 1.2 kg was distributed in a large paraffin block (neutron-slowing material) placed directly behind the beryllium target of the 60-inch cyclotron and was bombarded for 2 days with neutrons produced by the reaction of the deuteron beam with beryllium.

Figure 2.3 Emilio Segre, Berkeley, 1947.

The irradiated uranyl nitrate was placed in a continuously operating glass extraction apparatus, and the uranyl nitrate was extracted into diethylether. $^{239}_{93}$Np was isolated by use of the oxidation-reduction principle with lanthanum and cerium fluoride carriers (see below) and was reprecipitated six times in order to remove all uranium impurities. Measurement of the radiation from the $^{239}_{93}$Np made it possible to calculate that 0.5 µg was present to yield $^{239}_{94}$Pu upon decay. The resulting α-activity corresponded to a half-life of 30,000 years for the daughter $^{239}_{94}$Pu in demonstrable agreement with the present best value for the half-life of 24,110 years.

The group first demonstrated, on 28 March, 1941, with the sample containing 0.5 µg of $^{239}_{94}$Pu, that this isotope undergoes slow neutron-induced fission with a probability of reaction comparable to that of $^{235}_{92}$U (Ken 46). The samples were placed near the screened window of an ionization chamber that could detect the fission of $^{239}_{94}$Pu and $^{235}_{94}$U (present in natural uranium). Neutrons were then produced near the samples by bombarding a beryllium target with deuterons in the 37-inch cyclotron of

Berkeley's "Old Radiation Laboratory" (the name applied to the original wooden building, since torn down to make way for modern buildings). Paraffin around the samples slowed the neutrons down so they would be captured more readily by the plutonium and uranium. This experiment gave a small but detectable fission rate when a 6 μA beam of deuterons was used. To increase the accuracy of the measurement of the fission cross section, this sample, which had about 5 mg of rare earth carrier materials, was subjected to an oxidation-reduction chemical procedure (see below) that reduced the amount of carrier to a few tenths of a milligram. A slow neutron fission cross section for $^{239}_{94}$Pu, some 50% greater than that for $^{235}_{92}$U, was found, agreeing remarkably with the accurate values that were determined later. Once it was realized that plutonium, as $^{239}_{94}$Pu, might be used to make a nuclear weapon and that it might be created in quantity in a nuclear chain reactor followed by separation from uranium and the highly radioactive fission products, it became imperative to carry out chemical investigations of plutonium to develop large-scale chemical separation procedures. Once again, it was necessary to use code names to maintain secrecy, and a general numbering system—using the last numerals of the atomic number and mass number—was evolved for use in referring to any of a number of nuclides. Thus, in the case of $^{239}_{94}$Pu, that is, 94-239, the code number was "49"; those who worked on this project now refer to themselves as "forty-niners."

These investigations required that some of the work be done with weighable quantities, even though only microgram quantities could be produced using the cyclotron sources of neutrons available at that time. In August, 1942, B.B. Cunningham and L.B. Werner, working at the wartime Metallurgical Laboratory at the University of Chicago, succeeded in isolating about 1 μg of $^{239}_{94}$Pu which had been prepared by cyclotron irradiations (Cun 49). Thus, plutonium was the first man-made element to be obtained in visible quantity. The first weighing of this man-made element, using a larger sample (2.77 μg), was made by these investigators on 10 September, 1942.

Plutonium is now produced in much larger quantities than any other synthetic element. The large wartime chemical separation plant at Hanford, Washington, was constructed on the basis of investigations performed on the ultramicrochemical scale of investigation. The scale-up between ultramicrochemical experiments and the final Hanford plant corresponds to a factor of about 10^{10}, surely a scale-up of unique proportions (Sea 58).

The ultramicrochemical work on the Hanford chemical separation process was preceded by tracer work. As a result of tracer investigations during 1941 and early 1942 at the University of California, a great deal was learned about the chemical properties of plutonium. It was established that it had at least two oxidation states, the higher of which was not "carried" by lanthanum fluoride or cerium fluoride, while the lower state was quantitatively coprecipitated with these compounds.

The principle of the oxidation-reduction cycle was conceived and applied to the separation processes that were to become so useful later. This principle as applied to precipitation processes, involved the use of a carrier substance which would retain plutonium in one of its oxidation states but not in another. In practice, the Hanford chemical separation process, first worked out on the tracer scale at the Chicago wartime Metallurgical Laboratory, used a precipitate of bismuth phosphate to "carry" the plutonium in the IV oxidation state. The fission products, some of which were also "carried" and therefore accompanied the plutonium, were removed by dissolving the bismuth phosphate, oxidizing the plutonium to the VI oxidation state, and then reprecipitating the bismuth phosphate. Plutonium in the higher oxidation state was not carried by a bismuth phosphate precipitate, while the fission products behaved as previously and again were carried by the bismuth phosphate. The plutonium was then reduced to the IV oxidation state and the *decontamination* cycle repeated in order to effect still further separation. Since that time, a number of other plutonium production facilities such as the U.S. plant at Savannah River have gone into operation.

2.5 AMERICUM AND CURIUM

At the wartime Metallurgical Laboratory, after the completion of the most essential part of the chemical investigations involved in the production of plutonium, attention was turned to the synthesis and identification of the next transuranium elements. Seaborg, R.A. James, L.O. Morgan, and A. Ghiorso, were collaborators in this endeavor (Figure 2.4).

The first attempts to produce these elements ended in failure. Small amounts of $^{239}_{94}$Pu were irradiated with neutrons and deuterons but no new α-emitting products were found due to the use of insensitive detection techniques and because the experiments were based upon the premise that these elements should behave chemically like plutonium, that is, they could be oxidized to the VI oxidation state and chemically isolated. It was not until the summer of 1944, when it was first recognized that these elements were a part of an actinide transition series (with stable +3 oxidation states) that any progress was made. Success in their identification followed quickly.

Once it was realized that these elements could be oxidized above the

Figure 2.4 Codiscoverers of americium and curium. *Top*: left, Leon O. Morgan (1944); right, Ralph A. James (1945); *bottom*: Albert Ghiorso (in the Met Lab counting room, January, 1946).

III state only with difficulty, the use of a proper chemical procedure led quickly to the identification of an isotope of a transplutonium element. Thus, a new α-emitting nuclide, now known to be $^{242}_{96}$Cm (half-life 162.9 days), was produced in the summer of 1944 (Sea 45) by the bombardment of $^{239}_{94}$Pu with 32-MeV helium ions:

$$^{239}_{94}\text{Pu} + ^{4}_{2}\text{He} \rightarrow ^{242}_{96}\text{Cm} + ^{1}_{0}\text{n} \qquad (2.12)$$

The bombardment took place in the Berkeley 60-inch cyclotron after which the material was shipped to the Metallurgical Laboratory at Chicago for chemical separation and identification. The crucial step in the identification of the α-emitting nuclide as an isotope of element 96, $^{242}_{96}$Cm, was the identification of the known $^{238}_{94}$Pu as the α-decay daughter of the new nuclide.

The identification of an isotope of element 95, in late 1944 and early 1945, followed after the identification of this isotope of element 96 (^{242}Cm) as a result of the bombardment of $^{239}_{94}$Pu with neutrons in a nuclear reactor (Ghi 50). The production reactions, involving multiple neutron capture by plutonium are

$$^{239}_{94}\text{Pu} + ^{1}_{0}\text{n} \rightarrow ^{240}_{94}\text{Pu} + \gamma \qquad (2.13)$$

$$^{240}_{94}\text{Pu} + ^{1}_{0}\text{n} \rightarrow ^{241}_{94}\text{Pu} + \gamma \qquad (2.14)$$

$$^{241}_{94}\text{Pu} \xrightarrow[t_{1/2}=14.4\text{years}]{\beta^-} ^{241}_{95}\text{Am}(t_{1/2} = 432.7 \text{ years}) \qquad (2.15)$$

$$^{241}_{95}\text{Am} + ^{1}_{0}\text{n} \rightarrow ^{242}_{95}\text{Am} + \gamma \qquad (2.16)$$

$$^{242}_{95}\text{Am} \xrightarrow[t_{1/2}=16.0\text{ h}]{\beta^-} ^{242}_{96}\text{Cm} \qquad (2.17)$$

A confirmation of the identification of the nuclide $^{241}_{95}$Am involved the physical separation (based upon volatility) of $^{241}_{95}$Am from its parent $^{241}_{94}$Pu in a separated mass 241 sample.

Some comments should be made, at this point, concerning the similarity of these two elements to the rare earth elements. The hypothesis that elements 95 and 96 should have a stable III oxidation state and greatly resemble the rare earth elements in their chemical properties proved to be true. In fact, the near identity of their properties greatly hindered the efforts of the discovery team. The better part of a year was spent in trying, without success, to chemically separate the two elements from each other and from the fission product and carrier rare earth elements.

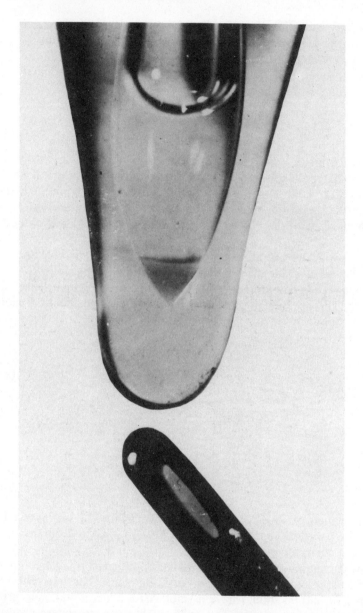

Figure 2.5 Photograph of an early hydroxide precipitation of americium, isolated in 1945 (magnified approximately 20 times). The americium, in the form of the isotope $^{241}_{95}$Am, can be seen at the bottom of a solution in the capillary tube. An ordinary sewing needle is shown below the capillary tube for purposes of comparison.

Although the discovery team was confident on the basis of their chemical and radioactive properties and the methods of production, that isotopes of elements 95 and 96 had been produced, the complete chemical proof still was lacking. The elements remained unnamed during this period of futile attempt at separation (although one of the group referred to them as "pandemonium" and "delirium" in recognition of their difficulties). The key to their chemical separation, which occurred later at Berkeley, and the technique which made feasible the separation and identification of subsequent transuranium elements was the ion-exchange technique.

The present names of these new elements were proposed on the basis of their chemical properties. The name "americium" was suggested for element 95, after the Americas, by analogy with the naming of its rare earth counterpart or homologue, europium, after Europe; and the name "curium" was suggested for element 96, after Pierre and Marie Curie, by analogy with the naming of its homologue, gadolinium, after the Finnish rare earth chemist J. Gadolin.

By chance, the discovery of these elements was revealed informally on a nationally broadcast radio program, the Quiz Kids, on which one of the authors (GTS) appeared as a guest on November 11, 1945. The discovery information had already been declassified (that is, removed from the "secret" category) for presentation at an American Chemical Society symposium at Northwestern University in Chicago the following Friday. Therefore, when one of the youngsters asked—during a session in which one of the authors was trying to answer their questions—if any additional new elements had been discovered in the course of research on nuclear weapons during the war, he was able to reveal the existence of the elements 95 and 96. Apparently many children in America told their teachers about it the next day, and judging from some of the letters which the author subsequently received from such youngsters, they were not entirely successful in convincing their teachers. The formal announcement of the discoveries was, of course, made later in the week, as planned.

Americium was first isolated by B.B. Cunningham, as the isotope ^{241}Am in the form of a pure compound, the hydroxide, in the fall of 1945 at the wartime Metallurgical Laboratory (Cun 45). A photograph of an early hydroxide precipitate is shown in Figure 2.5. Curium was first isolated in the form of a pure compound, the hydroxide, of $^{242}_{96}$Cm (produced by the neutron irradiation of $^{241}_{95}$Am) by L.B. Werner and I. Perlman at the University of California during the fall of 1947 (Wer 51).

2.6 BERKELIUM AND CALIFORNIUM

The story of the discovery of berkelium and californium began, in the words of Seaborg, shortly after the end of World War II (Sea 75).

I recall that we began planning for the possible synthesis and identification of transuranium elements as soon as, or even before, we returned to Berkeley from the Chicago Metallurgical Laboratory; that is, in late 1945 and in 1946. I thought that this would be a good Ph.D. thesis problem for Stan Thompson and it was, of course, natural that Al Ghiorso would participate on the radiation detection end of the problem as he had in the discovery of americium and curium in Chicago a year or two earlier.

On the basis of our confidence in the actinide concept we felt we could make the chemical identification, although we knew we would have to develop better chemical separation methods than were then available to us. And it seemed clear that we would use helium ion bombardments of americium and curium for our production reactions once these elements became available in sufficient quantity through production by prolonged neutron bombardment of plutonium, and we learned how to handle safely their intense radioactivity.

We knew these things but we didn't anticipate how long it would take to solve these simple problems. Actually, three years went by before we found ourselves ready to make our first realistic experiment.

The most important prerequisite to the process for making the transcurium elements was the manufacture of sufficiently large amounts of americium and curium to serve as target material. Because of the intense radioactivity of americium and curium, even in milligram or submilligram amounts, it was necessary to develop extremely efficient chemical separation methods to isolate the new elements from the target materials. This large degree of separation was necessary to detect the very small amounts of radioactivity due to the new elements produced in the presence of the highly radioactive starting materials. The dangerous radioactivity of the source material also made it necessary to institute complicated remote control methods of operation to keep health hazards to a minimum.

These problems were solved after three years work. Americium for target material was prepared in milligram amounts by intense neutron bombardment of plutonium over a long period of time, and curium target materials were prepared in microgram amounts as the result of the intense neutron bombardment of some of this americium. Both of these neutron bombardments took place in high-flux reactors (that is, reactors that deliver large concentrations of neutrons that can be used for transmutation purposes).

Element 97 was discovered by S.G. Thompson, Ghiorso and Seaborg, in December 1949, as the result of the bombardment of milligram quantities of $^{241}_{95}$Am with 35 MeV helium ions accelerated in the 60-inch cyclotron at Berkeley (Tho 50a). The nuclear reaction was

$$^{241}_{95}\text{Am} + ^{4}_{2}\text{He} \rightarrow ^{243}_{97}\text{Bk} + 2\,^{1}_{0}\text{n} \qquad (2.18)$$

BERKELIUM AND CALIFORNIUM   23

The new nuclide was expected to have a short half-life and thus relatively rapid chemical separation techniques had to be employed. For this purpose, cation-exchange was used. The original elution curve corresponding to the discovery of element 97 is shown in Figure 2.6.

The actual discovery experiments were not as simple as this description would indicate. Again, in the words of Seaborg (Sea 75)

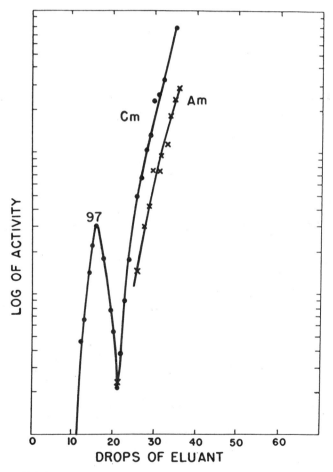

Figure 2.6 Original elution data corresponding to the discovery of berkelium, 19 December, 1949. Other activities were added for calibration purposes. Activity is due to conversion electrons (Bk) and alpha particles (Cm and Am). (Dowex 50 resin at 87°C with ammonium citrate as eluting agent.)

During the fall of 1949 we made a number of bombardments of americium with helium ions in the 60-inch cyclotron, with emphasis on looking for α-particle emitting isotopes of element 97, all with negative results. It was becoming clear that we should look for electron capture decay by detecting the accompanying conversion electrons and X-rays, so Ghiorso worked to improve the detection efficiency for such radiations.

The first successful experiment was performed on Monday, December 19, 1949. A target containing 7 milligrams of ^{241}Am was bombarded with helium ions in the 60-inch cyclotron, after which the chemical separation was started at 10:00 a.m. After the removal of the bulk of the americium by two oxidation cycles (utilizing oxidation to the hexapositive, fluoride-soluble, oxidation state of americium, which had just been discovered by Asprey, Stephanou and Penneman at Los Alamos), the 97, Cm and remaining Am were carried on lanthanum fluoride, dissolved and subjected to a group separation from fission product lanthanide elements (using a method of elution with concentrated HCl, just discovered by Ken Street), after which the actinide fraction was put through a cation exchange adsorption-elution procedure; this entire process was completed in seven hours. The prediction that element 97 would elute ahead of Cm and Am, in sequence, was of course the key to its successful chemical identification. Figure 2.6 shows a reproduction of the data obtained that afternoon. In this case, and especially in considering the data from following elution experiments, we were somewhat surprised to see the rather large gap between 97 and curium; we shouldn't have been surprised because there is a notably large gap between the elution peaks of the homologous lanthanide elements terbium and gadolinium.

Detected in the samples that eluted at the peak corresponding to element 97 were conversion electrons, X-rays of energy corresponding to decay by electron capture, and alpha particles at very low relative intensity (less than 1%). These radiations were found to decay with a half life of about 4.5 hours and it was immediately assumed that the isotope was ^{244}97 produced by the reaction: ^{241}Am (α, n) ^{244}97. Soon thereafter it was correctly surmised that the main isotope, that giving rise to the observed alpha particles, was actually ^{243}Bk produced by the reaction ^{241}Am$(\alpha, 2n)$ ^{243}Bk.

It is interesting to note that experiments as early as the first day, that is, Monday night, indicated that element 97 has two oxidation states, III and IV. The actinide concept provided the guidance to look for these two oxidation states, by analogy with the homologous element, terbium. In fact, the chemical identification procedure had been devised to accommodate either oxidation state and the large gap in the elution positions of element 97 and the curium was at first erroneously thought to be due to the fact that element 97 was in the IV oxidation state at that stage.

Element 98 was first produced and identified similarly by Thompson, K.

Street, Jr., Ghiorso and Seaborg (Figure 2.7), soon afterward in February of 1950, again at Berkeley (Tho 50b). The first isotope produced is now assigned the mass number 245 and decays by α-particle emission and orbital electron capture with a half-life of 44 min. This isotope was produced by the bombardment of microgram amounts of $^{242}_{96}$Cm with 35-MeV helium ions accelerated in the 60-inch cyclotron:

$$^{242}_{96}\text{Cm} + ^{4}_{2}\text{He} \rightarrow ^{245}_{98}\text{Cf}\,(t_{1/2} = 44\text{ min}) + ^{1}_{0}\text{n} \qquad (2.19)$$

It is interesting to note that this identification of element 98 was accomplished with a total of only some 5000 atoms; someone remarked at the time that this number was substantially smaller than the number of students attending the University of California.

The key to the discovery of element 98 was once again the use of ion-

Figure 2.7 The codiscoverers of berkelium and californium in Seaborg's office, Lawrence Berkeley Laboratory, January 20, 1975 (25th anniversary of discovery). Left to right: Kenneth Street, Jr., Stanley G. Thompson, Seaborg, Albert Ghiorso.

exchange techniques. On the basis of column calibration experiments, element 98 was expected to elute onto collection plate #13 in the 25th and 26th drops of eluant and this is exactly where it was found after a total elapsed chemical separation time of 2 h. The half life and α-particle energy were also in agreement with predictions. The original elution curve for the discovery of this element is shown in Figure 2.8.

Element 97 was called berkelium after the city of Berkeley, California, where it was discovered, just as its rare earth analogue, terbium, was given a name derived from Ytterby, Sweden, where so many of the early rare earth minerals were found. Element 98 was named californium, after the university and state where the work was done. This latter name, chosen for the reason given, does not reflect the observed chemical anal-

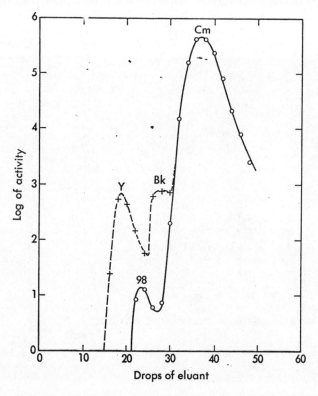

Figure 2.8 Original elution data corresponding to the discovery of californium, 9 February 1950. Other activities were added for calibration purposes. Solid curve indicates α-particle counts/minute; dashed curve, conversion electrons and beta particles. (Dowex 50 resin at 87°C with ammonium citrate as eluting agent.)

ogy of element 98 to dysprosium, as "americium," "curium," and "berkelium" signify that these elements are the chemical analogues of europium, gadolinium, and terbium, respectively. In their announcement of the discovery of element 98 in *Physical Review*, the authors commented, "The best we can do is point out, in recognition of the fact that dysprosium is named on the basis of a Greek word meaning 'difficult to get at,' that the searchers for another element (Au) a century ago found it difficult to get to California."

Upon learning about the naming of these elements, the "Talk of the Town" section of the *New Yorker* magazine had the following to say:

> New atoms are turning up with spectacular, if not downright alarming frequency nowadays, and the University of California at Berkeley, whose scientists have discovered elements 97 and 98, has christened them berkelium and californium, respectively. While unarguably suited to their place of birth, these names strike us as indicating a surprising lack of public relations foresight on the part of the university, located, as it is, in a state where publicity has flourished to a degree matched perhaps only by evangelism. California's busy scientists will undoubtedly come up with another atom or two one of these days, and the university might well have anticipated that. Now it has lost forever the chance of immortalizing itself in the atomic tables with some such sequence as universitium (97), offium (98), californium (99), berkelium (100).

The discoverers sent the following reply:

> 'Talk of the Town' has missed the point in their comments on naming of the elements 97 and 98. We may have shown lack of confidence but no lack of foresight in naming these elements 'berkelium' and 'californium.' By using these names first, we have forestalled the appalling possibility that after naming 97 and 98 'universitium' and 'offium,' some New Yorker might follow with the discovery of 99 and 100 and apply the names 'newium' and 'yorkium'.

The answer from the *New Yorker* staff was brief:

> We are already at work in our office laboratories on 'newium' and 'yorkium'! So far we have just the names.

In 1958, Cunningham and Thompson (Tho 59), at Berkeley, succeeded in isolating, for the first time, macroscopic amounts of berkelium (as the isotope $^{249}_{97}$Bk and californium (as a mixture of the isotopes ^{249,250,251,252}Cf). These were synthesized by the long-term irradiation of

$^{239}_{93}$Pu and its transmutation products with neutrons in the Materials Testing Reactor at the National Reactor Testing Station in Idaho.

The first compound of californium of proven molecular structure (by means of X-ray diffraction) was isolated in 1960 by Cunningham and J.C. Wallmann as 0.3 μg of californium ($^{249}_{98}$Cf) as oxychloride. The pure oxide and the trichloride were also prepared at that time. This experimental work was carried out on a *submicrogram* scale as the result of the further development of ultramicrochemical techniques.

The first compound of berkelium of proven molecular structure was isolated in 1962 by Cunningham and Wallmann. They isolated about 0.02 μg of berkelium ($^{249}_{97}$Bk) dioxide and used about one-tenth of this, about 0.002 μg (that is, 2 ng), for the determination of its molecular structure by means of the X-ray diffraction technique.

2.7 EINSTEINIUM AND FERMIUM

The discoveries of many of the transuranium elements were the result of careful planning, taking into account predictions of chemical and physical properties.

Elements 99 and 100, however, were unexpectedly discovered in debris from the "Mike" thermonuclear explosion which took place in the Pacific on 1 November, 1952. This was the first large test of a thermonuclear device. The device is shown in Figure 2.9. Its terrible power is best shown in the "before" and "after" pictures of a portion of the Eniwetok Atoll (Figure 2.10). Debris from the explosion was collected, first on filter papers attached to airplanes which flew through the clouds* and, later in more substantial quantity, gathered up as fall-out material from the surface of a neighboring atoll. This debris was brought to the United States for chemical investigation in a number of laboratories to establish the properties of the explosion.

Early analysis of the "Mike" debris by scientists at the Argonne National Laboratory near Chicago and the Los Alamos Scientific Laboratory in New Mexico showed the unexpected presence of new isotopes of plutonium, $^{244}_{94}$Pu and $^{246}_{94}$Pu. (At the time, the heaviest known isotope of plutonium was $^{243}_{94}$Pu.) The tenor of that work is best described in the words of one of the participants, Sherman Fried (Fri 79).

... we found 20 or 25% of plutonium-240 and 2% of plutonium-241. These

---

*This sampling effort cost the life of First Lieutenant Jimmy Robinson who waited too long before returning to his base, tried to land on Eniwetok and ditched about a mile short of the runway.

Figure 2.9 Closeup view of Mike device with its associated cryogenic equipment.

were incredible, incredible concentrations. That afternoon Tony Turkevich (a professor of chemistry at the University of Chicago) called up and said, 'Say look at these things you've got. You better look for plutonium-244.' And, sure enough, we looked at the mass spectrometric result, and there it was. The question then came up, what is the radiation of plutonium-244? That is easy to determine. We purified the plutonium and looked, and sure enough, there was β-activity in the plutonium fraction. That β-activity gave rise to and added to another β-activity which was in the americum fraction. And it had a 25-minute half-life, which pegged it immediately as ^{244}Am, as everybody knows. This meant that the plutonium-244 was a β-emitter, and

(a)

(b)

Figure 2.10 (a) The Eniwetok Atoll before the Mike explosion. Elugelab and the Mike device are at the top center. (b) After the Mike explosion. Elugelab has gone.

we were feeling pretty good about that. But Marty Studier said, 'Well, if you're so smart and if it's really a β-emitter, where's the curium-244?' (which would result from the decay sequence $^{244}\text{Pu} \xrightarrow{\beta^-} {}^{244}\text{Am} \xrightarrow{\beta^-} {}^{244}\text{Cm}$). When we looked, it wasn't there. We were feeling pretty sorry for ourselves, let me tell you. But one thing and another, we finally worked out that it was plutonium and that, indeed, it had to be plutonium-246, which we verified. Once having shown it was plutonium-246, it began to dawn on us that, my God, uranium has captured eight neutrons and we still, three weeks later, have enough stuff to see plutonium-246 easily. The way was open.

This observation led to the conclusion that the $^{238}_{92}\text{U}$ in the device had been subjected to an enormous neutron flux and had successively captured numerous neutrons. (Later calculations showed an integrated neutron fluence of $1-4 \times 10^{24}$ neutrons was delivered in a few nanoseconds — a few *moles* of neutrons!!)

How this conclusion was reached is described by another of the participants, Albert Ghiorso (Ghi 79).

On December 4, 1952, Glenn Seaborg received a (classified) teletype from James Beckerley, the Director of Office of Classification in Washington. It read: 'Radiochemical data on recent Eniwetok test indicates presence of some unique heavy element isotopes such as ^{244}Pu. We do not want to release any information on the properties of these isotopes, even their existence, at this time even though information is declassifiable under guide. Accordingly, you are requested to withhold publication of any information on the existence and properties of isotopes present in debris samples and consider such information as secret, restricted data. This prohibition applies even when information is disassociated from test. Please inform those in laboratory who might have access to these data.'

When Glenn showed the teletype to Stan Thompson and me, we were puzzled and began speculating as to what had happened and what we might do about it. Stan had been bombarding ^{239}Pu in the Materials Testing Reactor for a long time, and periodically searches had been made with our mass spectrograph for ^{244}Pu, which we expected to be made by successive neutron captures. Now suddenly we were told that it had been produced at one fell swoop in a nuclear explosion! Since we were not connected in any way with the bomb tests, we had no way of finding out what had happened, or when.

By the next morning, however, I had come up with a really wild idea and I got together with Stan and tried it out on him. First, I assumed that the curve of the yield at each mass number could be represented by a straight line on a semi-log plot (Figure 2.11). To get the slope of the line, I assumed that the starting material was ^{238}U and that the relative yield of ^{244}U (which

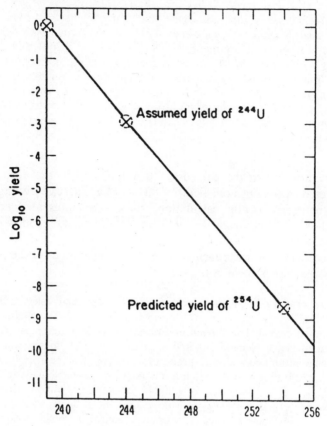

Figure 2.11 Ghiorso's original prediction of the yield of ^{254}U in the Mike explosion.

of course would β-decay to ^{244}Pu) was 10^{-3}. I felt that this number was the lowest yield that could be seen in the plutonium fraction with certainty on the mass spectrographs of that time. Extrapolating this line to the mass ^{254}U (ten neutrons heavier), which I assumed would β-decay all the way up to element 100, indicated a yield of roughly 10^{-8} or 10^{-9}. For some reason, which I don't recall now, I assumed that the bomb fraction of 10^{14} atoms was obtainable. This meant that we might get approximately a count per minute of α-activity from ^{254}100 if it had a half-life of about a month. It didn't take much persuasion to convince Stan that the possibility was real, although pretty far fetched, and that we should do something about it.

(The actual yield curve for this test is shown in Figure 2.12 and shows the remarkable accuracy of the Ghiorso prediction.)

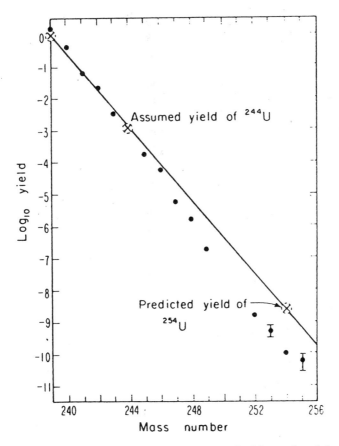

Figure 2.12 A comparison of Ghiorso's prediction (solid curve) and the measured yields of the explosion (data points).

Armed with the knowledge of the multineutron capture by ^{238}U, scientists at the University of California immediately began a search for transcalifornium isotopes in the bomb debris. Ion-exchange experiments of the type previously mentioned in the case of berkelium and californium immediately demonstrated the existence of a new element and within a few weeks, of a second new element. A typical early elution curve is shown in Figure 2.13. The first identification of element 100 was made with only about 200 atoms. To secure a larger amount of source material, it was necessary later to process many hundreds of pounds of coral from one of the atolls adjoining the explosion area. Eventually, such coral was processed by the ton, using bismuth phosphate as the carrier for the

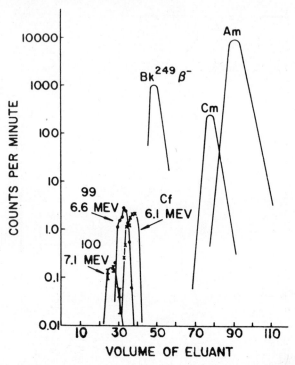

Figure 2.13 Elution in original experiments of elements einsteinium (99) and fermium (100) relative to other actinide elements with a citrate eluant.

tripositive actinide elements, in a pilot-plant operation which went under the name of "Paydirt."

Without going into the details, it may be pointed out that such experiments involving the groups at the three laboratories led to the positive identification of isotopes of elements 99 and 100. A twenty-day activity emitting alpha particles of 6.6 MeV energy was identified as an isotope of element 99 (with the mass number 253), and a 7.1 MeV α-activity with a half-life of 22 h was identified as an isotope of element 100 (with the mass number 255).

The path of successive neuron captures by $^{238}_{92}U$ and subsequent β^- decay of the capture products is shown in Figure 2.14. The β^- decay chains for each A value end in the first β-stable nuclide. Thus the first isotopes of elements 99 and 100 produced in such a device are those with $A = 253$ and 255, respectively.

The large group of scientists who contributed to the discovery of elements 99 and 100 included A. Ghiorso, S.G. Thompson, G.H. Higgins,

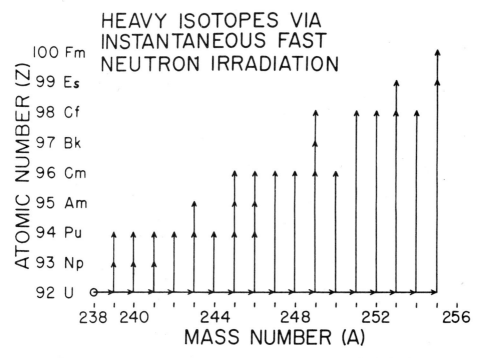

Figure 2.14 Production of uranium isotopes in the November, 1952 Mike thermonuclear device, and their decay to β-stable nuclei.

and G.T. Seaborg, of the Radiation Laboratory and Department of Chemistry of the University of California; M.H. Studier, P.R. Fields, S.M. Fried, H. Diamond, J.F. Mech, G.L. Pyle, J.R. Huizenga, A. Hirsch, and W.M. Manning of the Argonne National Laboratory; and C.I. Browne, H.L. Smith, and R.W. Spence of the Los Alamos Scientific Laboratory (Ghi 55a) (Figure 2.15). These researchers suggested the name einsteinium (symbol E) for element 99 in honor of the great physicist, Albert Einstein; and for element 100, the name fermium (symbol Fm) in honor of the father of the atomic age, Enrico Fermi, making these the first in a series of elements named after eminent scientists. The chemical symbols Es and Fm were adopted subsequently for these elements. The choice of name of fermium for element 100 has proven to be prescient since it is the last element to be synthesized using neutron capture reactions (which were extensively studied by Fermi).

Before removal of the "secret" label from this information and the subsequent announcement of the original discovery experiments could be

Figure 2.15 Codiscoverers of elements einsteinium and fermium at symposium commemorating the 25th anniversary of their discovery held at Lawrence Berkeley Laboratory, January 23, 1978. Left to right: front row: Louise Smith, Sherman Fried, Gary Higgins. Left to right: back row: Al Ghiorso, Rod Spence, Seaborg, Paul Fields, and John Huizenga.

accomplished, isotopes of elements 99 and 100 were produced by other, more conventional methods. Chief among these was that of successive neutron capture as the result of intense neutron irradiation of plutonium in the high-flux Materials Testing Reactor (MTR) at the National Reactor Testing Station in Idaho (see Figure 2.16). The difference between this method of production and that of the "Mike" thermonuclear explosion is

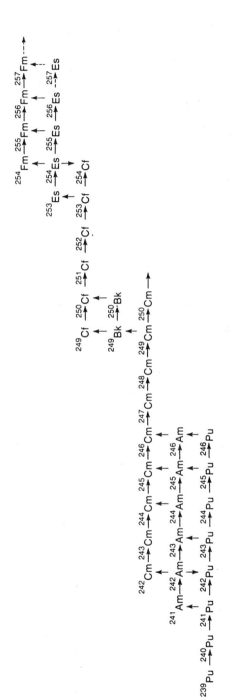

Figure 2.16 The sequence of nuclides produced in a high-flux reactor neutron irradiation of a ^{239}Pu target. The horizontal arrows represent neutron capture, vertical arrows up represent β^- decay, and vertical arrows down represent electron capture decay.

one of time as well as of starting material. In a reactor, it is necessary to bombard gram quantities of plutonium for two or three years; thus, the short-lived, intermediate isotopes of the various elements have an opportunity to decay. The path of element production proceeds up the valley of β-stability. In the thermonuclear device, larger amounts of uranium were subjected to an extremely high neutron flux for a period of nanoseconds; the subsequent β-decay of the ultraheavy isotopes of uranium led to the nuclides found in the debris.

It was not until 1961 that sufficient einsteinium had been produced through intense neutron bombardments of $^{239}_{94}$Pu in the Materials Testing Reactor to permit separation of a macroscopic and weighable amount. Cunningham, Wallmann, L. Phillips, and R.C. Gatti (Cun 61), working at Berkeley on the submicrogram scale, were able to separate a small fraction of pure $^{253}_{99}$Es. This was a remarkable feat, since the total amount of material involved was only a few hundredths of a microgram of einsteinium. As was true with all other isolations of weighable quantities of the transuranium elements, it was possible to observe a macroscopic property—in this case, the magnetic susceptibility of einsteinium. A macroscopic property of the "metallic" zero-valent state of fermium, the magnetic moment, was measured in 1971 at the Argonne National Laboratory by L.S. Goodman, H. Diamond, H.E. Stanton, and M.S. Fred (Goo 71) using the 3.2-hour ^{254}Fm in a modified atomic beam magnetic resonance apparatus.

2.8 MENDELEVIUM

The discovery of mendelevium was one of the most dramatic in the sequence of transuranium element syntheses. It marked the first time in which a new element was produced and identified one atom at a time.

By 1955, scientists at Berkeley had prepared an equilibrium amount of $\sim 10^9$ atoms of $^{253}_{99}$Es by neutron irradiation of plutonium in the Materials Testing Reactor. As the result of a "back of the envelope" calculation done by Ghiorso during an airplane flight, they thought it might be possible to prepare element 101 using the reaction

$$^{253}_{99}\text{Es} + ^{4}_{2}\text{He} \rightarrow ^{256}_{101}\text{Md} + ^{1}_{0}\text{n} \tag{2.20}$$

The amount of element 101 expected to be produced in an experiment can be calculated using the formula

$$N_{101} = \frac{N_{\text{Es}}\sigma\phi(1 - e^{-\lambda t})}{\lambda} \tag{2.21}$$

where N_{101} and N_{Es} are the number of element 101 atoms produced and the number of $^{253}_{99}\text{Es}$ target atoms, respectively, σ is the reaction cross section (estimated to be $\sim 10^{-27}$ cm^2), ϕ the helium ion flux ($\sim 10^{14}$ particles/s), λ the decay constant of $^{256}_{101}\text{Md}$ (estimated to be $\sim 10^{-4}\,\text{s}^{-1}$) and t the length of each bombardment ($\sim 10^4$ s).

$$N_{101} \approx \frac{(10^9)(10^{-27})(10^{14})(1 - e^{-(10^{-4})(10^{+4})})}{(10^{-4})} \approx 1 \text{ atom} \qquad (2.22)$$

Thus the production of only one atom of element 101 per experiment could be expected!

Adding immeasurably to the complexity of the experiment was the absolute necessity for the chemical separation of the one atom of element 101 from the 10^9 atoms of einsteinium in the target and its ultimate, complete chemical identification by separation with the ion-exchange method. This separation and identification would presumably have to take place in a period of hours, or perhaps even one hour or less, because the expected half-life was of this order of magnitude or less. Furthermore the target material had a 20 day half-life and one needed a nondestructive technique of using the target material over and over again.

These requirements indicated the desperate need for new techniques, together with some luck. Fortunately, both were forthcoming. The first new technique involved separation of the element 101 by *the recoil method* from the einsteinium in the target. The einsteinium was placed on a gold foil in an invisibly thin layer. The helium-ion beam was sent through the back of the foil so that the atoms of element 101, recoiling through a vacuum due to the momentum of the impinging helium ions, could be caught on a second thin gold catcher foil—as shown in Figure 2.17. This

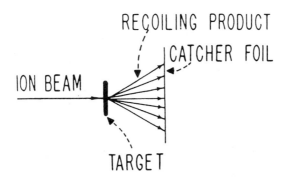

Figure 2.17 The recoil technique.

second gold foil, which contained recoil atoms and was relatively free of the einsteinium target material, was dissolved and was used for later chemical operations.

The preparation of the ^{253}Es target for this recoil experiment was another technical *tour de force*. After approximately five failures to prepare the target by vaporization of ^{253}Es from a hot filament, the essentially weightless deposit of ^{253}Es was electroplated onto the Au foil within a very small area.

An extremely reliable ion-exchange separation scheme had to be developed to unambiguously chemically identify atoms of a new element that were made. It took several months, involving hundreds of column elutions to develop the appropriate procedure. The final choice was the use of a Dowex 50 ion-exchange column run at an elevated temperature (87°C) with an α-hydroxyisobutyrate eluant. The procedure was so well developed that the discovery team could tell exactly in which drops of eluant the interesting activities would appear. Finally the 60-inch cyclotron ^{4}He^{2+} beam was increased by an order of magnitude in intensity to 100 μA per cm^2.

The earliest experiments were confined to a search for short-lived, α-emitting isotopes that might be due to element 101. For this purpose, it was sufficient to look quickly at the actinide chemical fraction as separated by the ion-exchange method. No alpha activity was observed that could be attributed to element 101, even when the time between the end of bombardment and the beginning of the alpha particle analyses had been reduced to five minutes. The experiments were continued and, in one of the subsequent overnight bombardments, two large pulses in the electronic detection apparatus due to spontaneous fission were observed. With probably unjustified self-confidence, it was thought that this might be a significant result. Although such an attitude might ordinarily have been considered foolish, it must be recalled that rapid decay by spontaneous fission was — up until that time — confined to only a few nuclides, none of which should have been introduced spuriously into the experiment. In addition, background counts due to this mode of decay should be zero in proper equipment.

The story of what happened next is best expressed by the words of one member of the discovery team, A. Ghiorso (Ghi 80).

> Just these two events in all the bombardments that we had made — what did they mean? I was very bold and proposed that we had produced an electron-capturing isotope of element 101 decaying to element 100 which then underwent spontaneous fission decay with a half-life of a few hours. The mass number would be 256.

This hypothesis completely changed our whole course of action. Up to this point we had assumed that the isotopes of element 101 that we would make would be short-lived alpha emitters. There was no way of knowing that they would be highly hindered for alpha decay. And now we had the possibility of finding an electron-capturing nuclide decaying to one undergoing spontaneous fission with both nuclides having reasonable half-lives. This was a really wild lucky guess considering that our only evidence were the two big kicks.

The major question, of course, was whether the experiment could be repeated. In a number of subsequent bombardments, one or two spontaneous fission events were observed in some, while none was observed in other experiments. This, of course, was to be expected, because of the statistical fluctuation inherent in the production of the order of one atom per bombardment. Furthermore, more advanced chemical experiments seemed to indicate that spontaneous fission counts, when they did appear, came in about the element 100 or 101 chemical fractions.

The definitive experiments were performed in a memorable, all-night session, 18 February, 1955. To increase the number of events that might be observed at one time, three successive three-hour bombardments were made, and, in turn, their transmutation products were quickly and completely separated by the ion-exchange method. Some of the nuclide $^{253}_{99}$Es was present in each case so that, together with the $^{246}_{98}$Cf produced from $^{244}_{96}$Cm also present in the target (via the ^{244}Cm (^{4}He, 2n) reaction), it was possible to define the positions in which the elements came off the column used to contain the ion-exchange resin. Five spontaneous fission counters then were used to count simultaneously the corresponding drops of solution from the three runs.

A total of five spontaneous fission counts were observed in the element 101 position, while a total of eight spontaneous fission counts were also observed in the element 100 position. No such counts were observed in any other position. The original data are presented in Figure 2.18.

A description of events of that night in the words of some members (Albert Ghiorso, Bernard Harvey, Gregory Choppin and Stanley Thompson, Figure 2.19) of the discovery team gives the reader a sense of one of those rare moments in scientific history (Sea 58a).

> We can see the powerful stream of helium nuclei as a thin blue beam of light if we allow it to escape through the target area out into the air. It can be photographed through the five-foot tank of water which acts as a viewing window into the room housing the cyclotron. This is the beam which will strike the target and, with luck, add the 2 protons of the helium to the 99 protons of the einsteinium in the target to make mendelevium.

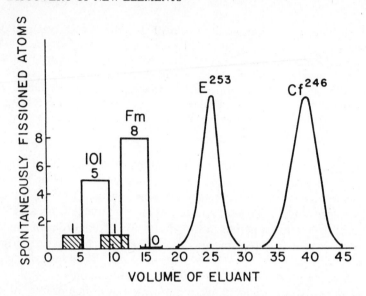

Figure 2.18 Original elution data corresponding to the discovery of mendelevium, February 18, 1955. The curves for einsteinium-253 (given the old symbol E^{253}) and californium-246 are for α-particle emission. (Dowex 50 ion exchange resin was used, and the eluting agent was ammonium α-hydroxyisobutyrate).

As soon as we got the all-clear signal, Bernie Harvey and Al Ghiorso pulled open the water-filled door and ran inside.

Al quickly took the "holder" from the machine. Bernie pulled off the second gold foil with a pair of tweezers and shoved it into a test tube.

Then he raced out through the narrow corridors and up a flight of steps to a temporary laboratory. In this makeshift lab, Bernie gave the foil to Gregory Choppin, who heated it in a solution to dissolve the gold.

The result was a liquid containing the gold, some other miscellaneous elements, and perhaps a few atoms of the mendelevium we were after.

The rest of the necessary chemical steps had to be done a mile away, up on the "hill" at the Radiation Laboratory.

For the mad dash up the hill, Al was ready at the wheel of a supercharged Volkswagen just outside the cyclotron building.

We had—we hoped—a few atoms of element 101, and our job was to isolate and identify them before they all decayed.

The precious drops were driven up to the nuclear chemistry building on the

Figure 2.19 Codiscoverers of mendelevium, Lawrence Berkeley Laboratory, March 28, 1980 (25th anniversary of discovery). Left to right: Gregory R. Choppin, Seaborg, Bernard G. Harvey, Albert Ghiorso.

hill. Greg and Bernie raced into the lab where Stanley Thompson was waiting with the apparatus with which to separate the element 101 from the einsteinium, gold, and anything else that might be present.

First the liquid was forced through an ion exchange column to get rid of the gold. The gold is held back in the column while a solution containing the mendelevium drips off at the bottom.

These drops were dried and redissolved. Then Stan forced them through a second column to isolate element 101 from any other elements still present.

The drops from the bottom of the column were caught, one by one, on little platinum plates which were then put under a heat lamp and dried.

The plates were taken into a counting room where Al put them in special counters, one to each counter.

If there were any mendelevium present in the particular drop being tested, it would show up as it decayed. Whenever an atom of the new element disintegrates, the fission fragments cause a large burst of ionization. This pulse of current makes a pen on the recording chart jump.

It is typical of these elusive heavy elements that we cannot positively identify an atom until the moment that it ceases to be that element and disintegrates into something else. It is rather like the man who only counts his money as he spends it.

In the first experiment, we waited more than an hour before the pen shot to mid-scale and dropped back, marking a line that related to the decay of the first known atom of mendelevium.

Since this was quite an event at the Rad Lab, we connected a fire bell in the hallway to the counters so that the alarm would go off after every time an atom of element 101 decayed. This was a most effective way of signaling the occurrence of a nuclear event, but quieter means of communication were soon substituted, following a suggestion put forward by the fire department.

We found only about one atom of mendelevium in each of our first experiments. We repeated the experiment perhaps a dozen times, and our grand total was 17 atoms of the new element.

The recording chart from the first complete experiment (Figure 2.20) shows the disintegrations of the first atoms of mendelevium ever made and completely identified.

The rate of spontaneous fission in both the element 101 and 100 fractions decayed with a half-life of about 3 h (later determined to be 160 min). This and other evidence led to the hypothesis that this isotope of element 101 has the mass number 256 and decays, by electron capture (designated by the symbol EC), with a half-life of the order of 1.5 h, to the isotope $^{256}_{100}$Fm, which is responsible for the spontaneous fission decay. The discovery reactions were

$$^{253}_{99}\text{Es} + ^{4}_{2}\text{He} \rightarrow ^{256}_{101}\text{Md} + ^{1}_{0}\text{n} \qquad (2.23)$$

$$^{256}_{101}\text{Md} \xrightarrow[t_{1/2}=1.3\,\text{h}]{\text{EC}} ^{256}_{100}\text{Fm} \qquad (2.24)$$

$$^{256}_{100}\text{Fm} \xrightarrow[t_{1/2}=2.63\,\text{h}]{} \text{spontaneous fission} \qquad (2.25)$$

On the basis of this evidence and the experiments which led to the production of 17 atoms of element 101, Ghiorso, Harvey, Choppin, Thompson, and Seaborg (Figure 2.19) announced the discovery of element 101 (Ghi 55b). The name mendelevium (symbol Mv) was suggested for the element, in recognition of the role of the great Russian chemist,

MENDELEVIUM 45

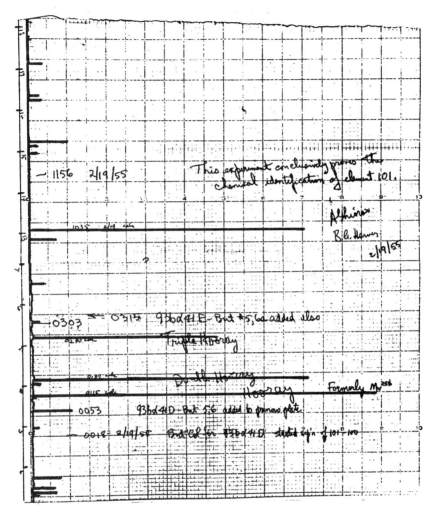

Figure 2.20 The ionization recording chart showing the first four events of the disintegration of mendelevium. The ordinate is the event time while the abscissa is the intensity of the ionization. The four pulses occurred at 1:15 a.m., 1:37 a.m., 2:40 a.m., and 10:35 a.m. on February 19, 1955. At 11:56 a.m., Ghiorso and Harvey made a note directly in the chart: "This experiment conclusively proves the chemical identification of element 101."

Dmitri Mendeleev, who was the first to use the periodic system of the elements to predict the chemical properties of undiscovered elements, a principle which was used in nearly all the transuranium element discovery experiments. The chemical symbol, Md, was later adopted for this element.

It is comforting to be able to record that subsequent experiments using larger amounts of einsteinium in the target led to the production of thousands of atoms of mendelevium, lending confirmation to the sparse evidence on which the original conclusions were made. The indications are clear that, as expected, mendelevium is a typical tripositive actinide element.

2.9 NOBELIUM—THE FIRST OF THE "CONTROVERSIAL" ELEMENTS

In 1957, a team of scientists from the Argonne National Laboratory in the United States, the Atomic Energy Research Establishment at Harwell in England and the Nobel Institute for Physics in Stockholm announced the discovery of element 102 based upon work done at the Nobel Institute (Fie 57). The group reported that in irradiations of $^{244}_{96}$Cm with ^{13}C ions accelerated at the Nobel Institute Cyclotron, they found an 8.5 MeV α-emitter with a half-life of about 10 min, presumably due to the $^{244}_{96}$Cm ($^{13}_{6}$C, 4n) or $^{244}_{96}$Cm ($^{13}_{6}$C, 6n) reactions. They claimed that this activity had been identified as a new element on the basis of ion-exchange chromatography in which the 8.5 MeV activity appeared in the "expected" element 102 position when eluted from a cation exchange column with α-hydroxyisobutyrate. The name of nobelium (chemical symbol No) was suggested for the new element in recognition of Alfred Nobel's contributions to the advancement of science.

However, neither experiments at Berkeley (Ghi 58) nor related experiments at the Kurchatov Institute in Moscow, U.S.S.R. (Fle 58a) confirmed this Stockholm work. In fact, subsequent experiments done in Berkeley have shown that the most stable oxidation state of element 102 in solution is +2; thus it would not appear in the "expected" element 102 tripositive position in a cation-exchange column.

In 1958, Ghiorso, T. Sikkeland, J.R. Walton, and Seaborg (Ghi 58) (Figure 2.21) announced the positive identification of $^{254}_{102}$No produced using the Berkeley heavy ion linear accelerator (HILAC) which they attributed to the reactions

NOBELIUM—THE FIRST OF THE "CONTROVERSIAL" ELEMENTS

Figure 2.21 The codiscoverers of nobelium, HILAC Building, Lawrence Berkeley Laboratory, 1958. Left to right: Albert Ghiorso, Torbjorn Sikkeland, John R. Walton (Seaborg absent).

$$^{246}_{96}\text{Cm} + ^{12}_{6}\text{C} \rightarrow ^{254}_{102}\text{No} + 4^{1}_{0}\text{n} \tag{2.26}$$

$$^{254}_{102}\text{No} \xrightarrow[t_{1/2} \sim 3\,\text{s}]{} ^{250}_{100}\text{Fm} + ^{4}_{2}\text{He} \tag{2.27}$$

The $^{250}_{100}$Fm daughter of the new element was collected using recoil techniques, one atom at a time. Eleven atoms of the $^{250}_{100}$Fm daughter were identified by their position in a cation exchange elution curve (see Figure 2.22). A half life of ~3 s was assigned to $^{254}_{102}$No on the basis of many recoil experiments in which an apparent α-emitting daughter of $^{254}_{102}$No was produced and direct counting of an 8.3 MeV α-emitting nuclide with a 3-s half-life. The 8.3 MeV α-emitter was found also to decay by spontaneous fission in 30% of its decays. It is now known that the 3-s activity originally assigned to $^{254}_{102}$No in the direct counting experiments was, in fact, $^{252}_{102}$No ($t_{1/2}$ = 2.3 s, E_α = 8.4 MeV) produced by a $^{244}_{96}$Cm (^{12}C, 4n) reaction in that the "^{246}Cm target" used by Ghiorso et al. had 20 × more ^{244}Cm than ^{246}Cm in it. $^{254}_{102}$No is now known to have a 55-s half-life.

The experimental claim for discovery of a new element, element 102,

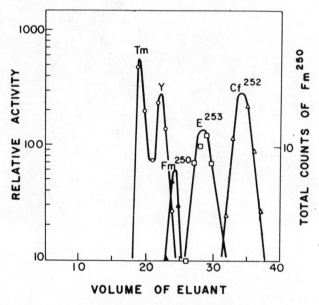

Figure 2.22 Elution data obtained in the discovery of element 102, showing the presence of $^{250}_{100}$Fm, the decay product of ^{254}102. Other activities were added for calibration purposes. (Dowex 50 ion-exchange resin was used, and the eluting agent was ammonium α-hydroxyisobutyrate).

however, must be judged upon the observation of the $^{250}_{100}$Fm daughter of $^{254}_{102}$No because this is the only evidence that establishes the atomic number of the new element. The recoil studies which assigned a 3-s half-life to $^{254}_{102}$No by observation of what was thought to be $^{250}_{100}$Fm were probably erroneous. What was probably being observed was the sequence

$$^{252}_{102}\text{No} \xrightarrow[t_{1/2}=2.3\,\text{s}]{\alpha} {}^{248}_{100}\text{Fm} \xrightarrow[t_{1/2}=36\,\text{s}]{\alpha} {}^{244}_{98}\text{Cf} \xrightarrow[t_{1/2}=19\,\text{min}]{} \quad (2.28)$$

in which the 19-min $^{244}_{98}$Cf granddaughter ($E_\alpha = 7.21$ MeV) was mistaken for the 30 min $^{250}_{100}$Fm ($E_\alpha = 7.43$ MeV). The ion-exchange elution curve (Figure 2.22) showing 11 atoms of $^{250}_{100}$Fm to appear in the proper position remains as the first definitive evidence for the production of element 102. The errors in this experiment indicate the difficulties associated with "one-atom-at-a-time" studies.

A parallel line of research on element 102 was carried out by G.N. Flerov and co-workers in the Kurchatov Institute in the U.S.S.R. In an experiment reported in 1958, Flerov et al. studied the reaction of $^{239}_{94}$Pu with $^{16}_{8}$O ions, reporting an α-emitter with $E_\alpha = 8.9$ MeV and $2 < t_{1/2} < 40$ s (Fle 58b). In 1964, E.D. Donets, V.A. Schegolev, and V.A. Ermakov (Don 64) of the Dubna laboratory reported the production of the new isotope $^{256}_{102}$No using recoil techniques with *chemical identification* of the α-emitting daughter $^{252}_{100}$Fm. The first correct identification of the half-life of $^{254}_{102}$No was in 1966–1967 by groups working at Dubna.

In summary one can say that the Berkeley group was the first group to clearly identify the atomic number of element 102 (i.e., to "discover" it) but important contributions to the definitive establishment of the existence of element 102 were made by the Soviet research scientists. Since the name of nobelium and symbol (No) for this element are in common use, the Berkeley scientists have suggested retention of the name suggested in the original, incorrect Stockholm experiment.

2.10 LAWRENCIUM—THE LAST ACTINIDE ELEMENT

The first identification of an isotope of element 103 was by Ghiorso, Sikkeland, A.E. Larsh, and R.M. Latimer in 1961 (Ghi 61) (Figure 2.23). Three micrograms of a mixture of californium isotopes ($A = 249, 250, 251, 252$) were bombarded with heavy ion beams of ^{10}B and ^{11}B at the Berkeley HILAC. Atoms recoiling from the target were caught by a long metallized mylar tape which was moved past a series of α-particle detectors. A new α-emitting nuclide with $E_\alpha = 8.6$ MeV and $t_{1/2} \sim 8$ s was ob-

Figure 2.23 The codiscoverers of lawrencium, HILAC Building, Lawrence Berkeley Laboratory, 1961. Left to right: Torbjorn Sikkeland, Albert Ghirso, Almon E. Larsh, Robert M. Latimer.

served and assigned to $^{257}_{103}\text{Lr}$. Later experiments indicated that this activity was due to $^{258}_{103}\text{Lr}$ ($E_\alpha = 8.6$ MeV, $t_{1/2} = 4.3$ s).

A subsequent identification of the atomic number of element 103 was made by Donets, Schegolev, and Ermakov at Dubna in 1965 (Don 65). The nuclear reaction used was

$$^{243}_{95}\text{Am} + ^{18}_{8}\text{O} \rightarrow ^{256}_{103}\text{Lr} + 5^{1}_{0}\text{n} \qquad (2.29)$$

Using a double recoil technique, they identified the α-emitter $^{256}_{103}\text{Lr}$ ($t_{1/2} \sim 45$ s) and linked it genetically to its granddaughter, the known $^{252}_{100}\text{Fm}$ via the decay sequence

$$^{256}_{103}\text{Lr} \xrightarrow{\alpha} ^{252}_{101}\text{Md} \xrightarrow{\text{EC}} ^{252}_{100}\text{Fm} \xrightarrow{\alpha} \qquad (2.30)$$

The relatively long half-life of $^{256}_{103}\text{Lr}$ (now known to be ~ 30 s) enabled

Silva, Sikkeland, Nurmia, and Ghiorso (Sil 70a) to establish in 1970 that element 103 exhibits a stable +3 oxidation state in solution, as expected by the actinide concept.

In the report of the original experiments of Ghiorso et al. (Ghi 61), they suggested the name lawrencium (subsequently accepted by the IUPAC) for element 103 in honor of E.O. Lawrence, the inventor of the cyclotron and founder of the Radiation Laboratory at Berkeley where so much of the transuranium research has been carried out. The accepted chemical symbol is Lr.

2.11 RUTHERFORDIUM AND HAHNIUM (Hyd 87)

There is considerable controversy, frequently punctuated by acerbic comments, over the discovery of elements 104, 105, and 106. In 1964, Flerov and co-workers (Fle 64) bombarded $^{242}_{94}$Pu with $^{22}_{10}$Ne from the Dubna cyclotron and reported finding a nuclide that decayed by spontaneous fission with $t_{1/2} \sim 0.3$ s. This nuclide was assigned to be ^{260}104 on the basis of nuclear reaction systematics. The name of kurchatovium (Ku) in honor of the Soviet nuclear physicist Igor Kurchatov was suggested later for element 104. Subsequently this group suggested that the half-life of this nuclide was 0.1 s, then 80 ms and most recently 28 ms. The identification of the atomic number of the new species on the basis of thermochromatography of the chlorides of this element (in a glass column without packing material) was claimed by I. Zvara, K.T. Chuburkov, R. Tsaletka, T.S. Zvarova, M.R. Shalaevskii, and B.V. Shilov in 1966 (Zva 66). However, if the half-life of the ^{260}104 nuclide was 28 ms, it is impossible that it could have survived passage through the apparatus of Zvara et al. which involved a 1.2-s transit time for the volatile chlorides. Furthermore, an important part of the interpretation of the thermochromatography experiment was the assumption of a 0.3 s $t_{1/2}$ for the species being detected. Much later Zvara and co-workers have claimed, in retrospect, that their original experiment probably measured the chemical behavior of 3-s ^{259}104 which would have survived transit through their apparatus. However in the description of the original thermochromatography experiments, Zvara et al. stated "positively that the half-life could not be 3.7 s". Because of questions about these thermochromatography experiments, many (in particular, see the work of Hyde et al. (Hyd 87)) have regarded these experiments as not providing a definitive characterization of the atomic number of the new species.

There is little doubt that Ghiorso, M. Nurmia, J. Harris, K. Eskola, and P. Eskola (Figure 2.24) did definitely produce isotopes of element

Figure 2.24 The codiscoverers of rutherfordium and hahnium with Seaborg, HILAC Building, Lawrence Berkeley Laboratory, 1970. Left to right: Matti Nurmia, James Harris, Kari Eskola, Seaborg, Pirrko Eskola, Albert Ghiorso

104 and identify their atomic number in experiments at Berkeley in 1969 (Ghi 69). The nuclear reactions involved were

$$^{249}_{98}\text{Cf} + ^{12}_{6}\text{C} \rightarrow ^{257}_{104}\text{Rf}(t_{1/2} \sim 3.8\,\text{s}) + 4^{1}_{0}\text{n} \qquad (2.31)$$

$$^{249}_{98}\text{Cf} + ^{13}_{6}\text{C} \rightarrow ^{259}_{104}\text{Rf}(t_{1/2} \sim 3.4\,\text{s}) + 3^{1}_{0}\text{n} \qquad (2.32)$$

The atomic numbers of the isotopes of element 104 were identified by detecting the known daughters of $^{257}_{104}\text{Rf}$ and $^{259}_{104}\text{Rf}$, $^{253}_{102}\text{No}$ and $^{255}_{102}\text{No}$. This group later suggested the name of rutherfordium (chemical symbol Rf) for element 104 in honor of Lord Ernest Rutherford. These results were confirmed in subsequent work by C.E. Bemis et al. at Oak Ridge National Laboratory (Bem 73).

Studies at Berkeley of the aqueous chemistry of rutherfordium have shown it to behave differently than the heavy actinides. Its solution chemistry resembles that of hafnium and zirconium, in agreement with the idea that rutherfordium is not an actinide but a Group IV element (Sil 70b).

Unfortunately, controversy also exists over the discovery of element 105. In 1968, Flerov and co-workers (Fle 68) in Dubna reported production of two new α-emitters, assigned to be $^{260}105$ and $^{261}105$ in the reaction of $^{243}_{95}$Am with $^{22}_{10}$Ne ions. The element 105 radioactivities were identified by detection of events in which the initial α-particles (9.7 and 9.4 MeV) emitted by the element 105 activities were found to be correlated in time with the α-particles emitted by the daughter (element 103) nuclides. A small number of such events (~10) was observed and the two element 105 nuclides were said to have half-lives in the range 0.1–3 and >0.01 s, respectively. The international groups who compile and certify nuclear data have generally considered this work to be inconclusive or possibly wrong because of the small number of observed events and the discrepancy between the reported element 105 α-particle energies of 9.7 and 9.4 MeV and those now known to be correct, that is, 9.1 and 8.9 MeV, respectively.

In 1970, A. Ghiorso, M. Nurmia, K. Eskola, J. Harris, and P. Eskola (Ghi 70) reported the observation of an isotope of element 105 with mass number 260 produced in the following reaction:

$$^{249}_{98}\text{Cf} + ^{15}_{7}\text{N} \longrightarrow ^{260}_{105}\text{Ha}(t_{1/2} = 1.5 \text{ s}) + 4^{1}_{0}\text{n} \qquad (2.33)$$

The Z and A of the element 105 nuclide were unambiguously identified in a manner similar to that used in the discovery of rutherfordium by observing the time correlation between α-particles emitted by the parent (element 105) and those of the known daughter (^{256}Lr). The Berkeley group's data combined more than ten times more events than were reported by Flerov et al. Their α-particle energies are in agreement with what is currently known about $^{260}_{105}$Ha. In honor of the German radiochemist Otto Hahn who discovered fission and developed many experimental techniques, the Berkeley group suggested the name of hahnium (symbol Ha) for this element. This work was subsequently confirmed by Bemis et al. (Bem 77).

At about the same time as the Berkeley work, Flerov, Y.T. Oganessian, Y.V. Lobanov, Y.A. Lasarev, S.P. Tretiakova, I.V. Kolesov, and V.M. Plotko (Fle 71) reported the observation of a nuclide with a half-life of 1.8 ± 0.6 s (which decayed by spontaneous fission) produced in the reaction $^{243}_{95}$Am with $^{22}_{10}$Ne. On the basis of nuclear reaction systematics and the angular distribution of the observed reaction products, those workers assigned this nuclide to $^{261}105$. This spontaneous fission activity was later (1975) reported to behave as if it were due to a group V element in a thermochromatography experiment although this conclusion has been criticized (Hyd 87). The Soviet group has suggested the name of nielsbohrium (symbol Ns) for element 105 in honor of the Danish physicist Niels Bohr.

2.12 ELEMENT 106

Experiments leading to competing claims for the discovery of element 106 were performed essentially simultaneously at Berkeley and Dubna in 1974. Ghiorso, J.M. Nitschke, J.R. Alonso, C.T. Alonso, M. Nurmia, G.T. Seaborg, E.K. Hulet, and R.W. Lougheed (Ghi 74) (Figure 2.25) reported

Figure 2.25 The codiscoverers of Element 106, HILAC Building, Lawrence Berkeley Laboratory, 1974. Left to right: Matti Nurmia, Jose R. Alonso, Albert Ghiorso, E. Kenneth Hulet, Carol T. Alonso, Ronald W. Lougheed, Seaborg, Joachim M. Nitschke.

the production of $^{263}106$ by the reaction

$$^{249}_{98}\text{Cf} + ^{18}_{8}\text{O} \longrightarrow ^{263}106 + 4^{1}_{0}\text{n} \qquad (2.34)$$

The new nuclide was shown to decay by α-emission with a half-life of 0.9 ± 0.2 s and a principal α-energy of 9.06 ± 0.04 MeV to previously known $^{259}_{104}\text{Rf}$ which in turn was shown to decay to the known nuclide $^{255}_{102}\text{No}$. Thus the atomic number of the new nuclide was firmly established by a genetic link to its daughters. Oganessian, Y.P. Tretyakov, A.S. Iljinov, A.G. Demin, A.S. Pleve, S.P. Tretyakova, V.M. Plotko, M.P. Ivanov, N.A. Danilov, Y.S. Korotkin, and G.N. Flerov (Oga 74) (Figure 2.26) reported the observation of a spontaneous fission activity with a half-life of 4–10 ms, produced by bombarding $^{207}_{82}\text{Pb}$ with $^{54}_{24}\text{Cr}$, which they assigned to $^{259}106$ on the basis of reaction systematics. We now know this assignment was erroneous in that the observed spontaneous fission activities were primarily due to the daughters of element 106, that is, $^{256,255}104$, and not element 106 (Dem 84). The isotope $^{260}106$ (which may have been produced also in the Oganessian et al. work) is now known to have a half-

Figure 2.26 At Laboratory of Nuclear Reactions, Dubna, USSR, September 23, 1975. Left to right: V.A. Druin, Seaborg, Albert Ghiorso, Georgiy N. Flerov, Yuri T. Oganessian, Ivo Zvara.

life of 4 ms with a partial half-life for spontaneous fission of ~7 ms. Neither group has suggested a name for element 106.

2.13 ELEMENT 107

In 1976, Oganessian and co-workers (Oga 76) reported the production of a spontaneous fission activity with a half-life of ~2 ms from the reaction of $^{209}_{83}$Bi with $^{54}_{24}$Cr which they attributed to ^{261}107. In 1981, G. Münzenberg, S. Hofmann, F.P. Hessberger, W. Reisdorf, K.H. Schmidt, J.R.H. Schneider, W.F.W. Schneider, P. Armbruster, C.C. Sahm, and B. Thuma (Mün 81) working at the Gesellschaft für Schwerionenforschung (GSI) at Darmstadt, West Germany, identified the nuclide ^{262}107 produced in the "cold fusion" reaction

$$^{209}_{83}\text{Bi} + ^{54}_{24}\text{Cr} \longrightarrow ^{262}107 + ^{1}_{0}\text{n} \qquad (2.35)$$

The recoiling product nuclei from the nuclear reaction were passed through a velocity separator (called SHIP) which guaranteed that they had the characteristic velocity of the product of complete fusion of projectile and target nuclei. The mass number of the velocity-separated product nuclei was roughly determined using a time-of-flight spectrometer and the atomic number and mass number were determined by observing the time correlated α-decay of ^{262}107 to its decay products (see Figure 2.27). One

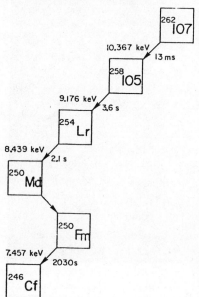

Figure 2.27 Sequence of time-correlated decay chain observed by Münzenberg et al. to identify the product of the ^{209}Bi (^{54}Cr,n) ^{262}107 reaction.

sequence of correlated decays ended in the known nucleus $^{254}_{103}$Lr, one ended in $^{246}_{98}$Cf, two ended in $^{250}_{100}$Fm decay, and one ended in $^{250}_{101}$Md. Five decays of ^{262}107 were observed with $E_\alpha = 10.4$ MeV and $t_{1/2} \sim 5$ ms. The cross section for producing these nuclei was $\sim 2 \times 10^{-34}$ cm^2 (approximately 1/5,000,000 of the production cross section assumed in the first one-atom-at-a-time experiments with Md!). It is a remarkable tribute to the quality of this experiment that the results of this experiment have found rapid, universal acceptance despite the exceedingly low production rate involved. By 1988, a total of 38 atoms had been observed. Subsequent experiments (Mün 89) identified three 107 species, ^{261}107 ($t_{1/2} = 11.8^{+5.3}_{-2.8}$ ms; $E_\alpha \sim 10.2$ MeV), ^{262}107 ($t_{1/2} = 102 \pm 26$ ms; $E_\alpha \sim 9.9$ MeV) and ^{262m}107 ($t_{1/2} \approx 8.0 \pm 2.1$ ms; $E_\alpha \sim 10.3$ MeV). Contrary to the initial observations of the Dubna group, no spontaneous fission activities with $t_{1/2} = 1-2$ ms were observed. No name has been suggested for element 107.

2.14 ELEMENT 108

In 1984, two reports of the successful synthesis of element 108 appeared. The Darmstadt group (G. Münzenberg, P. Armbruster, H. Folger, F.P. Hessberger, S. Hofmann, J. Keller, K. Poppensieker, W. Reisdorf, K.H. Schmidt, H.-J. Schott, M.E. Leino, and R. Hingmann, Figure 2.28) used the velocity separator SHIP to identify 3 atoms of element 108 (Mün 84). The nuclear reaction used was ^{208}Pb (^{58}Fe, 1n) ^{265}108 at a ^{58}Fe energy of 5.02 MeV/nucleon which should lead to an excitation energy of 18 ± 2 MeV for the compound system. The cross section for production of these nuclei was 1/10 that observed for the production of element 107 ($\sigma(108) \sim 2 \times 10^{-36}$ cm^2). Three time-correlated α-decay chains that clearly led to known nuclei ^{261}106 and ^{257}104 were observed (Figure 2.29). The observed species ^{265}108 appears to have a $t_{1/2} \sim 1.8$ ms and decays by the emission of a 10.36 MeV α-particle. In a second experiment (Mün 86), one atom of the even-even nuclide ^{264}108 ($t_{1/2} \sim 80$ μs) was produced in the ^{207}Pb (^{58}Fe, n) reaction. The observation of α-decay by element 108 is taken as a sign that spontaneous fission lifetimes are unexpectedly long for these nuclei, possibly portending the synthesis of still heavier nuclei.

At approximately the same time as the Münzenberg et al. report, Oganessian et al. (Oga 84) (Y.T. Oganessian, A.G. Demin, M. Hussonnois, S.P. Tretyakova, Y.P. Kharitonov, V.K. Utyonkov, I.V. Shirokovsky, O. Constantinescu, H. Bruchertseifer, and I. Korotkin) reported the observation of the possible decay of ^{263,264,265}108 produced in the reactions of 5.5 MeV/nucleon ^{55}Mn + ^{209}Bi ($\to ^{263}$108) and ^{58}Fe + ^{207,208}Pb ($\to ^{264,265}$108). The production cross sections reported by this Dubna group

Figure 2.28 Codiscoverers of element 108, GSI Laboratory, 1984. Left to right: Sigurd Hofmann, Karl Heinz Schmidt, Peter Armbruster, Willi Reisdorf, H.J. Schött, Gottfried Münzenberg.

were from one-tenth to one-quarter those observed by the Darmstadt group. None of the α-particle decays of these nuclei were observed directly. In the case of ^{264}108, a 8 ms and a 6 ms fission activity were observed and attributed to the granddaughter of ^{264}108, ^{256}104, a known 9 ms spontaneous fission activity. Similarly ^{263}108 was identified on the observation of a 1.1 s spontaneous fission activity attributed to its granddaughter, ^{255}104, a known 2 s spontaneous fission activity. The nuclide ^{265}108 was said to have been detected because of the observation of α-emitting ^{253}Es, a possible great-great-great-granddaughter of ^{265}108. Interesting as the observations of the Dubna group are, they are not sufficient by themselves to be a claim for the discovery of element 108 or to be a confirmation of the work of the Darmstadt group. No name has been suggested for element 108.

ELEMENT 109

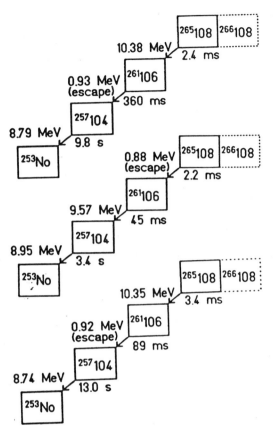

Figure 2.29 The three observed decay sequences associated with element 108. The nucleus ^{266}108 is assumed to have been the compound nucleus which emitted 1 neutron to form ^{265}108.

2.15 ELEMENT 109

In 1982, G. Münzenburg, P. Armbruster, F.P. Hessberger, S. Hofmann, K. Poppensieker, W. Reisdorf, K. Schneider, K.H. Schmidt, C. Sahm, and D. Vermeulen reported the observation (Mün 82), after about 2 weeks of bombardment, of one unusual time-correlated decay sequence that occurred for a reaction product that had been velocity-separated by SHIP from the $^{209}_{83}$Bi + $^{58}_{26}$Fe reaction. After implantation of the complete fusion reaction product in a detector, an 11.1 MeV α-particle decay was detected, followed 22 ms later by a detected α-particle of 1.14 MeV en-

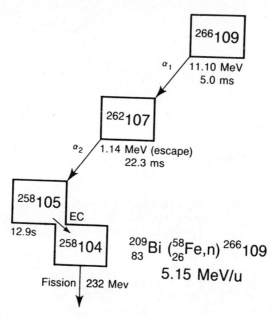

Figure 2.30 A possible decay sequence of an event attributed to the reaction ^{209}Bi (^{58}Fe,n) ^{266}109.

ergy, followed 13 s later by a spontaneous fission. A possible sequence for this decay is shown in Figure 2.30. The 1.14 MeV α-particle is assumed to result from a decay in which only part of the α-particle energy was deposited in the detector. Such a yield corresponds to a formation cross section of $\approx 10^{-35}$ cm². In a second experiment in 1988, two more time-correlated decay sequences similar to the first event were found (Mün 88). The combined results of both experiments give a value of the half-life of $3.4^{+6.1}_{-1.3}$ ms for ^{266}109 and a production cross section of 10^{+10}_{-6} pb. No name has been suggested for element 109.

According to the criteria for the discovery of a chemical element, no names *should* be suggested for elements 106, 107, 108, and 109 "until the initial discovery is confirmed."

2.16 ELEMENT 110

In a study (1986) of the reaction of ^{44}Ca with ^{232}Th, Oganessian and co-workers (Fle 87) reported the production of two spontaneously fissioning species with half-lives of 0.8 ± 0.3 ms and $8.6^{+4.0}_{-2.4}$ ms. The former activity was assigned to ^{240}Am^f ($t_{1/2} = 0.9 \pm 0.1$ ms). The latter activity (with $\sigma_{prod} \sim 8$ pb) was thought to be due to an isotope of element 110 ($t_{1/2} \sim 10$ ms) and was also produced (1987) in the reaction of 210 MeV ^{40}Ar with ^{235,236}U (at what would be a similar excitation energy of the ^{276}110 composite). This evidence does not meet the criteria for the discovery of a chemical element.

An attempt by the GSI team to observe the 9 ms spontaneous fission activity was made utilizing the ^{235}U + ^{40}Ar reaction. No activity was observed and an upper limit of 8×10^{-36} cm^2 was set for the formation cross section.

REFERENCES

Arm 85   P. Armbruster, Annu. Rev. Nucl. Part. Sci. **35**, 169 (1985).

Bem 73   C.E. Bemis, Jr., R.J. Silva, D.C. Hensley, O.L. Keller, Jr., J.R. Tarrant, L.D. Hunt, P.F. Dither, R.L. Hahn, and C.D. Goodman, Phys. Rev. Lett. **31**, 647 (1973).

Bem 77   C.E. Bemis, Jr., P.F. Dither, R.J. Silva, R.L. Hahn, J.R. Tarrant, L.D. Hunt, and D.C. Hensley, Phys. Rev. **C16**, 1146 (1977).

Cun 45   B.B. Cunningham, Metallurgical Laboratory Report CS-3312, (1945) pp. 5–6.

Cun 49   B.B. Cunningham and L.B. Werner, J. Am. Chem. Soc. **71**, 1521 (1949).

Cun 61   B.B. Cunningham, private communication.

Dem 84   A.G. Demin, S.P. Tretyakova, V.K. Utyonkov, and I.V. Shirokovsky, Z. Phys. **A315**, 197 (1984).

Don 64   E.D. Donets, V.A. Schegolev, and V.A. Ermakov, Atomnaya Energiya **16**, 195 (1964); English translation, Soviet Journal Atomic Energy **16**, 233 (1964).

Don 65   E.D. Donets, V.A. Schegolev, and V.A. Ermakov, Atomnaya Energiya **19**, 109 (1965); English translation, Soviet Journal Atomic Energy **19**, 995 (1965).

Fie 57   P.R. Fields, A.M. Friedman, J. Milsted, H. Atterling, W. Forsling, L.W. Holm, and B. Åström, Phys. Rev. **107**, 1460 (1957).

Fle 58a   G.N. Flerov et al., Sov. Phys. Dokl. **3**, 546 (1958).

Fle 58b   G.N. Flerov, P/2299, Proceedings of the Second United Nations International Conference on the Peaceful Uses of Atomic Energy, Geneva, 1958, vol. 14, (United Nations, New York, 1958), pp. 151–157.

Fle 64   G.N. Flerov, Y.T. Oganessian, Y.V. Lobanov, V.I. Kuznetsov, V.A.

	Druin, V.P. Perelygin, K.A. Gavrilov, S.P. Tret'yakova, and V. M. Plotko, Phys. Lett. **13**, 73 (1964).
Fle 68	G. N. Flerov, V.A. Druin, A.G. Demin, Y.V. Lobanov, N.K. Skokelev, G.N. Akap'ev, B.V. Fefilov, I.V. Kolesov, K.A. Gavrilov, Y.P. Kharitonov, and L.P. Chelnokov, Preprint JINR P7-3808, Dubna (1968).
Fle 71	G.N. Flerov, Y.T. Oganessian, Y.V. Lobanov, Y.A. Lasarev, S.P. Tretiakova, I.V. Kolesov, and V.M. Plotko, Nucl. Phys. **A160**, 181 (1971).
Fle 87	G.N. Flerov and G.M. Ter-Akopian in *Progress in Particle and Nuclear Physics*, Vol. 19, A. Falssler, Ed. (Pergamon, Oxford, 1987) pp. 197–239; see also Y.T. Oganessian et al. Dubna Preprint D7-87-392, June 1987.
Fri 79	S. Fried, in Lawrence Berkeley Laboratory Report LBL-7701, April 1979.
Ghi 50	A. Ghiorso, R.A. James, L.O. Morgan, and G.T. Seaborg, Phys. Rev. **78**, 472 (1950).
Ghi 55a	A. Ghiorso et al. Phys. Rev. **99**, 1048 (1955).
Ghi 55b	A. Ghiorso, B.G. Harvey, G.R. Choppin, S.G. Thomson, and G.T. Seaborg, Phys. Rev. **98**, 1518 (1955).
Ghi 58	A. Ghiorso, T. Sikkeland, J.R. Walton, and G.T. Seaborg, Phys. Rev. Lett **1**, 18 (1958).
Ghi 61	A. Ghiorso, T. Sikkeland, A.E. Larsh, and R.M. Latimer, Phys. Rev. Lett **6**, 473 (1961).
Ghi 69	A. Ghiorso, M. Nurmia, J. Harris, K. Eskola, and P. Eskola, Phys. Rev. Lett. **22**, 1317 (1969).
Ghi 70	A. Ghiorso, M. Nurmia, K. Eskola, J. Harris, and P. Eskola, Phys. Rev. Lett. **24**, 1498 (1970).
Ghi 74	A. Ghiorso, J.M. Nitschke, J.R. Alonso, C.T. Alonso, M. Nurmia, G.T. Seaborg, E.K. Hulet and R.W. Lougheed, Phys. Rev. Lett. **33**, 1490 (1974).
Ghi 79	A. Ghiorso, in Lawrence Berkeley Laboratory Report LBL-7701, April 1979.
Ghi 80	A. Ghiorso, in Lawrence Berkeley Laboratory Report LBL-11599, March 1980.
Ghi 82	A. Ghiorso in *Actinides in Perspective*, N.M. Edelstein, Ed. (Pergamon, Oxford, 1982) p. 23. An original account of the discovery of the transplutonium elements by someone who participated in the discovery of almost all of these elements.
Goo 71	L.S. Goodman, H. Diamond, H.E. Stanton, and M. S. Fred, Phys. Rev. **A4**, 473 (1971).
Har 76	B.G. Harvey, G. Herrmann, R.W. Hoff, D.C. Hoffman, E.K. Hyde,

REFERENCES

	J.J. Katz, O.L. Keller, Jr., M. Lefort, and G.T. Seaborg, Science 193, 1271 (1976). In this paper an international group of scientists identifies the criteria for discovery of a new element.
Hui 79	J.R. Huizenga, Lawrence Berkeley Laboratory Report LBL-7701, April 1979, p. 17.
Hyd 87	E.K. Hyde, D.C. Hoffman, and O.L. Keller, Radiochimica Acta **42**, 57 (1987).
Ken 46	J.W. Kennedy, G.T. Seaborg, E. Segre, and A.C. Wahl, Phys. Rev. **70**, 555 (1946).
Mag 48	L.B. Magnusson and T.J. LaChapelle, J. Am. Chem. Soc. **70**, 3534 (1948).
McM 40	E.M. McMillan and P.A. Abelson, Phys. Rev. **57**, 1185 (1940).
Mün 81	G. Münzenberg et al., Z. Phys. **A300**, 107 (1981).
Mün 82	G. Münzenberg et al., Z. Phys. **A309**, 89 (1982), see also G. Münzenberg et al., Z. Phys. **A315**, 145 (1984).
Mün 84	G. Münzenberg et al., Z. Phys. **A317**, 235 (1984), see also G. Münzenberg et al., Z. Phys. **A328**, 49 (1987).
Mün 86	G. Münzenberg et al., Z. Phys. **A324**, 489 (1986).
Mün 88	G. Münzenberg et al., Z Phys. **A330**, 435 (1988).
Mün 89	G. Münzenberg et al., Z Phys. **A333**, 163 (1989).
Oga 74	Y.T. Oganessian Y.P. Tret'yakov, A.S. Iljinov, A.G. Demin, A.A. Pleve, S.P. Tret'yakova, V.M. Plotko, M.P. Ivanov, N.A. Danilov, Y.S. Korotkin, and G.N. Flerov, JETP Lett. **20**, 265 (1974).
Oga 76	Y.T. Oganessian A.G. Demin, N.A. Danilov, M.P. Ivanov, A.S. Iljinov, N.N. Kolesnikov, B.N. Markov, V.M. Plotko, S.P. Tret'yakova, and G.N. Flerov, JETP Lett. **23**, 277 (1976); Y.T. Oganessian et al., Nucl. Phys. **A273**, 505 (1976).
Oga 84	Y.T. Oganessian et al., Z. Phys. **A319**, 215 (1984).
Sea 45	G.T. Seaborg, Chem. Eng. News **23**, 2190 (1945).
Sea 46a	G.T. Seaborg, E.M. McMillan, J.W. Kennedy, and A.C. Wahl, Phys. Rev. 69, 366 (1946).
Sea 46b	G.T. Seaborg, A.C. Wahl, and J.W. Kennedy, Phys. Rev. 69, 367 (1946).
Sea 48	G.T. Seaborg and A.C. Wahl, J. Am. Chem. Soc. **70**, 1128 (1948).
Sea 58	G.T. Seaborg, *The Transuranium Elements* (Yale University Press, New Haven, 1958).
Sea 58a	G. T. Seaborg and E. G. Valens, *Elements of the Universe* (Dutton, New York, 1958).
Sea 63	G.T. Seaborg, *Man-Made Transuranium Elements* (Prentice-Hall, Kennedy, and A.C. Wahl, Phys. Rev. **69**, 366 (1946).
Sea 72	G.T. Seaborg, *Nuclear Milestones* (W.H. Freeman and Company, San

	Francisco, 1972) A collection of speeches recognizing historic discoveries and landmarks of nuclear science.
Sea 75	G.T. Seaborg in Lawrence Berkeley Laboratory Report LBL-4366, January, 1975.
Sea 78	G.T. Seaborg, Ed., *Transuranium Elements: Products of Modern Alchemy* (Dowden, Hutchinson; Ross, Stroudsberg, 1978). Annotated collection of the original papers describing the discovery of the transuranium elements, Np-106, and related work.
Sil 70a	R.S. Silva, T. Sikkeland, M. Nurmia, and A. Ghiorso, Inorg. Nucl. Chem. Lett. **6**, 733 (1970).
Sil 70b	R. Silva, J. Harris, M. Nurmia, K. Eskola, and A. Ghiorso, Inorg. Nucl. Chem. Lett. **6**, 871 (1970).
Tho 50a	S.G. Thompson, A. Ghiorso, and G.T. Seaborg, Phys. Rev. **77**, 838 (1950).
Tho 50b	S.G. Thompson, K. Street, Jr., A. Ghiorso, and G.T. Seaborg, Phys. Rev. **78**, 298 (1950).
Tho 59	S.G. Thompson, Lawrence Radiation Laboratory Report UCRL-8615, April 1959.
Wer 51	L.B. Werner and I. Perlman, J. Am. Chem. Soc. **73**, 5215 (1951).
Zva 66	I. Zvara, Y.T. Chuburkov, R. Tsaletka, T.S. Zvarova, M.R. Shalaevskii, and B.V. Shilov, Sov. At. Energy **21**, 709 (1966).

Retrospective accounts of the discovery of elements 99–101 can be found in Lawrence Berkeley Laboratory Reports LBL-4366, LBL-7701 and LBL-11599 available from the National Technical Information Service, U.S. Dept. of Commerce, 5285 Port Royal Road, Springfield, VA 22161. Many of the first person narratives of the element discoveries come from these reports.

3

CHEMICAL PROPERTIES

The elements beyond uranium represent a large, diverse group of artificial chemical substances. The placing of these man-made elements in the periodic table represents one of the few significant alterations of the periodic table of Dimitri Mendeleev. The transuranium elements constitute a unique group of chemical species with which we can test our understanding of the most fundamental aspects of chemistry. Since little is known about the chemistry of the transactinide elements, they represent a unique opportunity for chemists to predict an elemental chemistry before the relevant experiments are performed. In this chapter, we shall systematically discuss the principal physical and chemical properties of the transuranium elements. Our emphasis will not be on a detailed description of the properties of each element but instead we shall try to concentrate on the common (and not so common) characteristics of the entire group of heavy, man-made species. Many excellent descriptions of the chemistry of the transuranium elements have been published (KSM 86, Kel 71, FK 86, Ede 82, Sea 63, Cot 88, RA 83) and the reader is referred to them and the references contained therein for the details of the chemistry discussed here.

3.1 THE EVOLUTION OF THE PERIODIC TABLE

During the eighteenth century, about a dozen new chemical elements were discovered, and the atomic theory of matter was born. About 60 more elements were identified in the nineteenth century. In the same period Mendeleev, the great Russian chemist, brought order out of the chaos about the elements with the perfection of his periodic table, giving us, at the same time, the tremendous advantage of being able to predict the properties of the then undiscovered elements.

The periodic table was soon elaborated to show positions for 92 elements. By the middle of the third decade of the present century, all 92 of these elements had been discovered, with the exception of those with the atomic numbers 43, 61, 85, and 87. The state of the understanding of the atomic nucleus was such in the 1930s that it could be shown that the missing elements were all radioactive, with such short half-lives that their existence in appreciable concentrations on the earth was not possible. Figure 3.1 shows the periodic table as it looked before World War II, when scientists first tried to produce elements beyond uranium.

In the 1920s, lanthanum and the rare earth elements were fitted between barium and hafnium, as indeed they are today. Even up until World War II, however, the three heaviest known elements, thorium, protactinium, and uranium, were believed to be related to hafnium, tantalum, and tungsten, respectively. The next element, number 93, was thus expected to have chemical properties resembling those of rhenium. Similarly, elements 94 to 104 were expected to fit into the periodic table in the manner shown in Figure 3.1.

The discovery of an element with atomic number higher than 92 came in 1940 as the result of the work of E.M. McMillan and P.H. Abelson (see Chapter 2). This was followed shortly by the discovery of plutonium by G.T. Seaborg, E.M. McMillan, J.W. Kennedy, and A.C. Wahl, in late 1940. The tracer chemical experiments with neptunium and plutonium showed that their chemical properties were much like those of uranium and not at all like those of rhenium and osmium.

For a few years following this, uranium, neptunium, and plutonium, were considered to be sort of "cousins" in the periodic table, but the family relationship was not clear. It was thought that elements 95 and 96 should be much like them in their chemical properties. Thus it was thought that these elements formed a "uranide" (chemically similar to uranium) group.

The periodic table of 1944 shown in Figure 3.2, therefore implied that the chemical properties of elements 95 and 96 should be very much like those of neptunium and plutonium. These assumptions proved to be wrong, and the experiments directed toward the discovery of elements 95 and 96 failed (see Chapter 2). Again, the undiscovered elements 95 and 96 apparently refused to fit the pattern indicated by the periodic table of 1944.

Then, in 1944, Seaborg conceived the idea that perhaps all the known

1 H																	2 He
3 Li	4 Be											5 B	6 C	7 N	8 O	9 F	10 Ne
11 Na	12 Mg											13 Al	14 Si	15 P	16 S	17 Cl	18 Ar
19 K	20 Ca	21 Sc	22 Ti	23 V	24 Cr	25 Mn	26 Fe	27 Co	28 Ni	29 Cu	30 Zn	31 Ga	32 Ge	33 As	34 Se	35 Br	36 Kr
37 Rb	38 Sr	39 Y	40 Zr	41 Nb	42 Mo	43	44 Ru	45 Rh	46 Pd	47 Ag	48 Cd	49 In	50 Sn	51 Sb	52 Te	53 I	54 Xe
55 Cs	56 Ba	57–71 La–Lu*	72 Hf	73 Ta	74 W	75 Re	76 Os	77 Ir	78 Pt	79 Au	80 Hg	81 Tl	82 Pb	83 Bi	84 Po	85	86 Rn
87	88 Ra	89 Ac	90 Th	91 Pa	92 U	93	94	95	96	97	98	99	100	101	102	103	104

*Lanthanides	57 La	58 Ce	59 Pr	60 Nd	61	62 Sm	63 Eu	64 Gd	65 Tb	66 Dy	67 Ho	68 Er	69 Tm	70 Yb	71 Lu

Figure 3.1 The periodic table of the 1930s; atomic numbers of then undiscovered elements are in shaded squares.

68 CHEMICAL PROPERTIES

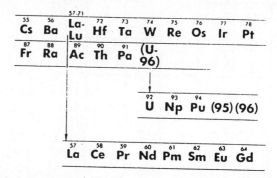

Figure 3.2 The position of the heavy elements in the periodic table in 1944. Atomic numbers of the undiscovered elements are in parentheses.

elements heavier than actinium were misplaced in the periodic table. The theory advanced was that these elements heavier than actinium might constitute a second series similar to the series of rare earth or lanthanide elements. The lanthanides are chemically very similar to each other and usually are listed in a separate row below the main part of the periodic table. This would mean that all these heavier elements really belonged with actinium directly after radium in the periodic table—just as the known lanthanides fit in with lanthanum between barium (Ba) and hafnium (Hf).

The revised periodic table, then listed the heaviest elements as a second rare earth series. These heaviest elements (including undiscovered elements), with the name "actinide" elements, were paired off with those in the already known lanthanide rare earth series, as in Figure 3.3.

The new concept meant that elements 95 and 96 should have some properties in common with actinium and some in common with their rare earth sisters, europium and gadolinium, especially with respect to the difficulty of oxidation above the III state. When experiments were designed according to this new concept, elements 95 and 96 were soon discovered that is, they were synthesized and chemically identified.

Since the elements beyond actinium (through lawrencium, element 103) belong to the actinide group, the elements thorium, protactinium, and uranium, have been removed from the positions they occupied in the periodic table before World War II and placed in this transition family. In the modern periodic table (Figure 3.4 and overleaf) elements 104, 105, and 106 take over the positions previously held by thorium, protactinium, and uranium. Thus we have the interesting result that the newcomers have affected the face of the periodic table, and a change has been made after many years even though it seemed to have assumed its final form.

1 H																1 H	2 He
3 Li	4 Be											5 B	6 C	7 N	8 O	9 F	10 Ne
11 Na	12 Mg	13 Al										13 Al	14 Si	15 P	16 S	17 Cl	18 A
19 K	20 Ca	21 Sc	22 Ti	23 V	24 Cr	25 Mn	26 Fe	27 Co	28 Ni	29 Cu	30 Zn	31 Ga	32 Ge	33 As	34 Se	35 Br	36 Kr
37 Rb	38 Sr	39 Y	40 Zr	41 Cb	42 Mo	43	44 Ru	45 Rh	46 Pd	47 Ag	48 Cd	49 In	50 Sn	51 Sb	52 Te	53 I	54 Xe
55 Cs	56 Ba	57-71 La–Lu*	72 Hf	73 Ta	74 W	75 Re	76 Os	77 Ir	78 Pt	79 Au	80 Hg	81 Tl	82 Pb	83 Bi	84 Po	85	86 Rn
87	88 Ra	89-103 Ac–†	90 Th	91 Pa	92 U	93 Np	94 Pu	95	96								

*Lanthanides	57 La	58 Ce	59 Pr	60 Nd	61	62 Sm	63 Eu	64 Gd	65 Tb	66 Dy	67 Ho	68 Er	69 Tm	70 Yb	71 Lu
†Actinides	89 Ac	90 Th	91 Pa	92 U	93 Np	94 Pu	95	96							

Figure 3.3 The periodic table, as published by G.T. Seaborg, in 1945, shows the heaviest elements as members of an actinide series with atomic numbers of then undiscovered elements in shaded squares.

1 H																	2 He
3 Li	4 Be											5 B	6 C	7 N	8 O	9 F	10 Ne
11 Na	12 Mg											13 Al	14 Si	15 P	16 S	17 Cl	18 Ar
19 K	20 Ca	21 Sc	22 Ti	23 V	24 Cr	25 Mn	26 Fe	27 Co	28 Ni	29 Cu	30 Zn	31 Ga	32 Ge	33 As	34 Se	35 Br	36 Kr
37 Rb	38 Sr	39 Y	40 Zr	41 Nb	42 Mo	43 Tc	44 Ru	45 Rh	46 Pd	47 Ag	48 Cd	49 In	50 Sn	51 Sb	52 Te	53 I	54 Xe
55 Cs	56 Ba	57 La	72 Hf	73 Ta	74 W	75 Re	76 Os	77 Ir	78 Pt	79 Au	80 Hg	81 Tl	82 Pb	83 Bi	84 Po	85 At	86 Rn
87 Fr	88 Ra	89 Ac	104 Rf	105 Ha	106	107	108	109	(110)	(111)	(112)	(113)	(114)	(115)	(116)	(117)	(118)
(119)	(120)	(121)	(154)	(155)	(156)	(157)	(158)	(159)	(160)	(161)	(162)	(163)	(164)	(165)	(166)	(167)	(168)

LANTHANIDES	58 Ce	59 Pr	60 Nd	61 Pm	62 Sm	63 Eu	64 Gd	65 Tb	66 Dy	67 Ho	68 Er	69 Tm	70 Yb	71 Lu	
ACTINIDES	90 Th	91 Pa	92 U	93 Np	94 Pu	95 Am	96 Cm	97 Bk	98 Cf	99 Es	100 Fm	101 Md	102 No	103 Lr	
SUPER-ACTINIDES	(122)	(123)	(124)	(125)	(126)										(153)

XBL 751-2036A

Figure 3.4 The modern periodic table with atomic numbers of the undiscovered elements in parentheses.

3.2 ELECTRONIC STRUCTURES OF THE GASEOUS TRANSURANIUM ATOMS: f ELECTRON CHEMISTRY

3.2.1 Nonrelativistic Orbitals

The actinide and the known transactinide elements are transition elements, that is, they have partly filled f or d electronic orbitals. As such, they are metals (being mostly known or predicted to be hard, strong, high melting and high boiling materials that conduct electricity and heat well). Like other transition metals, most of them are sufficiently electropositive to dissolve in mineral acids. However, there is an important distinction that separates the actinide elements from the other transition elements, including the transactinide elements. The partially filled d orbitals of most transition elements extend out to the boundary of the atoms and are influenced greatly by (or can influence) the chemical environment of the atom or ion. Thus the chemical properties of elements with partly filled d shells are highly complex and seem to vary somewhat irregularly as one passes from element to element. But the 5f orbitals of the actinides are better screened from the chemical environment of the atom or ion by the higher lying s and p shell electrons and thus there is a greater similarity in chemical properties among the actinides compared to other transition elements. (Correspondingly, the 4f orbitals of the lanthanides are even better screened than the 5f actinide orbitals and the chemical behavior of the lanthanides is even more homologous). Because we know so little about the chemistry of the transactinide elements, we shall focus our attention in this chapter largely on the chemistry of the actinide elements. Thus it is important to understand as much as possible about the f electrons in atoms.

While the shapes of the s, p, and d orbitals, as calculated by nonrelativistic quantum mechanics, and their properties are familiar to most students in chemistry, the same cannot be said to be true about the nonrelativistic f orbitals. Therefore, we tabulate in Table 3.1 the angular portions of nonrelativistic f orbital wave functions and show their general shapes in Figure 3.5. The f orbitals are unusual in that no single set of wave functions is useful in all situations. In Table 3.1, we show two commonly used sets: (a) the general set (which simply arises from solving the Schrodinger equation for the hydrogen-like atom) and (b) the cubic set which is derived from three orbitals from the original general set (f_{z^3}, $f_{z(x^2-y^2)}$ and f_{xyz}) and linear combinations of the remaining four orbitals. The cubic set has the desirable property of giving the crystal field splittings in cubic, tetrahedral, and octahedral symmetry (that is, symmetries involving triply degenerate orbitals) while the general set is

TABLE 3.1 Angular Parts of the f Orbitals[a] **(Fri 64)**

a. The general set

True Polynomial	Simplified Polynomial	$Y_{lm}(\theta, \phi)$
$z(5z^2 - 3r^2)$	z^3	$\frac{1}{4}\left(\frac{\sqrt{7}}{\pi}\right)^{1/2} (5\cos^3\theta - 3\cos\theta)$
$x(5z^2 - r^2)$	xz^2	$\frac{1}{8}\left(\frac{\sqrt{42}}{\pi}\right) \sin\theta (5\cos^2\theta - 1)\cos\phi$
$y(5z^2 - r^2)$	yz^2	$\frac{1}{8}\left(\frac{\sqrt{42}}{\pi}\right) \sin\theta (5\cos^2\theta - 1)\sin\phi$
$z(x^2 - y^2)$	–	$\frac{1}{4}\left(\frac{\sqrt{105}}{\pi}\right) \sin^2\theta \cos\theta \cos 2\phi$
xyz	–	$\frac{1}{4}\left(\frac{\sqrt{105}}{\pi}\right) \sin^2\theta \cos\theta \cos 2\phi$
$x(x^2 - 3y^2)$	–	$\frac{1}{8}\left(\frac{\sqrt{70}}{\pi}\right) \sin^3\theta \cos 3\phi$
$y(3x^2 - y^2)$	–	$\frac{1}{8}\left(\frac{\sqrt{70}}{\pi}\right) \sin^3\theta \sin 3\phi$

b. The cubic set

True Polynomial	Simplified Polynomial	$Y_{lm}(\theta, \phi)$
$x(5x^2 - 3r^2)$	x^3	$\frac{1}{4}\left(\frac{\sqrt{7}}{\pi}\right) \sin\theta \cos\phi (5\sin^2\theta\cos^2\phi - 3)$
$y(5y^2 - 3r^2)$	y^3	$\frac{1}{4}\left(\frac{\sqrt{7}}{\pi}\right) \sin\theta \sin\phi (5\sin^2\theta\sin^2\phi - 3)$
$z(5z^2 - 3r^2)$	z^3	$\frac{1}{4}\left(\frac{\sqrt{7}}{\pi}\right) 5\cos^3\theta - 3\cos\theta)$
xyz	–	$\frac{1}{4}\left(\frac{\sqrt{105}}{\pi}\right) \sin^2\theta \cos\theta \sin 2\phi$
$x(z^2 - y^2)$	–	$\frac{1}{4}\left(\frac{\sqrt{105}}{\pi}\right) \sin\theta \cos\phi (\cos^2\theta - \sin^2\theta\sin^2\phi)$
$y(z^2 - x^2)$	–	$\frac{1}{4}\left(\frac{\sqrt{105}}{\pi}\right) \sin\theta \sin\phi (\cos^2\theta - \sin^2\theta\cos^2\phi)$
$z(x^2 - y^2)$	–	$\frac{1}{4}\left(\frac{\sqrt{105}}{\pi}\right) \sin^2\theta \cos\theta \cos 2\phi$

[a] The angular part and the radial part of the wave functions have been separately normalized to 1.

ELECTRONIC STRUCTURES OF GASEOUS TRANSURANIUM ATOMS

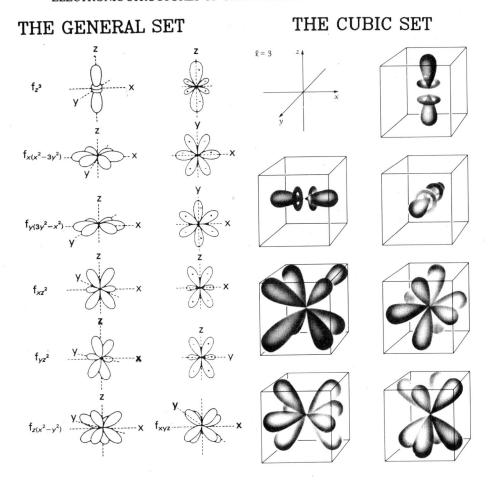

Figure 3.5 Pictorial representation of the nonrelativistic f orbital shapes.

most useful in treating tetragonal trigonal, and other symmetries involving doubly degenerate orbitals.

In the cubic set, the f_{z^3}, f_{x^3}, f_{y^3} are identical orbitals except for their orientation in space (Figure 3.5). The f_{xyz} orbital has eight identical lobes which point towards the corners of a cube centered about the nucleus. The $f_{x(z^2-y^2)}$, $f_{y(z^2-x^2)}$, and $f_{z(x^2-y^2)}$ orbitals are the same except the lobes have been rotated 45° about the x, y and z axes, respectively.

The f_{z^3} orbital of the general set is the same as in the cubic set as are the f_{xyz} and $f_{z(x^2-y^2)}$ orbitals. The "new" orbitals are the f_{xz^2} and f_{yz^2}

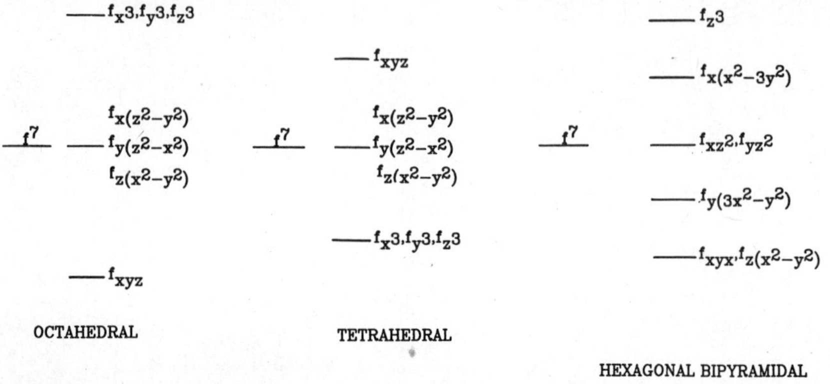

74 CHEMICAL PROPERTIES

```
                    —— f_{x^3}, f_{y^3}, f_{z^3}
                                            —— f_{xyz}                              —— f_{z^3}
                                                                                    —— f_{x(x^2-3y^2)}
          f_{x(z^2-y^2)}                     f_{x(z^2-y^2)}
    f^7 ——f_{y(z^2-x^2)}            f^7 ——f_{y(z^2-x^2)}           f^7             —— f_{xz^2}, f_{yz^2}
          f_{z(x^2-y^2)}                     f_{z(x^2-y^2)}
                                                                                    —— f_{y(3x^2-y^2)}
                                            —— f_{x^3}, f_{y^3}, f_{z^3}
                                                                                    —— f_{xyx}, f_{z(x^2-y^2)}
          —— f_{xyz}

          OCTAHEDRAL                         TETRAHEDRAL
                                                                          HEXAGONAL BIPYRAMIDAL
```

Figure 3.6 Crystal field splitting of the f orbitals in octahedral, tetrahedral, and hexagonal bipyramidal ligand fields.

orbitals which are the same except for the direction of the small lobes. The $f_{x(x^2-3y^2)}$ and $f_{y(3x^2-y^2)}$ orbitals are identical except for their spatial orientation in the xy plane.

The crystal field splittings of the f orbitals in octahedral, tetrahedral, and hexagonal bipyramidal ligand fields are shown in Figure 3.6. In an octahedral field, the $f_{x^3}, f_{y^3}, f_{z^3}$ orbitals point directly towards the ligands and are of highest energy, the $f_{x(z^2-y^2)}, f_{y(z^2-x^2)}$ and $f_{z(x^2-y^2)}$ orbitals point partly towards the ligands and are of next lowest energy while the f_{xyz} orbital lobes do not point directly at any ligand and have the lowest energy. Similar arguments can be used to predict the tetrahedral field splittings with the cubic orbitals. The splittings for the hexagonal bipyramidal field are derived using the general set of orbitals.

The radial probability distributions for the f orbitals are frequently compared to the distributions for other outer orbitals to explain chemical behavior. In Figure 3.7, we show the radial probability distributions for the hydrogen-like atom for the 5f, 6d, 7s, and 7p electrons. The primary f orbital electron density is concentrated well within the principal parts of the 6d, 7s, and 7p electron distributions but there is a significant tail to the density distribution that extends into the outer electron regions. The extent of this tail changes with atomic number for the actinides and is an important feature of their chemistry. The 5f orbitals extend further out relative to the 7s and 7p orbitals than do the lanthanide 4f orbitals relative to the 6s and 6p orbitals. This allows a small amount of covalency from the 5f orbitals whereas this is not observed with the 4f orbitals.

ELECTRONIC STRUCTURES OF GASEOUS TRANSURANIUM ATOMS

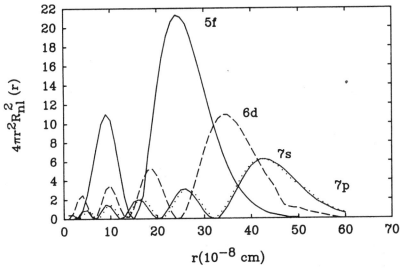

Figure 3.7 Radial probability distributions for the outer valence orbitals of heavy hydrogenlike atoms.

3.2.2 Relativistic Orbitals

While the above discussion allows one to understand the basic properties of the f orbitals in the same nonrelativistic terms that we use to describe the familiar s, p, and d orbitals, it is not a sufficient description to understand the chemistry of the heaviest elements. As the atomic number Z of the nucleus increases, the electrons become more tightly bound. As the binding energy of the electrons increases, so does their velocity with the result that for $Z \sim 90$, the electron velocities exceed $\sim c/2$, i.e., clearly in the relativistic domain. Instead of using the Schrödinger equation, we must use the fully relativistic treatment of Dirac. (Simple reviews of relativistic quantum chemistry are found in Whi 31, Pow 68, Sza 69, McK 83 while more advanced treatments include Bet 57, Pit 79, and Pyy 79).

The solution of the Dirac equation for the hydrogen-like atom leads to eigenfunctions that are products of radial and angular factors although the eigenfunction itself is a four-component column vector. Each state is specified by four quantum numbers which do not have the same meaning as in the Schrödinger atom. They are: n, the principal quantum number taking on values 1, 2, 3, 4 ...; l, the azimuthal quantum number taking on values $0, \ldots, n-1$, denoted by s, p, d, f (BUT l no longer represents the orbital angular momentum); j, the angular momentum quantum number taking on the two values of $l \pm \frac{1}{2}$ (usually denoted as a subscript

76 CHEMICAL PROPERTIES

to l); and m, the magnetic quantum number, taking on half-integer values from $-j$ to $+j$. Thus the three p, five d, and seven f levels are no longer degenerate and split into one $p_{1/2}$, two $p_{3/2}$, two $d_{3/2}$, three $d_{5/2}$, three $f_{5/2}$, and four $f_{7/2}$ levels, respectively, with occupancy of each level being $2j + 1$ (the so-called spin-orbit splitting). The orbital shapes are governed by j and m and orbitals with the same j and m have the same shape. Figure 3.8 shows the predicted shapes of the angular probability distributions associated with the complex eigenfunctions. (Note that no attempt has been made to form linear combinations of the complex eigenfunctions to get real eigenfunctions with useful symmetry properties, as is traditionally done with the Schrödinger orbitals). What is most surprising to the traditional chemist is that the p orbitals are shaped like a sphere ($p_{1/2}$), a doughnut ($p_{3/2}$, $m = 3/2$), and a dog-bone ($p_{3/2}$, $m = 1/2$). Another key fact is that none of the orbitals have nodes in the wave function in that none of the four components of the column vector Ψ ever have nodes in the same place. One also notices that the state with the highest m value

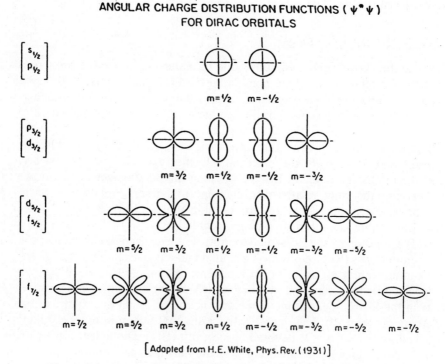

[Adapted from H.E. White, Phys. Rev. (1931)]

Figure 3.8 Pictorial representation of the relativistic orbital shapes.

for a given j always has a doughnut shaped distribution while the lowest value of m corresponds to a distribution stretched along the z-axis with no nodes. States of intermediate m are multi-lobed toroids.

The calculated (Des 73) Dirac–Fock orbital binding energies are presented in Figure 3.9 for the valence electrons of elements 92–110. The 5f and 6d orbitals are sufficiently similar that the exact electron configuration of these elements is determined by interelectronic forces that are not included in the calculations. Thus while the known electron configurations of these elements cannot be predicted straightforwardly from the energy levels shown, the known configurations are at least plausible within the uncertainties (or errors) in the predicted energy eigenvalues. Also, it should be noted that some Dirac–Slater calculations (Fri 77) for $Z \geq 100$ place the 5f orbitals much closer to the 7s, allowing a simple filling of the electronic levels to lead to the known electron configuration. Some authors have also suggested that the $7p_{1/2}$ level may be involved in the valence shell as early as lawrencium (Des 80). In considering the properties of the actinide metals, the overlap between the 5f, 6d, and 7s electrons which is greatest in the light actinides leads to an itinerancy of the 5f electrons as part of the 6d-7s conduction band.

The comparable energies of the 5f, 6d, 7s, and 7p orbitals and their spatial overlap will lead to bonding involving any or all of them. Thus

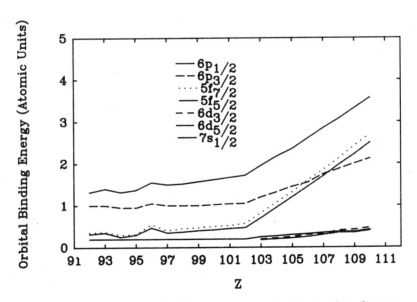

Figure 3.9 Relativistic orbital binding energies for the heaviest elements.

complex formation is an important feature of actinide chemistry. Also since the differences in energy of the electronic levels are similar to chemical bond energies, the most stable oxidation states of the actinides will change from one chemical compound to another and the solution chemistry will be sensitive to the ligands present.

There are three "relativistic effects" seen in the predicted atomic orbitals from the Dirac approach compared to the more familiar Schrödinger solutions. They are: (a) a contraction of the radii of the $s_{1/2}$, and $p_{1/2}$ orbitals (with a concomitant increase in stability) due to the relativistic increase in electron mass, (b) the spin-orbit splitting (which produces two levels for each l), (c) the expansion (and destabilization) of the d and f orbitals due to increased screening effect of the s and p orbitals because of effect (a). Figure 3.10 shows the magnitude of these effects for the uranium atom.

These effects are observable in the chemical behavior of other heavier elements. For example the differences between silver and gold are largely due to relativistic effects. From simple electrostatic arguments, one might expect it to be easier to remove the 6s electron from gold than to remove the 5s electron from silver due to the difference in principal quantum numbers of the electrons. Yet one finds the first ionization potential of gold to be larger than that of silver due to the relativistic contraction of

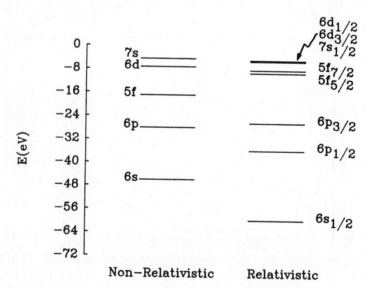

Figure 3.10 A comparison of the predicted (nonrelativistic and relativistic) energies of the valence electronic levels for the uranium atom.

the 6s orbital. (This effect also explains the nobility of gold and the existence of the Au⁻ ion). However, due to the relativistic expansion of the 5d shell, the second ionization potential of gold is less than that of silver. Another effect thought to be largely relativistic in origin is the "inert pair" effect. The heaviest members of Groups IIIA–VIIA exhibit dominant oxidation states that are two less than the lighter members of each group due to "an inert pair" of s electrons. The inertness of this pair of s electrons is due primarily to their increased binding.

3.2.3 Electron Configurations

The atomic spectra of the heaviest elements are quite complex and it is difficult to identify levels in terms of quantum numbers and configurations.

Nonetheless, much work has been done in this area along with studies involving atomic beams and the resulting valence electron configurations of the ground state, gaseous neutral atoms of the actinide elements are shown in Table 3.2, as well as the configurations for the ionized species. For the lighter actinides, the 6d level lies relatively lower than the corresponding 5d level for the lanthanides with a resulting difference in electron configurations. For elements plutonium and beyond, the electron con-

TABLE 3.2 Valence Electron Configurations of f-Block Atoms and Ions[a]

Lanthanide Series			Actinide Series					
Element	Gaseous Atom	$M^{3+}(g)$	Element	Gaseous Atom	$M^+(g)$	$M^{2+}(g)$	$M^{3+}(g)$	$M^{4+}(g)$
La	$5d\,6s^2$		Ac	$6d\,7s^2$	$7s^2$	$7s$		
Ce	$4f\,5d\,6s^2$	$4f$	Th	$6d^2\,7s^2$	$6d\,7s^2$	$5f\,6d$	$5f$	
Pr	$4f^3\,6s^2$	$4f^2$	Pa	$5f^2\,6d\,7s^2$	$5f^2\,7s^2$	$5f^2\,6d$	$5f^2$	$5f$
Nd	$4f^4\,6s^2$	$4f^3$	U	$5f^3\,6d\,7s^2$	$5f^3\,7s^2$	$5f^3\,6d?$	$5f^3$	$5f^2$
Pm	$4f^5\,6s^2$	$4f^4$	Np	$5f^4\,6d\,7s^2$	$5f^5\,7s?$	$5f^5?$	$5f^4$	$5f^3$
Sm	$4f^6\,6s^2$	$4f^5$	Pu	$5f^6\,7s^2$	$5f^6\,7s$	$5f^6$	$5f^5$	$5f^4$
Eu	$4f^7\,6s^2$	$4f^6$	Am	$5f^7\,7s^2$	$5f^7\,7s$	$5f^7$	$5f^6$	$5f^5$
Gd	$4f^7\,5d\,6s^2$	$4f^7$	Cm	$5f^7\,6d\,7s^2$	$5f^7\,7s^2$	$5f^8$	$5f^7$	$5f^6$
Tb	$4f^9\,6s^2$	$4f^8$	Bk	$5f^9\,7s^2$	$5f^9\,7s$	$5f^9$	$5f^8$	$5f^7$
Dy	$4f^{10}\,6s^2$	$4f^9$	Cf	$5f^{10}\,7s^2$	$5f^{10}\,7s$	$5f^{10}$	$5f^9$	$5f^8$
Ho	$4f^{11}\,6s^2$	$4f^{10}$	Es	$5f^{11}\,7s^2$	$5f^{11}\,7s$	$5f^{11}$	$5f^{10}$	$5f^9$
Er	$4f^{12}\,6s^2$	$4f^{11}$	Fm	$5f^{12}\,7s^2$	$(5f^{12}\,7s)$	$(5f^{12})$	$(5f^{11})$	$(5f^{10})$
Tm	$4f^{13}\,6s^2$	$4f^{12}$	Md	$(5f^{13}\,7s^2)$	$(5f^{13}\,7s)$	$(5f^{13})$	$(5f^{12})$	$(5f^{11})$
Yb	$4f^{14}\,6s^2$	$4f^{13}$	No	$(5f^{14}\,7s^2)$	$(5f^{14}\,7s)$	$(5f^{14})$	$(5f^{13})$	$(5f^{12})$
Lu	$4f^{14}\,5d\,6s^2$	$4f^{14}$	Lr	$(5f^{14}\,6d^1\,7s^2)$ or $5f^{14}\,7s^2\,7p^1)$	$(5f^{14}\,7s^2)$	$(5f^{14}\,7s)$	$(5f^{14})$	$(5f^{13})$

[a]Predicted configurations in parentheses.

figurations of the lanthanide and actinide elements resemble each other. As mentioned previously, relativistic effects may cause the lowering of the $7p_{1/2}$ level to the point where lawrencium has an electron configuration $5f^{14} 7p_{1/2} 7s^2$ rather than $5f^{14} 6d 7s^2$. This might lead to the stabilization of oxidation states below the III state, but attempts to establish the existence of such states for lawrencium have not been successful (Jos 88). An upper limit for the reduction potential of $E_0 \leq -0.44$ V for the Lr^{3+}/Lr^{1+} half reaction has been determined (Sch 88b).

The expected gaseous atom electron configurations of the transactinide elements are shown in Table 3.3 (Fri 75). The crucial point is that the electron configurations (and thus, the chemistry) of rutherfordium, hahnium, 106, et cetera, are predicted to be different than that of the heaviest actinide elements due to the absence of valence f orbitals in the transactinides. One expects a 6d series from rutherfordium to 112 analogous to the 5d series from hafnium to mercury.

In analogy with the situation with lawrencium, rutherfordium is predicted due to relativistic effects to have the ground state electron configuration [Rn] $5f^{14} 6d 7s^2 7p$ (Gle 89). Experiments designed to search for this configuration have not yielded any evidence for element 104 being a p-block element (Zhu 89).

Despite the utility of the gaseous atom electron configurations in rationalizing the chemistry of the elements, one must remember that the electron configurations in metals, for example, are different from those in the free atom. (This difference might reflect the atomic geometry in metals, the need to bond to other atoms and resulting preference for configurations that are not spherically symmetric.) Brewer has calculated for example, the bonding electron configurations for the metals plutonium through lawrencium as $f^5 d^{1.5} sp^{0.5}$, $f^6 d^1 sp^1$, $f^7 d^1 sp^1$, $f^8 d^1 sp^1$, $f^9 d^1 sp^1$, $f^{10} d^1 sp^1$, $f^{11} d^1 sp^1$, $f^{13} sp^1$, $f^{14} sp^1$, and $f^{14} d^1 s^1 p^1$ (Bre 71). Thus the 6d and 7p orbitals may be playing important roles in the hybrid orbitals being used in bonding in metals although their role would not be clear from the gaseous atom electron configurations.

3.3 IONIC RADII

One of the most famous examples of the actinide concept in describing heavy element chemistry is the *actinide contraction* analogous to the lanthanide contraction. The radii of the M^{3+} and M^{4+} ions (Table 3.4) are observed to decrease with increasing positive charge of the nucleus. This contraction is a consequence of the addition of successive electrons to an inner f electron shell, so that the imperfect screening of the increasing

TABLE 3.3 Dirac–Fock Ground-State Configurations of Free Neutral Atoms of Elements 104–168 (Fri 75)

	Element Rn "core" + 5f¹⁴ +										Element 118 "core" +						
	5g	6d	6f	7s	7p$_{1/2}$	7p$_{3/2}$	7d	8s	8p$_{1/2}$		5g	6f	7d	8s	8p$_{1/2}$	9s	9p$_{1/2}$
Rf		2		2						137	11	3	1	2	2		
Ha		3		2						138	12	3	1	2	2		
106		4		2						139	13	2	2	2	2		
107		5		2						140	14	3	1	2	2		
108		6		2						141	15	2	2	2	2		
109		7		2						142	16	2	2	2	2		
110		8		2						143	17	2	2	2	2		
111		9		2						144	18	1	3	2	2		
112		10		2						145	18	3	2	2	2		
113		10		2	1					146	18	4	2	2	2		
114		10		2	2					147	18	5	2	2	2		
115		10		2	2	1				148	18	6	2	2	2		
116		10		2	2	2				149	18	6	3	2	2		
117		10		2	2	3				150	18	6	4	2	2		
118		10		2	2	4				151	18	8	3	2	2		
119		10		2	2	4		1		152	18	9	3	2	2		
120		10		2	2	4		2		153	18	11	2	2	2		
121		10	1	2	2	4		2		154	18	12	2	2	2		
122		10	1	2	2	4	1	2		155	18	13	2	2	2		
123		10	3	2	2	4	1	2		156	18	14	2	2	2		
124		10	3	2	2	4		2		157	18	14	3	2	2		
125	1	10	2	2	2	4		2	1	158	18	14	4	2	2		
126	2	10	2	2	2	4		2	1	159	18	14	4	2	2	1	
127	3	10	2	2	2	4		2	1	160	18	14	6	2	2	1	
128	4	10	2	2	2	4	1	2	1	161	18	14	5	2	2	1	
129	5	10	2	2	2	4		2	2	162	18	14	6	2	2		
130	6	10	2	2	2	4		2	2	163	18	14	8	2	2		
131	7	10	2	2	2	4		2	2	164	18	14	9	2	2		
132	8	10	2	2	2	4		2	2	165	18	14	10	2	2	1	
133	8	10	3	2	2	4		2	2	166	18	14	10	2	2	2	
134	8	10	4	2	2	4		2	2	167	18	14	10	2	2	2	1
135	9	10	4	2	2	4		2	2	168	18	14	10	2	2	2	2
136	10	10	4	2	2	4		2	2								

CHEMICAL PROPERTIES

TABLE 3.4 Ionic Radii of Actinide and Lanthanide Elements (KSM 86)

No. of 4f or 5f Electrons	Lanthanide Series				Actinide Series			
	Element	Radius (nm)	Element	Radius (nm)	Element	Radius (nm)	Element	Radius (nm)
0	La^{3+}	0.1061			Ac^{3+}	0.1119	Th^{4+}	0.0972
1	Ce^{3+}	0.1034	Ce^{4+}	0.092	(Th^{3+})	(0.108)	Pa^{4+}	0.0935
2	Pr^{3+}	0.1013	Pr^{4+}	0.090	(Pa^{3+})	(0.105)	U^{4+}	0.0918
3	Nd^{3+}	0.0995			U^{3+}	0.1041	Np^{4+}	0.0903
4	Pm^{3+}	(0.0979)			Np^{3+}	0.1017	Pu^{4+}	0.0887
5	Sm^{3+}	0.0964			Pu^{3+}	0.0997	Am^{4+}	0.0878
6	Eu^{3+}	0.0950			Am^{3+}	0.0982	Cm^{4+}	0.0871
7	Gd^{3+}	0.0938			Cm^{3+}	0.0970	Bk^{4+}	0.0860
8	Tb^{3+}	0.0923	Tb^{4+}	0.084	Bk^{3+}	0.0949	Cf^{4+}	0.0851
9	Dy^{3+}	0.0908			Cf^{3+}	0.0934		
10	Ho^{3+}	0.0894			Es^{3+}	0.0925		
11	Er^{3+}	0.0881						
12	Tm^{3+}	0.0869			Md^{3+}	0.0896^a		
13	Yb^{3+}	0.0858						
14	Lu^{3+}	0.0848			Lr^{3+}	0.0882^a		

^{a}Brü 88.

nuclear charge by the additional f electron results in a contraction of the outer or valence orbital. Since the ionic radii are generally similar for ions of the same oxidation state, one expects the ionic compounds to be isostructural. The observed decrease in actinide ionic radii with increasing atomic number is in good agreement with recent, multidimensional Dirac–Fock calculations (Des 84).

3.4 OPTICAL AND MAGNETIC PROPERTIES (KSM 86, Hes 80)

As mentioned previously, the atomic emission spectra of excited heavy atoms are very complex; 100,000 lines have been measured for uranium and 5000–20,000 lines for each of the elements from plutonium to berkelium. Of these only 2500 have been assigned for uranium and about 100 for curium. This complexity is due to the interaction of the valence electrons with the 5f electrons and with each other and the resulting splitting of the energy levels.

The characteristic colors of the common aqueous actinide cations under "average" conditions are shown in Table 3.5. A knowledge of these colors will aid the chemist in diagnosing what is happening in a complex chemical

TABLE 3.5 Ion Types and Colors[a] for Actinide Ions

Element	M^{3+}	M^{4+}	MO_2^+	MO_2^{2+}	MO_5^{3-}
Actinium	Colorless				
Thorium		Colorless			
Protactinium		Colorless	Colorless		
Uranium	Red	Green	Color unknown	Yellow	
Neptunium	Blue to purple	Yellow-green	Green	Pink to red	Dark green
Plutonium	Blue to violet	Tan to orange-brown	Reddish-purple	Yellow to pink-orange	Dark green
Americium	Pink or yellow	Color unknown	Yellow	Rum-colored	Color unknown
Curium	Pale green	Color unknown			
Berkelium	Green	Yellow			
Californium	Green				

[a] For a photograph showing many of these colors, see G.T. Seaborg, The Transuranium Elements (Yale University Press, New Haven, CT, 1958) p. 122.

84 CHEMICAL PROPERTIES

procedure involving the actinides. The wide range of observed colors is typical of transition metals such as the actinides.

The magnetic properties of the actinide ions and compounds, which arise from the effects of the spin and orbital angular momenta of the unpaired electrons, are difficult to interpret. One does observe magnetic moments that are less than those predicted from Russell–Saunders coupling, giving evidence for the expected importance of j–j coupling in these heavy atoms.

3.5 OXIDATION STATES

The actinides show an unusually broad range of oxidation states, ranging from +2 to +7 in solution. The known oxidation states are shown in Table 3.6. The most common oxidation state is +3 for the transplutonium actinide elements similar to the lanthanides with +2 oxidation states being observed for the heaviest species. A stable +4 state is observed for the elements thorium through plutonium and berkelium. The +5 state is well established for protactinium through americium and the +6 state for uranium through americium. Following the normal trend for polyvalent cations, lower oxidation states are stabilized by acid conditions while the higher oxidation states are more stable in basic solutions. Complexing, can, of course, change this general trend.

In compounds of the +2, +3 and +4 oxidation states, the elements are present as simple M^{+2}, M^{+3} or M^{4+} cations but for higher oxidation

TABLE 3.6 The Oxidation States of the Actinide Elements (KSM 86)[a]

Atomic Number:	89	90	91	92	93	94	95	96	97	98	99	100	101	102	103
Element:	Ac	Th	Pa	U	Np	Pu	Am	Cm	Bk	Cf	Es	Fm	Md	No	Lr
Oxidation States															
														1?	
							(2)	(2)		(2)	(2)	2	2	2	
	3	(3)	(3)	3	3	3	3	3	3	3	3	3	3	3	3
		4	4	4	4	4	4	4	4	4	(4) 4?				
			5	5	5	5	5	5?		5?					
				6	6	6	6	6?							
					7	(7)	7?								

[a] The most common oxidation states are underlined, unstable oxidation states are shown in parentheses. Question marks indicate species that have been claimed but not substantiated.

states, the most common forms in compounds and in solution are the oxygenated actinyl ions MO_2^+ and MO_2^{2+}.

It is interesting to consider why, in view of the ground state gaseous atom electron configurations of $5f^n 7s^2$, the most stable oxidation state for the heavier actinides is +3. We can understand the issues involved by considering the stability of the +3 ion relative to the +2 and +4 ions. The heats of formation of the ions, ΔH_f, are given as

$$M(s) \xrightarrow{\Delta H_{sub}} M(g) \xrightarrow{I_1+I_2} M(g)^{2+} \xrightarrow{\Delta H_2} M^{2+}(aq)$$
$$\Delta H_f^{2+} = I_1 + I_2 + \Delta H_2 + \Delta H_{sub}$$

$$M(s) \xrightarrow{\Delta H_{sub}} M(g) \xrightarrow{I_1+I_2+I_3} M(g)^{3+} \xrightarrow{\Delta H_3} M^{3+}(aq)$$
$$\Delta H_f^{3+} = I_1 + I_2 + I_3 + \Delta H_3 + \Delta H_{sub}$$

$$M(s) \xrightarrow{\Delta H_{sub}} M(g) \xrightarrow{I_1+I_2+I_3+I_4} M(g)^{4+} \xrightarrow{\Delta H_4} M^{4+}(aq)$$
$$\Delta H_f^{4+} = I_1 + I_2 + I_3 + I_4 + \Delta H_4 + \Delta H_{sub}$$

where I_n, ΔH_n are the nth ionization potential and the heat of hydration of the $+n$ ion, respectively and ΔH_{sub} is the heat of sublimation. If the +3 ion is to be stable with respect to the +2 ion or +4 ion, the following ΔH values must be negative (neglecting the entropy term $T\Delta S$ as ~3% of ΔH).

$$\Delta H^{2 \to 3} = \Delta H_f^{3+} - \Delta H_f^{2+} = I_3 + \Delta H_3 - \Delta H_2$$
$$\Delta H^{4 \to 3} = \Delta H_f^{3+} - \Delta H_f^{4+} = \Delta H_3 - \Delta H_4 - I_4$$

Thus the oxidation ($2 \to 3$) is exothermic if $|\Delta H_3 - \Delta H_2| > |I_3|$. Because the M^{3+} cations would be very much smaller than the M^{2+} cations, their heats of hydration will be much larger, leading to the fulfillment of the requirement. Similarly the difference $|\Delta H_4 - \Delta H_3|$ is expected to be such as to cause $\Delta H_3 - \Delta H_4 - I_4$ to be negative.

Disproportionation reactions are an important part of the chemistry of the heaviest elements, especially for the +4 and +5 oxidation states. A few comments on the various oxidation states found in solution follow.

M(I)   The existence of Md^+ has been reported by Mikheev et al. (Mik 73) but studies by other groups have failed to confirm these observations (Hul 80).

$M(II)$ Both No^{2+} and Md^{2+} form readily in solution. No^{2+} is the most stable oxidation state of nobelium, a result of the stability of the completely filled $5f^{14}$ shell. Pulse radiolysis of acid solutions has been used to make unstable Am^{2+}, Cm^{2+}, Bk^{+2}, Cf^{2+}, and Es^{2+}. Fm^{2+} is more stable than these ions but less stable than Fm^{3+}.

$M(III)$ As mentioned previously, the +3 oxidation state is the most stable oxidation state for americium through mendelevium and for lawrencium. It is easy to produce Pu^{3+} and Np^{3+} but U^{3+} is such a strong reducing agent that it is difficult to keep in solution.

$M(IV)$ The most stable oxidation state of thorium is +4. Pa^{4+}, U^{4+}, and Np^{4+} are stable but are oxidized by O_2. Pu^{4+} is stable in acid solutions with low plutonium concentration. Americium, curium, and californium can be oxidized to the +4 state with strong oxidizing agents like persulfate, pulse radiolysis or other strong oxidation and complexation techniques. Bk^{4+} is relatively stable as a result of the half-filled shell, $5f^7$.

$M(V)$ The actinides protoactinium through americium form MO_2^+ ions in solution. PuO_2^+ may be the dominant species in solution in low concentrations in natural waters that are relatively free of organic materials.

$M(VI)$ UO_2^{2+} is the most stable oxidation state of uranium. Neptunium, plutonium and americium form MO_2^{2+} ions in solution with the stability ordering being $U > Pu > Np > Am$.

$M(VII)$ M(VII) species, in oxygenated form such as MO_5^{3-} have been reported for neptunium, plutonium, and americium but are unstable.

The redox potentials of the actinide elements (at zero ionic strength) are shown in Figure 3.11 (KSM 86). The overall stability of the +3 oxidation state is well summarized in the plots of $nE°(= -\Delta G°/RT)$ in Figure 3.12. With increasing Z, one observes that the electropositive character of the metals increases and the stability of higher oxidation states decreases. The reactions which do not involve the making or breaking of chemical bonds, that is, $M^{3+} \rightarrow M^{4+}$, $MO_2^+ \rightarrow MO_2^{2+}$, are fast and reversible while the reactions that involve chemical bond formation, $M^{3+} \rightarrow MO_2^+$, $M^{3+} \rightarrow MO_2^{2+}$, $M^{4+} \rightarrow MO_2^+$, $M^{4+} \rightarrow MO_2^{2+}$ are slow and are not reversible.

The oxidation states of plutonium deserve some special discussion, because the oxidation-reduction relationships are among the most intricate known to chemistry. The proximity to each other of the oxidation-reduction potentials in acidic solution signifies that appreciable amounts of three or four oxidation states can exist together at equilibrium. The most

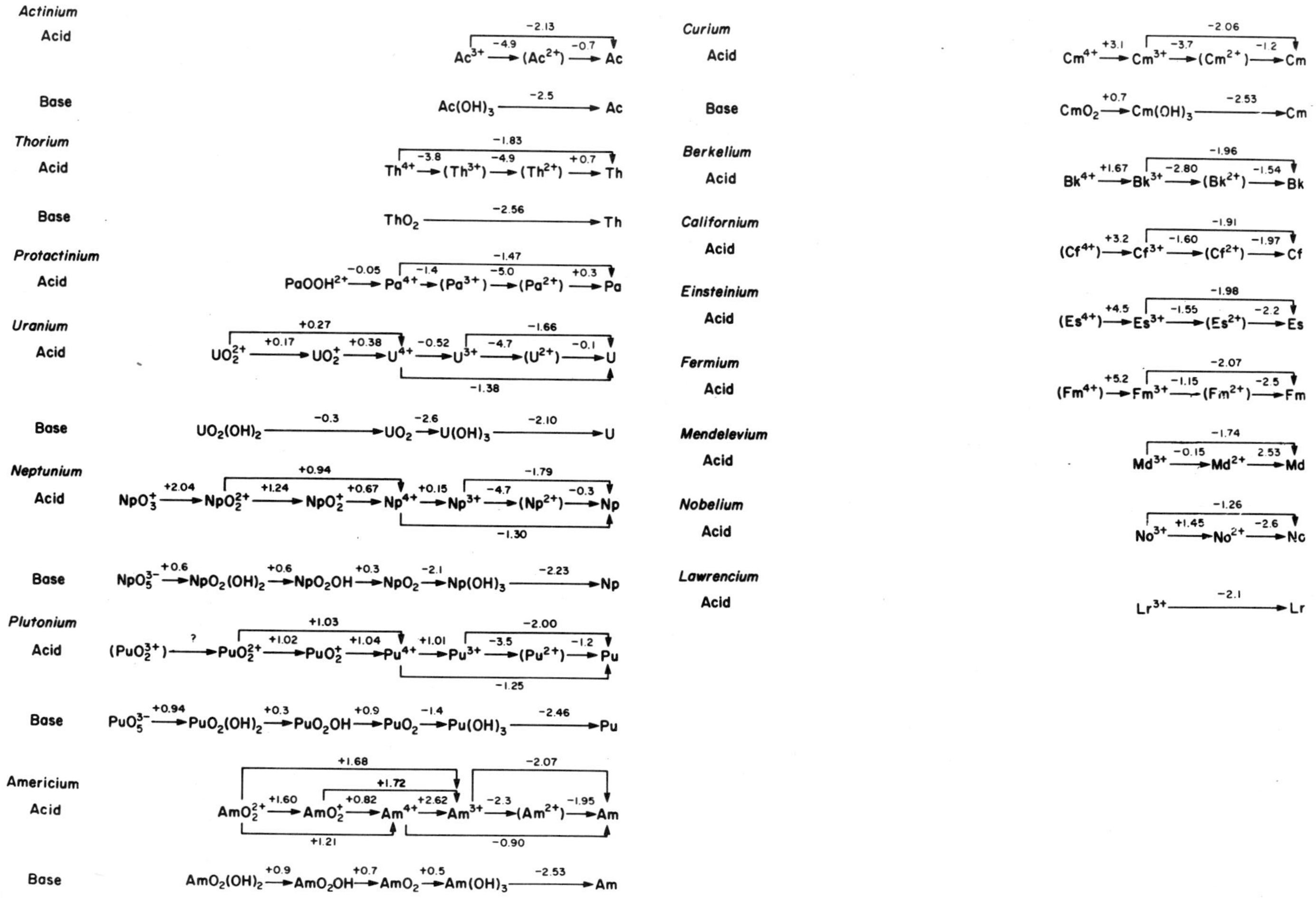

Figure 3.11 Standard reduction potentials for actinium and the actinide ions in acidic (pH = 0) and basic (pH = 14) aqueous solutions. Potentials are given in volts versus the standard hydrogen electrode (KSM 86).

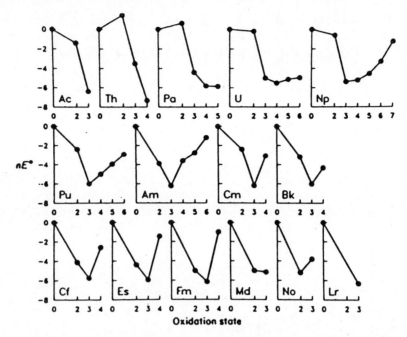

Figure 3.12 Comparative stability of the actinide aquo ions (KSM 86).

important equilibrium is that represented by the reaction:

$$3\,\text{Pu (IV)} = 2\,\text{Pu (III)} + \text{Pu (VI)}$$

For example, in 0.5 M HCl at 25°C the equilibrium percentages of plutonium in the various oxidation states are found to be 27.2% Pu (III), 58.4% Pu (IV), 13.6% Pu (VI), and 0.75% Pu (V). For 0.1 M perchloric acid at 25°C the value of K defined by

$$K = \frac{[\text{Pu (VI)}][\text{Pu (III)}]}{[\text{Pu (V)}][\text{Pu (IV)}]}$$

has been found to be 10.7, illustrating the existence of appreciable concentrations of all four ionic species. The reaction apparently may be written as

$$\text{Pu}^{4+} + \text{PuO}_2^+ = \text{PuO}_2^{2+} + \text{Pu}^{3+}$$

indicating the absence of any acidity effect on the equilibrium. Equilibrium is established rapidly, since only electron transfer and no breaking of plutonium-oxygen bonds is involved. Under conditions where this equilibrium involving four oxidation states is established rapidly, the equilibrium

$$3\ \text{Pu (IV)} = 2\ \text{Pu (III)} + \text{Pu (VI)}$$

is established only slowly because this involves the breaking of plutonium-oxygen bonds. Thus the rather striking condition can exist in which all four of the oxidation states are in equilibrium with respect to one reaction, whereas three oxidation states are not in equilibrium with respect to another reaction. The complexity of the oxidation-reduction relationships is further illustrated by a consideration of the disproportionation mechanisms of PuO_2^+ and Pu^{4+}, which are slow reactions due to the breaking of plutonium-oxygen bonds. Both disproportionations apparently involve the reaction

$$\text{Pu (IV)} + \text{Pu (IV)} \rightleftarrows \text{Pu (III)} + \text{Pu (V)} \qquad (1)$$

At low concentrations of Pu (III) the disproportionation of Pu (V) apparently proceeds through the reaction

$$\text{Pu (V)} + \text{Pu (V)} \rightleftarrows \text{Pu (IV)} + \text{Pu (VI)} \qquad (2)$$

However, in the presence of Pu (III), which may appear, for example, as the product of the oxidation of Pu (V) by Pu (IV), reaction 1, proceeding as written to the left, becomes the rate-determining step. The mechanism for the disproportionation of Pu (IV) also apparently involves reaction 1 proceeding as written to the right, with the additional rapid oxidation of Pu (V) to Pu (VI) by Pu (IV).

3.6 HYDROLYSIS

All metal cations in aqueous solution undergo hydrolysis and exist to some extent as hydrated species. The more highly charged the metal cation, the stronger the interaction with water. Aquo cations, especially the +3 and +4 ions, tend to act as Bronsted acids in solution:

$$M(H_2O)_x^{n+} = M(H_2O)_{x-1}(OH)^{(n-1)+} + H^+$$

The acidity of such aquo cations increases with increasing charge on the metal cation and therefore it is not surprising that the transuranium elements undergo extensive hydrolysis. The +4 cations, having the greatest charge/radius ratio undergo hydrolysis most readily. Uranium (IV) undergoes hydrolysis in solutions with pH > 2.9 with $U(OH)_3^+$ being the principal hydrolyzed species. Pu (IV) hydrolyzes extensively in moderately acid solutions and may form polymers of molecular weights as high as 10^{10}. These positively charged polymers of colloidal dimensions have created serious problems for the plutonium processing industry. (Their capricious behavior and their slow approach to equilibrium make it difficult to determine the concentration and nature of species present). Immediately after formation, such polymers can be readily decomposed to simpler species by acidification or oxidation. However, if the polymers are allowed to "age", then irreversible changes take place which makes it very difficult to decompose the resulting polymeric species. It is thought that this aging process involves the replacement of hydroxy bridges between the Pu atoms by oxo bridges.

The actinyl ions, MO_2^+ and MO_2^{2+}, are less acidic than the +4 ion and therefore have less tendency to undergo hydrolysis (hydrolysis decreases in the order $M^{4+} > MO_2^{2+} > M^{3+} > MO_2^+$).

3.7 COMPLEX IONS

Hydrolysis is a special type of complex ion formation. The large positive charge associated with transuranium cations that leads to hydrolysis is also the driving force for the interaction of nucleophiles with the transuranium cations. Water is only one example of a nucleophilic ligand. Other nucleophilic ligands present in solution may replace water molecules directly bound to the metal cation to form inner sphere complexes or alternatively, they may displace water molecules only from the outer hydrate shell to form outer sphere complexes.

In view of the competition between water and other ligands for positions in the inner coordination sphere of the central transuranium atom, it is not surprising that the stability of the complexes formed with a given ligand, decreases, in the order $M^{+4} > MO_2^{2+} > M^{+3} > MO_2^+$. (Note that the strength of complexation does not depend simply on the net cation charge, but rather the charge density seen by the anion or ligand as it approaches the metal. In the case of MO_2^{2+}, the effective charge is about +3.3 rather than +2.) Although there is some variation within the given cationic types, the general order of complexing power of different anions is $F^- > NO_3^- > Cl^- > ClO_4^-$ for singly charged anions and $CO_3^{2-} > C_2O_4^{2-} > SO_4^{2-}$ for doubly charged anions.

The actinide cations are "hard acids", that is, their bonding to ligands is described in terms of electrostatic interactions, and they prefer to interact with hard bases such as oxygen or fluorine rather than softer bases such as nitrogen or phosphorus or sulfur. The actinide cations do form complexes with the soft bases but only in non-aqueous solvents.

As typical hard acids, the stabilities of the actinide complexes are due to favorable entropy effects. The enthalpy terms are either endothermic or very weakly exothermic and are of little importance in determining the overall position of the equilibrium in complex formation.

The formation of complexes could be thought to be a three step process:

$$M(aq) + X(aq) \rightleftarrows [M(H_2O)_nX](aq) \rightleftarrows [M(H_2O)X]aq \rightleftarrows MX(aq)$$

The first step is diffusion controlled while in the second step an "outer sphere" complex is formed with at least one solvent water molecule intervening between the ligand and the metal atom. In the third or rate-determining step, a direct connection between the metal and ligand is established with the formation of an "inner sphere" complex. The process could terminate after the second step if the ligand cannot displace the solvent water. Actinides form both inner and outer sphere complexes although in most cases, the stronger inner sphere complexes are formed. The halide, nitrate, sulfonate, and trichloroacetate form outer sphere complexes of the trivalent actinides while fluoride, iodate, sulfate, and acetate form inner sphere complexes.

The actinide-ligand bonds in complexes are predominantly ionic in character. Because of the ionic nature of the bonding, the stereochemistry of actinide complexes is different than the stereochemistry of d block transition elements where covalent bonding dominates. For the +3 oxidation state, octahedral coordination is often found, but higher coordination numbers are also observed. The +4 oxidation state is frequently associated with eight-coordination. The MO_2^{2+} ion is linear with the structure $[O{=}M{=}O]^{2+}$. Planar 5-coordination appears to be the proper description of a number of MO_2^{2+} complexes with the ligand atoms lying in the equatorial plane of the O—M—O group.

The phosphate anion PO_4^{3-} and organic phosphates are powerful complexing agents for actinide ions. The phosphate anion acts as a bridge between metal ions to form aggregates that are insoluble in water. The M^{4+} and MO_2^{2+} ions form complexes with many organic phosphates, either neutral or anionic, that are preferentially soluble in nonpolar aliphatic hydrocarbons. Typical of such ligands are tributyl phosphate (TBP) and dibutyl phosphate (DBP). Phosphine oxides are also potent coordinating ligands. Oxygen-containing donor compounds such as the ketones, diisopropyl ketone or methyl isobutyl ketone, and the ethers, diethyl ether, ethyleneglycol diethyl ether or diethyleneglycol dibutyl ether, act likewise and are good complexing agents for actinide ions. All of these ligands have oxygen atoms with lone electron pairs not otherwise engaged in chemical bonding that can act as electron donors in coordination interactions.

Chelating ligands form strong complexes with actinide ions. Examples of such are the β-diketones, 8-hydroxyquinoline and its derivatives, and ethylenediaminetetraacetic acid (EDTA), among many others. In its enol form, acetylacetone, a typical diketone, forms very strong metal complexes with M^{4+} ions. Even though these complexes have significant water solubility, they are easily and completely extracted by benzene, carbon tetrachloride or similar nonpolar solvents. The acetylacetone complexes of the MO_2^{2+} actinyl ions form weaker complexes that show little preference for a nonpolar organic phase. The structure of the diketone can be modified to enhance the preferential solubility of the metal complex for the organic phase. The most important of such modified β-diketone chelating agents is 2-thenoyltrifluoroacetone (TTA,

$$\underset{\diagdown\diagup}{S}\text{—C—CH}_2\text{—}\underset{\|}{\overset{\overset{O}{\|}}{C}}\text{—CF}_3)$$

which has been widely used to extract plutonium from aqueous solutions into nonpolar solvents. Ethylenediaminetetraacetic acid (EDTA) and diethylenetriaminepentaacetic acid (DTPA) are effective sequestering agents for the actinide ions in aqueous solutions. The strongest complexes are formed by M^{4+}. The strength of the EDTA complexes with M^{3+} ions increases steadily from Pu^{3+} to Cf^{3+}. Possibly for steric reasons, EDTA interacts in a different way with actinyl (VI) ions; in these systems EDTA is a bridging ligand, and gives rise to linear polynuclear complexes. The stability constants for the DTPA complexes are about four orders of magnitude greater than those of the EDTA complexes. DTPA is the reagent of choice for preferentially removing actinide ions from biological systems although the linear sulfonated catechoylamide-3,4,3-LICAMS has also proved effective in this regard.

3.8 CHEMICAL SEPARATIONS OF THE TRANSURANIUM ELEMENTS

Frequently it is necessary to chemically separate the transuranium elements from each other and from the intensely radioactive fission products which frequently accompany the transuranium elements during their production. The procedures designed to effect these separations have served as unique means of identifying the atomic number of the transuranium nuclei, an important criterion for the discovery of a new element. These chemical separation procedures have been of great practical importance in the development of nuclear reactors and nuclear weapons. They have continuing importance in reducing the hazards associated with the radioactive waste from nuclear power plants, although it should be noted that at the time of this writing (1989), no chemical reprocessing of spent fuel from commercial nuclear power plants takes place in the United States.

Procedures designed to separate one transuranium element from all others and the fission products such as the tripositive lanthanide elements generally take advantage of the ability to place the desired element in a unique oxidation state which allows selective extraction, complexation, et cetera. The multiple oxidation states and complex chemistry of the actinide elements are, in this sense, blessings in disguise in that the chemical separation of these elements would be far more difficult without the existence of these "complications."

3.8.1 The Bismuth Phosphate Process (Sea 58)

The first chemical separation procedures for the transuranium elements were laboratory procedures involving the use of carriers and precipitation.

The first large-scale actinide separation process, the bismuth phosphate process, was developed during World War II. The problem confronting people at that time was to find a means of separating plutonium in high yield and purity from the many tons of uranium in which plutonium was present at a maximum concentration of 250 ppm. Because of this low concentration, compounds of plutonium could not be precipitated directly, and any precipitation-separation process had to be based upon coprecipitation phenomena, that is the use of so-called "carriers" for plutonium. At the same time, the radioactive fission products produced along with plutonium in the uranium (as a result of the fission of ^{235}U) had to be separated so that less than one part in 10^7 parts originally present with the plutonium would exist with the final product from the process. This requirement was necessary in order to make it safe to handle the plutonium, for without a separation of the fission products, the plutonium from each ton of uranium would have more than 10^5 Ci of energetic gamma radiation associated with it.

S. G. Thompson is largely responsible for the conception and early development of the process which was finally chosen. The key to the process is the quantitative carrying of plutonium (IV) from acid solution by bismuth (III) phosphate, an unexpected phenomenon, and the expected noncarrying of plutonium (VI) by the same carrier material. This method, operates as follows: neutron-irradiated uranium is dissolved in nitric acid, and, after the addition of sulfuric acid to prevent the precipitation of uranium, plutonium (IV) is coprecipitated with bismuth phosphate. The precipitate is dissolved in nitric acid, the plutonium (IV) is oxidized to plutonium (VI), and a by-product precipitate of bismuth phosphate is formed and removed, the plutonium (VI) remaining in the solution. After the reduction of plutonium (VI) to plutonium (IV), the latter is again coprecipitated with bismuth phosphate, and the whole "decontamination cycle" is repeated. At this point the carrier is changed to lanthanum fluoride, and a similar "oxidation-reduction cycle" is performed, using this carrier, thereby achieving further decontamination and concentration. The plutonium at this point is sufficiently concentrated that final purification can be accomplished without the use of carrier compounds and plutonium peroxide is precipitated from acid solution.

The overall recovery of plutonium in this process exceeded 95% and the overall decontamination from fission products was 10^7. This process was developed at the wartime Metallurgical Laboratory of the University of Chicago, demonstrated at the X-10 pilot plant at Oak Ridge in 1944 and put in operation for the large-scale recovery of plutonium at Hanford in early 1945. The process suffers from its batch nature, the large amounts of chemicals used and waste generated and the inability to recover uranium.

3.8.2 Solvent Extraction (Ahr 73)

The next general type of chemical separation process applied to the problem of actinide separations was that of solvent extraction, a technique that is still of interest, both on a laboratory and an industrial scale. In solvent extraction, the species to be separated is caused to transfer between two immiscible or partially miscible phases, such as water and a nonpolar organic phase. The transfer is effected by selectively complexing the species of interest causing its solubility in water to decrease with a concomitant increase in its solubility in the organic phase.

The hydrated actinide metal ion (M^{Z+}) will always prefer the aqueous phase to the organic phase due to hydrogen bonding and dipole action in the organic phase. To get the metal ion to extract, some or all of the inner hydration sphere must be removed. The resulting complex must be electrically neutral and organophilic, that is, have an organic "surface" that interacts with the organic solvent. This can be done by:

  a. forming a neutral complex MA_Z by coordination with organic anions A^-

b. or replacing water in the inner coordination sphere by large organic molecules B such that one forms MB_N^{Z+} which is extracted into the organic phase as an ion-association complex $(MB_N)^{Z+}L_Z^{Z-}$

c. or forming metal complexes of form ML_N^{Z-N} with ligands (L) such that they combine with large organic cations RB^+ to form ion pair complexes $(RB^+)_{N-Z}(ML_N)^{N-Z}$

The extracting agents are thus divided into three classes, polydentate organic anions A^-, neutral organic molecules B or large organic cations RB^+.

Polydentate organic anions, which form chelates (ring structures of 4–7 atoms) with the actinides, are important extracting agents. Among these are the β-diketonates, such as acetylacetonate, the pyrazolones, benzoylacetonate, and thenoyltrifluoroacetone (TTA), with the extraction increasing strongly through this sequence. Representing the organic chelating agent as HA, the overall reaction involved in the chelate extraction of a metal ion, M^{n+}, is

$$M^{n+}(aq) + nHA(o) \rightleftarrows MA_n(o) + nH^+(aq)$$

When an aqueous solution containing extractable metal ions is brought into contact with an organic phase containing chelating agent, the chelating agent dissolves in the water phase, ionizes, complexes the metal ion, and the metal chelate dissolves in the organic phase. The low solubility of the metal complexes and their slow rates of formation limit the industrial use of this type of anionic extraction.

However, a number of organophosphorus compounds are efficient extractants as they and their complexes are very soluble in organic solvents. The most important of these are monobasic diethylhexylphosphoric acid (HDEHP) and dibutylphosphoric acid (HDBP). The actinide MO_2^{2+} ions are very effectively extracted by these reagents as are the actinide (IV) ions.

Among the neutral extractants, alcohols, ethers, and ketones have enjoyed long use. The most famous example of these is the extraction of uranyl nitrate into diethyl ether, the process used in the Manhattan Project to purify the uranium used in the first reactors. In one of the early large-scale processes (the Redox process) to recover uranium and plutonium from irradiated fuel, methyl isobutyl ketone was used to extract the actinides as nitrates.

The most widely used neutral extractants, however, are the organophosphorus compounds, of which the ester, tributylphosphate (TBP), is the most important. TBP forms complexes with the actinide elements thorium, uranium, neptunium, and plutonium by bonding to the central metal atom via the phosphoryl oxygen in the structure

$$(C_4H_9O)_3P^+ \text{———} O^-$$

The overall reactions are

$$MO_2^{2+}(aq) + 2NO_3^-(aq) + 2\,TBP(o) \rightleftarrows MO_2(NO_3)_2 \cdot 2\,TBP(o)$$

or

$$M^{4+}(aq) + 4\,NO_3^-(aq) + 2\,TBP(o) \rightleftarrows M(NO_3)_4(TBP)_2(o)$$

These equilibria can be shifted to the right, increasing the degree of extraction by increasing the concentration of uncombined TBP in the organic phase or by increasing the concentration of $[NO_3^-(aq)]$. The latter increase is done by adding a salting agent such as HNO_3 or $Al(NO_3)_3$. These extraction equilibria are the basis of the Purex process, used almost exclusively in all modern reprocessing of spent nuclear fuel. In the Purex process, irradiated UO_2 fuel is dissolved in HNO_3 with the uranium being oxidized to $UO_2(NO_3)_2$ and the plutonium oxidized to $Pu(NO_3)_4$. A solution of TBP in a high-boiling hydrocarbon, such as n-dodecane, is used to selectively extract the hexavalent $UO_2(NO_3)_2$ and the tetravalent $Pu(NO_3)_4$ from the other actinide and fission product nitrates in the aqueous phase. In a second extraction apparatus, the TBP solution is contacted with a dilute nitric acid solution of a reducing agent, such as ferrous sulfamate, which reduces plutonium to the trivalent state but leaves uranium in the hexavalent state. Plutonium will then transfer to the aqueous phase leaving uranium in the organic phase. The uranium is stripped from the organic phase using water.

An important modern addition to the Purex process is the solvent extraction procedure known as TRUEX (*Trans Uranium Ex*traction). This process is used for the removal of transuranium elements from waste solutions generated during the reprocessing of spent fuel. The process is based on using a bifunctional extractant drawn from the class of carbamoylmethyl- phosphoryl (CMPO) compounds. A typical member of this class of compounds is octyl(phenyl)-N,N-diisobutylcarbamoylmethylphosphine oxide.

Like TBP, these compounds extract actinides at high acidity and can be stripped at low acidity or with complexing agents. Because of this, these solvents are mixed with Purex solvents, in a combined Purex-Truex process.

A third group of extractants (the cationic extractants) are the amines, especially the tertiary or quarternary amines. These strong bases form complexes with actinide metal cations. The efficiency of the extraction is improved when the alkyl groups have long carbon chains, such as occurs for trioctylamine or triisooctylamine. The extraction is conventionally thought of as a "liquid anion exchange" in that the reaction for metal extraction can be written as an anion exchange, that is,

$$xRB^+L^-(o) + ML_n^{-x} \rightleftarrows (RB^+)_x ML_n^x(o) + xL^-$$

where ML_n^{-x} is the metal anion complex being extracted and RB^+ is the ammonium salt of the amine. Hexavalent and tetravalent actinides are efficiently extracted using this technique while trivalent actinides are not well extracted under ordinary conditions.

Detailed information about the appropriate conditions, yields, and so on, for transuranium solvent extraction procedures has been summarized (Ahr 73, Sho 80) and the reader is advised to consult these references for details. It should be noted also that mixtures of different types of extractants sometimes have distribution coefficients that are greater than the sum of the distribution coefficients of the individual extractants. Such mixtures are said to be *synergistic*. Such synergistic effects appear to be due to having one of the extractants replace coordinated water on the extracted metal complex making it more organophilic or simply adding to a coordinatively unsaturated complex, making it more stable. There are a number of actinide extraction schemes involving synergism (Sho 80).

A technique closely related to solvent extraction is extraction chromatography. Extraction chromatography is a solvent extraction system in which one of the liquid phases is made stationary by adsorption on a solid support. The other liquid phase is mobile. Either the aqueous or the organic phase can be mobile. The aqueous phase can be made stationary by adsorption on silica gel, diatomaceous earth, or microspheres of 5–10 μm silica. The same extracting agents that are used in ordinary solvent extraction can be used in extraction chromatography, although the most effective extracting agents appear to be HDEHP or Aliquat-336-S (trialkylmethylammonium chloride with alkyl = C_8H_{17} or $C_{10}H_{21}$). The organic phase can be adsorbed on plastic beads (50–200 μm in diameter). When the stationary phase is organic, the technique is referred to as reversed-phase high-performance liquid chromatography. The stationary

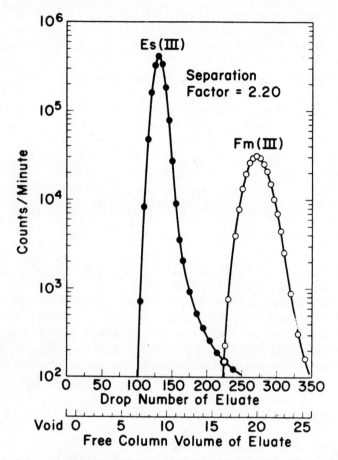

Figure 3.13 The elution of Es (III) and Fm (III) with 0.410 N HNO_3 from HDEHP on Celite (Hor 69). Reprinted with permission from J. Inorg. Nucl. Chem. 31, 1164 (1969).

phase is used in a column just as in ion-exchange chromatography. High-pressure pumps are usually used to force the liquid phase through these columns, just as in conventional high-performance liquid chromatography. Reversed-phase high-performance liquid chromatography (HPLC) has been used to separate Es (III) and Fm (III), and to effect other difficult separations (Figure 3.13).

3.8.3 Ion Exchange (Kor 89)

Ion exchange is a very important method of chemical separation for the transuranium elements. It is fast, efficient, and has the unique ability of permitting a determination of the atomic numbers of the elements being separated even for samples containing a few atoms. In ion exchange, anions or cations are portioned between a mobile aqueous phase and stationary solid phase. Most commonly the solid phase will be an organic polymer containing sulfonic acid groups (cation exchange resin) or quaternary ammonium groups (anion exchange resin). Actinide ions with oxidation states ranging from +3 to +6 can be sorbed by cation exchange resins and eluted with anions such as citrate, lactate or α-hydroxyisobutyrate. In anion exchange, the actinides are complexed with ligands like Cl^- to form anionic complexes which can then be adsorbed on anion exchange resins.

The order of elution of the actinide ions from a cation exchange column is generally in order of the radii of the hydrated ions with the largest hydrated ions eluting first; thus lawrencium is eluted first and americium last for the tripositive ions (see Figure 3.14) although this ordering may not be maintained in all situations. The exact position in which a given element will elute can be predicted ahead of time by careful comparison of the given column for the elution of actinides and lanthanides (Figure 3.14).

The general problems faced in the chemical separation of the actinide elements are (a) the separation of the actinide elements as a group from the lanthanide elements (which are present as fission products produced in the same irradiation that produced the actinides), and (b) the separation of the actinide elements from each other. The former problem can be addressed using a cation exchange separation. Both the lanthanides and actinides will be strongly adsorbed by a cation exchange resin. One then uses concentrated hydrochloric acid or 20% ethanol saturated with hydrochloric acid as an eluant. Because the actinides form chloride complexes more easily than the lanthanides (KSM 86), they will be eluted first *as a group* leaving the lanthanides on the column. Alternatively, one can use anion exchange with a strong ($\sim$12 M) chloride eluant, in which case the actinides will be most strongly held on the column and the lanthanides will be eluted as a group.

The second problem is usually solved by separating the actinides from one another using cation exchange with a citrate, lactate or α-hydroxyisobutyrate eluant. When one compares these actinide separations with those achieved with the lanthanides (Figure 3.14) one sees a strikingly analogous behavior. For example, the break in peak position between Gd and Tb is

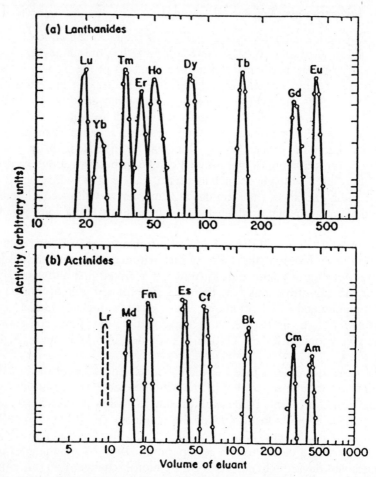

Figure 3.14 Elution of tripositive lanthanide and actinide ions on Dowex 50 cation-exchange resin using ammonium α-hydroxyisobutyrate as an eluant (KSM 86).

analogous to the break between Cm and Bk, presumably reflecting effects due to half-filled 4f and 5f orbitals. Such separation of tripositive actinide elements by adsorption-elution was the key to the discovery of the elements with atomic numbers 97–102, berkelium to nobelium (see Chapter 2).

For cation exchange the strength of adsorption goes as $M^{4+} > M^{3+} > MO_2^{2+} > M^{2+} > MO_2^{+}$, in order of the ionic charge if one allows that MO_2^{2+} has a higher charge on the central metal atom than M^{2+}.

These differences in adsorption are sufficiently large to allow separation of the different types of ions. As mentioned previously for the M^{3+} type, the larger hydrated cations of the M^{3+} type elute first. These effects can be enhanced by the use of appropriate oxycarboxylates of which α-hydroxyisobutyrate appears to be the most selective.

There is an extensive literature concerning the anion exchange behavior of various elements. An especially extensively studied system involves hydrochloric acid eluant and the Dowex 1 resin (a copolymer of styrene and divinylbenzene containing quaternary ammonium groups). Typical distribution ratios for various metals as a function of $[Cl^-]$ concentration are shown in Figure 3.15. One usually sees a steep rise in the distribution coefficient, D, until a maximum is reached and then a gradual decrease in D is observed with further increases in eluant concentration. The maximum occurs when the number of ligands bonding to the metal atom, n, equals the initial charge on the ion. The decrease in D with further increases in eluant concentration is due to free anions from the eluant competing with the metal complexes for ion-exchange resin sites. Usually the tetravalent and hexavalent actinides are strongly absorbed while the trivalent actinides are weakly absorbed or not absorbed at all.

A "cookbook" of ion-exchange procedures exists (Kor 89) with one volume devoted to the actinide elements.

3.8.4 Fast Methods of Separation

Most of the chemical separation procedures we have discussed take hours to perform. However many interesting actinide nuclei and all of the transactinides have much shorter half-lives. Thus it is appropriate to discuss the principles of rapid chemical procedures (procedures that take of the order of seconds or minutes) and refer the reader to the fine reviews of this subject (Her 82, Mey 80) for details.

In most chemical separation procedures, one's goal is to selectively transfer the element of interest from one phase to another, leaving behind any unwanted species. The actual transfer from one phase to another is generally quite rapid but the chemical reactions, procedures to place the desired element in an appropriate chemical state for transfer, are slow. The goal of rapid radiochemical separation procedures is to utilize very fast chemical reactions or procedures or to speed up existing chemical procedures by clever techniques.

Two approaches are commonly used for rapid radiochemical separations, the continuous approach and the batch approach. In the former method, production of the desired element by means of a nuclear reaction is carried out continuously along with the steps to isolate and count the

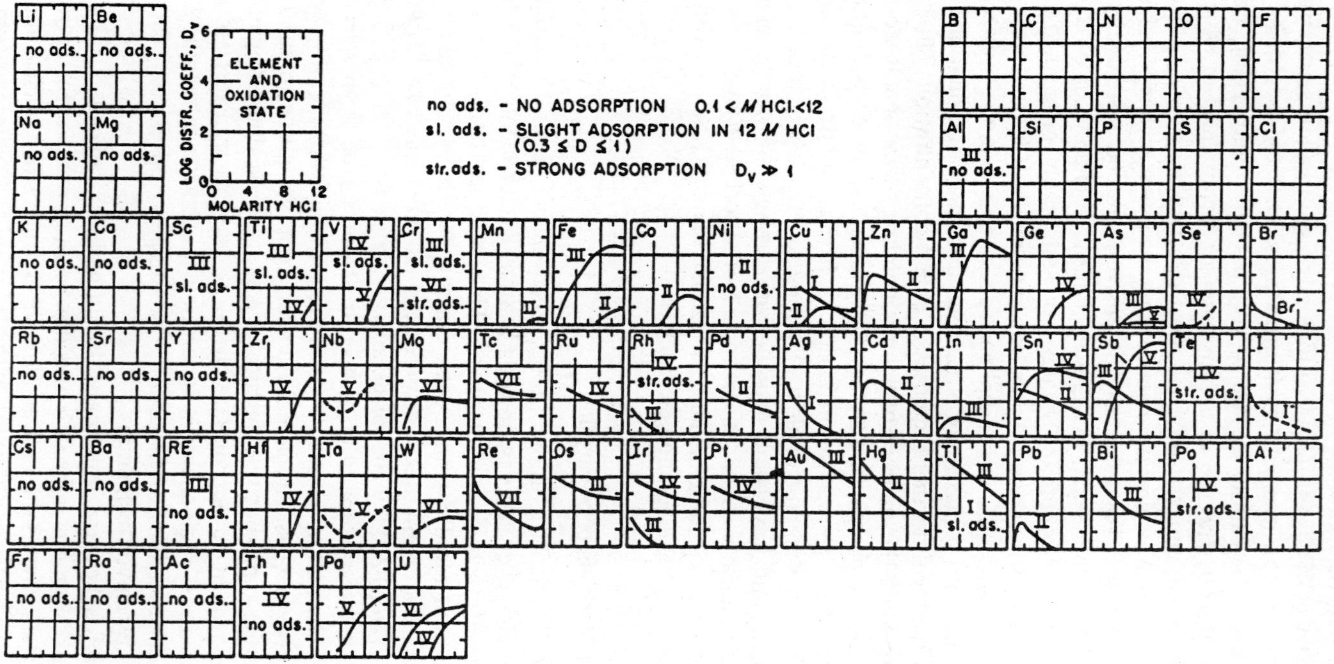

Figure 3.15 Elution of elements from anion exchange resin (Kra 56).

element of interest. In a batch procedure, the desired activities are produced in a short irradiation, separated, counted and the experiment repeated many times to reduce the statistical uncertainties in the data. Batch procedures are relatively easy to perform but must be repeated tens or hundreds or thousands of times to reduce the necessary statistical uncertainty in the final measurements.

Ion-exchange separations generally take hours to perform, but the use of high pressure techniques in conjunction with high-performance solid phases has allowed separation of the actinide elements from each other to be carried out within 20 min with manual operations (Sch 78). The use of an on-line helium jet transport system with a microcomputer controlled automated high-performance liquid chromatography system has allowed these separations to be carried out in a few minutes (Sch 88a). Solvent extraction procedures are intrinsically fast with the separation of the two phases being the rate-determining step. One trick that is frequently used to speed up batch procedures is to fix the organic phase to a fine-grained carrier and filter the aqueous solution rapidly through a layer of this organic-coated carrier. Perhaps the most interesting use of solvent extraction for rapid separations has been the use of continuous procedures utilizing high-speed centrifuges for phase separation. Using these techniques, chemical separations can be achieved in fractions of a second, and very short-lived activities can be studied (Ska 80).

One of the most promising techniques for rapid chemical separation is that of gas chromatography which has been developed for use with the transuranium elements by Zvara and co-workers at the Dubna Laboratory. In gas chromatography, volatile elements or compounds are separated from one another due to differences in their distribution between a mobile gas phase and a stationary solid phase. The most well developed of these techniques is that of thermochromatography. In thermochromatography, gases are passed through a column whose temperature decreases continuously with distance from the entrance. In this thermal gradient, the less volatile species deposit on the column walls first with the more volatile species depositing last.

To illustrate this approach, we describe the thermochromatography of rutherfordium (Zva 72) (Figure 3.16). The 3-s isotope ^{259}Rf was produced by the reaction ^{242}Pu (^{22}Ne, 5n). The recoiling ^{259}Rf atoms and other reaction product atoms were stopped in the nitrogen carrier gas and swept towards the entrance of the chromatagraphic column. Just before entering the column, the atoms were chlorinated using $TiCl_4$ and $SOCl_2$. The protuberances in the first 30 cm of the column induced turbulent flow and helped to deposit the less volatile chlorides. The first and second sections were kept at a constant temperature while a temperature gradient was

104 CHEMICAL PROPERTIES

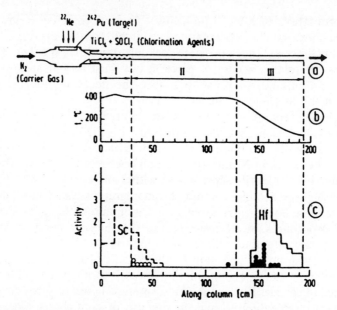

Figure 3.16 Schematic diagram of the thermochromatographic separation of element 105 (Zva 72).

established in the third section of the column. The column walls were lined with mica track detectors which are sensitive to spontaneous fission decay. Scandium tracer and some spontaneously fissioning heavy actinides deposit at the beginning of the column while hafnium and the rutherfordium activity deposit at the same position in the third section of the column. Thus, as expected, rutherfordium, the first transactinide element, forms a volatile chloride like hafnium and has a different chemistry than the heavy actinide elements.

3.9 THE METALLIC STATE

The actinide metals are highly electropositive and react with water vapor and oxygen of the air. Because of this electropositive character, preparation of the metals requires the use of strong reducing agents and vigorous conditions. The first preparation of many of the metals involved reduction of the anhydrous chlorides or fluorides with lithium or barium metal at high temperatures. This method is still used for submilligram quantities of the metals. Reduction of the oxides with lanthanum or thorium metal is used to prepare milligram to gram quantities of americium, curium, berkelium, californium, and einsteinium.

Actinide metals can be purified by volatilization of the impurities in a high vacuum, by volatilization of the metal or by electrodeposition. The van Arkel process, which consists of converting the crude metal to a volatile iodide by reaction with iodine at an elevated temperature, followed by thermal decomposition of the iodide is often used to obtain very pure metals.

The crystal structures and phase transformations of the actinide metals are listed in Table 3.7 along with the densities, melting points, and enthalpies of vaporization (KSM 86). The structures of the lighter actinides are different than those of their lanthanide congeners. The actinides uranium, neptunium, and plutonium have a bcc structure at the melting point. The elements americium through einsteinium show fcc structures at the melting point and double hexagonal close-packed (dhcp) structures below the melting point, a general similarity to the lanthanide metals. Their magnetic moments are similar to those of the lanthanides.

The differences between the lanthanide and actinide metals can be explained in terms of the behavior of the 4f and 5f electrons. In the lanthanide elements, the 4f electrons interact only weakly with the 5d electrons because they are strongly localized in the core of the atomic electron distribution. In the actinide elements, however, the situation is quite different. The delocalized 5f electrons in the first members of the actinide series are free to interact or mix with the outer valence electrons causing hybridization of these electrons in a large band with the 6d and/or 7s electrons. The 5f electrons are sufficiently localized to allow the structure to resemble that of the corresponding lanthanide only in americium and heavier elements.

The electronic properties of the 5f metals are intermediate between those of the familiar d-block transition elements where there is extensive mixing of the 5d electrons with outer s and p valence electrons and the lanthanide-elements where there is little or no mixing of the 4f electrons with the d, s, or p valence electrons. Furthermore, because the energy differences between localized and delocalized 5f electrons are small, relatively small perturbations can cause the mobilization of the 5f electrons. This delicate balance in the middle of the actinide series gives rise to an unusual set of crystal structures, magnetic properties, superconductivity, et cetera. The explanation of these complex properties is a challenge to modern theories of the metallic state.

One important feature of the metals is the onset of divalency for the higher actinides. At 25°C, divalency is seen in einsteinium and fermium metals and one expects mendelevium and nobelium will be divalent met-

TABLE 3.7 Properties of the Transuranium Metals (KSM 86)

Element	Melting Point (°C)	Enthalpy of Vaporization at 25°C (kJ mol^{-1})	Lattice Symmetry	Temp. Range of Stability (°C)	X-Ray Density (g cm^{-3})	Atoms per Unit Cell	Metallic Radius, CN12[a] (Å)
Actinium	1050	418	fcc		10.01	4	1.878
Thorium	1750	598	α, fcc	<~1360	11.724	4	1.798
			β, bcc	~1360–1750	11.10	2	1.80
Protactinium	1572	660	α, bc tetragonal	Below 1165	15.37	2	1.642
			β, bcc	1165–1572	13.87	2	1.775
Uranium	1133	536	α, orthorhombic	Below 668	19.16	4	1.542
			β, tetragonal	668–775	18.11	30	1.548
			γ, bcc	775–1133	18.06	2	1.548
Neptunium	637	465	α, orthorhombic	Below 280	20.45	8	1.503
			β, tetragonal	280–576	19.36	4	1.511
			γ, bcc	576–637	18.04	2	1.53
Plutonium	640	342	α, monoclinic	Below 122	19.86	16	1.623
			β, monoclinic	122–207	17.70	34	1.571
			γ, orthorhombic	207–315	17.14	8	1.588
			δ, fcc	315–457	15.92	4	1.640
			δ', bc tetragonal	457–479	16.00	2	1.640
			ε, bcc	479–640	16.51	2	1.592
Americium	1173	284	α, dhcp	Below 658	13.6	4	1.730
			β, fcc	793–1004	13.65	4	1.730
			γ, ?	~1050–1173			
Curium	1345	387	α, dhcp	Below ~1277	13.5	4	1.743
			β, fcc	1277–1345	12.9	4	1.782
Berkelium	1050	310	α, dhcp	Below 930	14.79	4	1.704
			β, fcc	930–986	13.24	4	1.767
Californium	900	196	α, dhcp	Below 900	15.10	4	1.694
			β, fcc		18.74	4	2.030
Einsteinium	860	133	α, dhcp	Below 860	?	4	
			β, fcc		8.84	4	2.03

[a]Radii are corrected to coordination number 12.

als (Nug 73, Sea 89). This has been confirmed in thermochromatography experiments (Hub 80).

Amongst the elements shown in Table 3.7, one notes that uranium, neptunium, and plutonium have complex crystal structures with many allotropes. Plutonium, in particular, shows six distinct allotropic forms between room temperature and its melting point. For no pair of phases does the coefficient of thermal expansion and the temperature coefficient of resistivity have the same sign (that is, normally a metal expands upon heating with an increase in electrical resistivity, but in plutonium if a phase expands upon heating, the resistance decreases). All of these facts are of great importance in nuclear technology of plutonium.

Superconductivity of the actinides seems to be relatively well understood on the basis of their electronic structure. Thorium and uranium are

TABLE 3.8a Single-Parameter Limits for Uniform Aqueous Solutions Containing Fissile Nuclides (Ben 81)

	Subcritical Limit for		
Parameter	^{235}U	^{233}U	^{239}Pu
Mass of fissile nuclide (kg)	0.76	0.55	0.51
Solution cylinder diameter (cm)	13.9	11.5	15.7
Solution slab thickness (cm)	4.6	3.0	5.8
Solution volume (liters)	5.8	3.5	7.7
Concentration of fissile nuclide (g/liter)	11.5	10.8	7.0
Areal density of fissile nuclide (g/cm^2)	0.40	0.35	0.25
Uranium enrichment (wt% ^{235}U)	1.00	–	–
Uranium enrichment in presence of two nitrate ions per uranium atom (wt% ^{235}U)	2.07	–	–

TABLE 3.8b Single-Parameter Limits for Metal Units

	Subcritical Limit for		
Parameter	^{235}U	^{233}U	^{239}Pu
Mass of fissile nuclide (kg)	20.1	6.7	4.9
Cylinder diameter (cm)	7.3	4.6	4.4
Slab thickness (cm)	1.3	0.54	0.65
Uranium enrichment (wt% ^{235}U)	5.0	–	–

known to be superconductors at low temperatures as are protactinium and americium.

The pure metals are generally silvery in color, lustrous, brittle, and quite reactive. They have high densities, with uranium being one of the densest metals (~19.0 g/cm^3), and combine directly with most elements. Upon standing in air the surface of the metal develops a green and subsequently a black, nonprotective film. Powdered uranium is often pyrophoric as is plutonium. This contributes to the difficulty of handling these metals. The metals are readily soluble in dilute acids unless acid attack is inhibited by passivation or formation of an oxide layer, as is frequently observed with concentrated HNO_3 and H_2SO_4. Uranium reacts with boiling water to produce UO_2 and hydrogen. Frequently the hydrogen formed will react further with the uranium to form a hydride, which is not mechanically stable. The actinide metals form extensive alloys with one another and do form intermetallic compounds with other metals.

Table 3.8 shows typical data for the critical masses of the "Big Three" fissile nuclides. Note the decrease in the critical mass for solutions due to the moderation and reflection of neutrons by water. Care must be taken to maintain the amounts and/or concentrations of the actinides below subcritical limits in all manipulations.

3.10 SOLID COMPOUNDS

There are literally thousands of compounds of the transuranium elements. Our purpose is to outline the general properties of these compounds and to refer the reader to more encyclopedic works for lists of their detailed properties, preparation, structure, et cetera (KSM 86, Kel 71). One aspect of a number of these compounds that deserves further comment is the nonstoichiometric character of many of them. The oxides of the actinide elements form a continuous spectrum of nonstoichiometric compounds (Table 3.9). Also the special stability of the MO_2 (fluorite) structure is noteworthy, leading to such compounds as AmO_2 and CmO_2 despite the fact that these elements are tripositive in their solution chemistry. The uranium-oxygen compounds are interesting and quite complex. There are close to a dozen distinct phases for the composition range UO_2 to UO_3. Also there are six modifications of UO_3. The main oxides are the orange-yellow UO_3, the green-black U_3O_8, and the brown-black UO_2. All the oxides dissolve readily in nitric acid to give UO_2^{2+} salts. The fusion of uranium oxides with alkali metal or alkaline earth carbonates gives orange-yellow materials known as uranates. The stoichiometry of these compounds is usually $M_2U_xO_{3x+1}$. The most famous of these compounds is

TABLE 3.9 Well-Characterized Binary Actinide Oxides (KSM 86)

An (III)	An (IV)	An (V)			An (VI)
$AcO_{1.5}$					
	ThO_2				
	PaO_2		$PaO_{2.5}$		
	UO_2	$UO_{2.25}$	$UO_{2.33}$	$UO_{2.67}$	$UO_3(\gamma)$
	NpO_2		$NpO_{2.5}$		
$PuO_{1.5}$	$PuO_{1.61}$	PuO_2			
$AmO_{1.5}$	$AmO_{1.6}(?)$	AmO_2			
$CmO_{1.5}$	$CmO_{1.714}$	CmO_2			
$BkO_{1.5}$	$BkO_{1.8}$	BkO_2			
$CfO_{1.5}$	$CfO_{1.714}$	CfO_2			

ammonium diuranate, made by adding aqueous ammonia to $UO_2(NO_3)_2$ solutions. These compounds feature octahedral U(VI) with unsymmetrical oxygen coordination.

The binary halides of the actinide elements have the formulas MX_3, MX_4, MX_5, and MX_6. Their properties, structure, preparation, and so on, have been summarized elsewhere (Bro 68). The lower fluorides are usually prepared by the action of hydrogen fluoride on the oxides, while the higher fluorides are generally prepared with the further addition of fluorine. The chlorides can be made by direct combination of chlorine with the metal oxyhydride, by hydrochlorination of the oxides, or by chlorination of the oxides with carbon tetrachloride, phosgene, sulfur chloride, bromine trichloride, or other chlorinating agents. The bromides may be prepared by direct combination of bromine with the metal or hydrobromination of the oxides. Except for the trifluorides, the halides are so hygroscopic as to require handling in an anhydrous atmosphere. Uranium hexafluoride gas is used for the separation of ^{235}U by the application of gaseous diffusion to a mixture of $^{235}UF_6$ and $^{238}UF_6$. Plutonium hexafluoride is very reactive, in that it will react with the inert gas xenon to form the ionic compound $Xe^+Pu\,F_6^-$.

The precipitation properties of the actinide M^{3+} ions are similar to those of the tripositive lanthanide ions while the behavior of the actinide M^{4+} ions closely resembles that of Ce^{4+}. Thus the fluorides and oxalates are insoluble in acid solution, whereas the nitrates, halides, sulfates, perchlorates, and sulfides are all soluble. The hydroxides of all four types of ion are insoluble.

3.11 ORGANOMETALLIC COMPOUNDS

One of the most active areas of actinide chemistry is the study of the organometallic compounds of the actinides. The unique properties of the actinides relative to those of the d-block transition elements offer hope of gaining new insight into the character of organometallic compounds. For example, the large size of the actinide ions permits coordination numbers and molecular geometries not seen in the d-block elements. The different symmetry properties and availability of the f orbitals as compared to the d-orbitals allow different bonding arrangements for the actinide organometallic compounds. For detailed information about this fast-growing field, the reader should consult any of a group of excellent review articles (KSM 86, Mar 86, Mar 82, Mar 79, Fag 82).

The first organoactinide (Rey 56) was a complex of the cyclopentadienyl ligand, UCp_3Cl, where Cp^- represents the cyclopentadienide ion, $C_5H_5^-$. The geometry of this compound is shown below with the arrangement of the ligands being approximately tetrahedral about the uranium atom.

The synthetic approach used to make this molecule is a general one: displacement of halide ligands from the actinide ion

$$AnX_4 + 3\,MCp \rightarrow U\,(Cp)_3X + 3MX$$

where M is an alkali metal. The organoactinide is extracted from the reaction mixture with an appropriate organic solvent and then purified by sublimation, if possible. Since the synthesis of the first organoactinide in 1956, several hundred organoactinides have been made. Some representative cyclopentadienyl derivatives are shown in Table 3.10. The stoichiometries of these compounds generally reflect the available oxidation states. Thus the higher actinides show tris(cyclopentadienyl)actinide (III) complexes while the tetravalent complexes are seen in the lighter actinides. The actinide (III) compounds are soluble in organic solvents, reasonably stable, relatively volatile and are air sensitive. The actinide (IV) complexes are soluble in organic solvents, air stable but not volatile.

A number of derivatives of these compounds with the formula MCp_3X where M is a tetravalent actinide and X is an anion have been prepared.

ORGANOMETALLIC COMPOUNDS

TABLE 3.10 Some Representative Actinide Cyclopentadienyl Complexes

Compound	Color	Melting Point[a] (°C)	Sublimation Temperature[b] (°C)
$Th(C_5H_5)_3$	Green		
$Th(C_5H_5)_3Cl$	Colorless		
$Th(C_5H_5)_4$	Colorless	170d	250–290
$Pa(C_5H_5)_4$	Orange-yellow	220d	220d
$U(C_5H_5)_3$	Brown	>200	
$U(C_5H_5)_3Cl$	Red-brown	260	260
$U(C_5H_5)_4$	Red	250d	200–220d
$Np(C_5H_5)_3Cl$	Brown		
$Np(C_5H_5)_4$	Red-brown	220d	200–220d
$Pu(C_5H_5)_3$	Green	180d	140–165
$Am(C_5H_5)_3$	Flesh	330d	160–200
$Cm(C_5H_5)_3$	Colorless		180
$Bk(C_5H_5)_3$	Amber		135–165
$Cf(C_5H_5)_3$	Red		135–320

[a] d indicates decomposition.
[b] At a pressure of 10^{-3}–10^{-4} Torr.

X can be a halide, sulfate, nitrate, and so on along with more exotic species such as BH_4^-, or R where R is an organic alkyl or aryl group.

The metal-ligand bonding in the actinide cyclopentadienyl compounds has been extensively studied. The cyclopentadienide ion, C_p^-, has π electrons in two conjugated double bonds capable of coordinating with vacant metal atom orbitals. In general one finds substantially more ionic character in the bonding in the organoactinides compared to the d-block transition elements. However, there is some evidence in the lighter actinides that there is some covalent bonding between the actinide and the ligand possibly involving 6d and 5f orbitals. For the heaviest actinides, the 5f orbitals are better shielded and the bonding is mostly ionic.

In 1968, the first cyclooctatetraene-actinide compound, uranocene, was prepared (Str 68) by reacting UCl_4 with the potassium salt of cyclooctatetraene (COT) as follows:

$$UCl_4 + 2K_2COT \rightarrow U(COT)_2 + 4KCl$$

Since then, analogous compounds of protactinium, thorium, neptunium and plutonium have been prepared. All of these compounds have a sandwich structure with the actinide metal ion sandwiched between two planar COT rings, as shown on the following page.

112 CHEMICAL PROPERTIES

Most of these compounds involve the actinide (IV) species although complexes of the form $M(COT)^-$ have been prepared as well as "piano stool" molecules of the form:

$$\begin{array}{c} \text{Cl} \diagup \overset{\displaystyle M}{} \diagdown \text{S} \\ \text{S} \quad \text{Cl} \end{array}$$

There is little or no barrier to rotation of the rings.

A qualitative molecular orbital picture of the bonding in uranocene has been prepared (which is analogous to the bonding in ferrocene). The uranium f_{xyz} and $f_{z(x^2-y^2)}$ orbitals have the same e_{2u} symmetry as a pair of the highest occupied molecular orbitals of the $C_8H_8^{2-}$ and therefore maximum overlap occurs. This explanation is consistent with a wide range of experimental observations such as that electron-donor substituents of the COT rings stabilize uranocene (by raising the energy of one of the e_{2u} orbitals and providing a better energy matching with the f orbitals). Also because the highest occupied and lowest vacant orbitals are metal orbitals, nucleophilic and electrophilic reagents will attack at the central metal atom.

Although searches were made for organoactinide compounds involving a direct actinide-carbon σ bond during the World War II Manhattan Project (because of their potential usefulness in isotope separation), it was three decades later that such molecules were synthesized. These molecules are synthesized via reactions such as

$$M(C_5H_5)_3X + RLi(\text{or } RMgX) \longrightarrow \text{[}M\text{-}R\text{]} + LiX(\text{or } MgX_2)$$

M = Th, Pa, U, Np
X = Cl, Br, I
R = alkyl, aryl, alkenyl, alkynyl

These molecules belong to a class of molecules called hydrocarbyls. The actinide hydrocarbyls are thermally stable due to the large bulky substituents that protect the central metal atom from reaction and the high coordinative saturation. A rich and exciting chemistry of the actinide-carbon σ bond has been uncovered in which one gains insight about many general reactions such as CO activation at metal surfaces as well as the chemistry of unusual species such as the organoactinide hydrides (Mar 82). The thermochemistry of actinide organometallic bonds can be predicted from electronegativity considerations (Mar 89). Organometallic compounds involving actinide hydrogen and actinide transition metal σ bonds are also known (KSM 86).

3.12 CHEMISTRY OF THE TRANSACTINIDES

The last member of the actinide series is lawrencium, element 103. The known elements beyond 103 (rutherfordium, hahnium, 106, 107, 108, and 109) should be members of a new fourth transition series extending from rutherfordium to element 112, in which the 6d electronic shell is being filled. The chemical properties of the members of this group should generally resemble those of their congeners hafnium to mercury in the third transition series (Dirac–Fock calculations of the ground state electron configurations of these elements have been discussed previously (Table 3.3)). The predicted chemical properties of these elements (Kel 77) are shown in Table 3.11. The predicted chemical properties are generally the result of a judicious extrapolation of periodic table trends for each group together with consideration of relativistic effects predicted for these elements (see Section 7.2). Of especial interest are the properties of 112, eka-Hg. Pitzer has suggested that relativistic effects will make 112 more noble than mercury and that 112 will be a volatile liquid or a gas (Pit 79).

Of the transactinide elements, only the chemistry of rutherfordium and hahnium has been studied. These elements all have short half-lives and study of their chemical properties must occur at the accelerators where they are produced. Since typical production rates are such that the elements are produced one atom at a time, the experiments to deduce the chemistry of these elements must be carried out many times with the results of the individual experiments being added together to produce a statistically significant result. The experience must be very reproducible and involve sensitive detection techniques such as high-resolution α-particle spectroscopy and fission track counting.

The chemistry of rutherfordium has been shown to be similar to the chemistry of hafnium rather than the chemistry of the heavier actinides, a clear demonstration of the expected end of the actinide series at lawren-

TABLE 3.11 Predicted Chemical Properties of Elements 104–112 [Kel 77]

	Rf	Ha	106	107	108	109	110	111	112
Stable oxidation states	III, IV	IV, V	IV, VI	III-VII	II-VIII	I-VI	I-VI	III, (−1)	I, II
first ionization energy (eV)	5.1	6.6	7.6	6.9	7.8	8.7	9.6	10.5	11.4
Standard electrode potential (V)	4→0	5→0	4→0	5→0	4→0	3→0	2→0	3→0	2→0
Ionic radius (Å)	−1.8	−0.8	−0.6	+0.1	+0.4	+0.8	+1.7	+1.9	+2.1
	(+4)	(+5)	(+4)	(+5)	(+4)	(+3)	(+2)	(+3)	(+2)
	0.71	0.68	0.86	0.83	0.80	0.83	0.80	0.76	0.75
Atomic radius (Å)	1.50	1.39	1.32	1.28	1.26				
Density (g cm^{-3})	23	29	35	37	41				
Heat of sublimation (kJ mol^{+1})	694	795	858	753	628	594	481	335	29
Boiling point (K)	5800								
Melting point (K)	2400								

cium. This demonstration involved both aqueous and gas phase chemistry. In the gas phase experiments (Zva 72), the 3-s isotope ^{259}Rf produced in the ^{242}Pu (^{22}Ne, 5n) reaction was used. Zvara and co-workers were able to use thermochromatography to show a difference in volatility of the RfCl$_4$ which condensed at ~220°C as compared to the chlorides of the heavier actinides which have much higher condensation temperatures. The details of this experiment are discussed in Section 3.8.4.

In the aqueous solution experiments, the 1-min isotope ^{261}Rf produced in the ^{248}Cm (^{18}O, 5n) reaction was used (Sil 70). Atoms of rutherfordium recoiling from the target were caught in an NH$_4$Cl layer sublimed onto platinum discs, dissolved with ammonium α-hydroxyisobutyrate solution and added to a heated Dowex 50 cation exchange resin column. The neutral and anionic complexes of hafnium, zirconium, and rutherfordium were not absorbed on the cation exchange column while actinides were strongly absorbed. Thus the hafnium, zirconium (tracers) and the rutherfordium atoms eluted within a few column volumes while the actinides eluted after several hundred column volumes. The time from end of bombardment to start of sample counting was less than one half-life of ^{261}Rf and after several hundred experiments Silva and co-workers were able to detect the decay of 17 atoms of ^{261}Rf in the eluant.

This work was extended by Hulet and co-workers (Hul 80) to the chloride complexes of rutherfordium. Computer automation was used to perform the chemical operations rapidly and reproducibly. An HCl solution containing ^{261}Rf was passed through an extraction chromatography column loaded with trioctylmethylammonium chloride which strongly extracts anionic chloride complexes. Such complexes are formed by the Group IVB elements such as rutherfordium while the actinides, and members of Groups IA and IIA form weaker complexes and are not extracted. Thus the actinide recoil products elute first and zirconium, hafnium, and rutherfordium were shown to elute in a second fraction as expected for Group IVB elements.

Hahnium (Ha) is expected to have the valence electron configuration $7s^2 6d^3$ and thus to be a homolog of Ta ($6s^2 5d^3$). Zvara et al. (Zva 76) have carried out a set of thermochromatography experiments similar to those done with rutherfordium. The isotope used for the study was the 1.8-s ^{261}Ha produced in the ^{243}Am (^{22}Ne, 4n) reaction, with detection by observation of its decay by spontaneous fission. The results of the experiments show the volatility of the chlorides and bromides of hahnium to be less than that of niobium (a $4d^3$ element which has the same deposition temperature in the chromatographic apparatus as tantalum) but relatively similar to that of hafnium. If this result is verified by additional gas phase or aqueous solution work, it could indicate an unexpected departure in the chemical behavior of hahnium from that expected for eka-Ta. However, capricious absorption behavior can lead to uncertainties in the interpretation of the results of such experiments. Gregorich et al. (Gre 88) have investigated some aqueous solution chemistry of hahnium, using the 35-s ^{262}Ha produced in the ^{249}Bk (^{18}O, 5n) reaction. With nearly a thousand batch experiments hahnium was found to hydrolyze in strong HNO_3 and adhere to glass surfaces. Such hydrolysis is characteristic of Group VB elements and different from IVB elements as verified in experiments with tantalum and niobium tracers under the same conditions. In other experiments, hahnium did not form extractable anionic fluoride complexes in HNO_3/HF solutions under conditions in which tantalum was extracted nearly quantitatively (Kra 89). This observation may be explained by an extrapolation of the Group VB properties in that the tendency to hydrolyze or form F^- complexes may be stronger for hahnium than tantalum, leading to a failure to observe extraction. In the pioneering work of Gregorich et al., the total study involved the identification of 47 atoms of ^{262}Ha on the basis of observation of decay by spontaneous fission and alpha emission, including the time correlation of alpha decays from ^{262}Ha and its 4-s daughter ^{258}Lr.

GENERAL REFERENCES

Ahr 73   S. Ahrland, J.O. Liljenzin, and J. Rydberg, *Comprehensive Inorganic Chemistry*, Vol. 5 (Pergamon, Oxford, 1973) pp. 465–635. Comprehensive review of actinide solution chemistry and solvent extraction.

Cot 88   F.A. Cotton and G. Wilkinson, *Advanced Inorganic Chemistry*, 5th Edn. (Wiley, New York, 1988). Still the "Bible" of inorganic chemistry textbooks.

Ede 82   N.M. Edelstein, Ed., *Actinides in Perspective* (Pergamon, Oxford, 1982) The proceedings of the Actinides-81 conference in which many excellent reviews of the actinides were presented.

Fre 74   A.J. Freeman and J.B. Darby, Jr., Eds., *The Actinides: Electronic Structure and Related Properties* (Academic Press, New York, 1974). Survey with emphasis on metallic properties.

FK 86   A.J. Freeman and C. Keller, Eds., *Handbook on the Physics and Chemistry of the Actinides* (North-Holland, Amsterdam, 1986). Detailed, comprehensive discussions

Her 82   G. Herrmann and N. Trautman, Annu. Rev. Nucl. Part. Sci. **32**, 117 (1982). Discussion of fast methods of separation.

Kel 71   C. Keller, *The Chemistry of the Transuranium Elements* (Verlag Chemie, Berlin, 1971). A comprehensive survey.

KSM 86   J.J. Katz, G.T. Seaborg, and L. Morss, *Chemistry of the Actinide Elements*, 2nd Edn. (Chapman and Hall, London, 1986). A very comprehensive, two volume survey.

RA 83   Radiochim. Acta, **32** (1983). A special issue devoted to the actinide elements and their properties.

Sea 58   G.T. Seaborg, *The Transuranium Elements* (Yale University Press, New Haven, Conn., Addison-Wesley Reading, 1958). Seaborg's Silliman Lectures contain much historical material and unique insights into the properties of the transuranium elements.

Sea 63   G. T. Seaborg, *Man-Made Transuranium Elements* (Prentice-Hall, Englewood Cliffs, 1963). A clear simple introduction to the nuclear and chemical properties of the transuranium elements.

REFERENCES

Ben 81   M. Benedict, T. Pigford and H. Levi, *Nuclear Chemical Engineering*, 2nd Edn., (McGraw-Hill, New York, 1981).

Bet 57   H.A. Bethe and E.E. Salpeter, *Quantum Mechanics of One- and Two-Electron Atoms* (Academic Press, New York, 1957).

Bre 71   L. Brewer, J. Opt. Soc. **61**, 1101 (1971).

Bro 68   D. Brown, *Halides of the Lanthanides and Actinides* (Wiley, London, 1968).

Brü 88   W. Brüchle et al., Inorg. Chim. Acta **146**, 267 (1988).

Cho 83   G.R. Choppin, Radiochim. Acta **32**, 43 (1983).

Des 73   J.P. Desclaux, Atom. Data Nucl. Data **12**, 311 (1973).

Des 80   J.P. Desclaux and B. Fricke, J. Phys. **41**, 943 (1980).

Des 84   J.P. Desclaux and A.J. Freeman in *Handbook on the Physics and Chemistry of the Actinides*, Vol. 1, A.J. Freeman and G.H. Lander, Eds. (North-Holland, Amsterdam, 1984) p. 1.

Fag 82   P.J. Fagan, E.A. Maatta, J.M. Manriquez, K.G. Moloy, A.M. Seyam, and T.J. Marks in *Actinides in Perspective*, N.M. Edelstein, Ed. (Pergamon, Oxford, 1982) p. 433.

Fri 64   H.G. Friedman, Jr., G.R. Choppin, and D.G. Fuerbachev, J. Chem. Ed. **41**, 354 (1964).

Fri 75   B. Fricke, Structure and Bonding **21**, 89 (1975).

Fri 77   B. Fricke and G. Soff, Atom. Data Nucl. Data Tables **19**, 83 (1977)

Gle 89   V.A. Glebor, L. Kasztura, V.S. Nefedor, and B.L. Zhuikov, Radiochim. Acta **46**, 117 (1989).

Gre 88   K.E. Gregorich, et al., Radiochim. Acta **43**, 223 (1988).

Hes 80   J.P. Hesler and W.T. Carnall in *Lanthanide and Actinide Chemistry and Spectroscopy*, N. Edelstein, Ed., (ACS, Washington, 1980).

Hor 69   E.P. Horwitz, C.A.A. Bloomquist and D.J. Henderson, J. Inorg. Nucl. Chem. **31**, 1149 (1969).

Hub 80   S. Hubener, Radiochem. Radioanal. Lett. **44**, 79 (1980); S.Hubener and I. Zvara, Radiochim. Acta 31, 89 (1982).

Hul 80   E.K. Hulet, R.W. Longheed, J.F. Wild, J.H. Landrum, J.M. Nitschke, and A. Ghiorso, J. Inorg. Nucl. Chem. **42**, 79 (1980).

Jos 88   D.T. Jost, H.W. Gäggeler, Ch. Vogel, M. Schädel, E. Jäger, B. Eichler, K.E. Gregorich, and D.C. Hoffman, Inorg. Chim. Acta **146**, 255 (1988).

Kel 77   O.L. Keller, Jr. and G.T. Seaborg, Annu. Rev. Nucl. Sci. **27**, 139 (1977).

Kor 89   J. Korkisch, *Handbook of Ion Exchange Resins: Their Application to Inorganic Analytical Chemistry*, Vol. I, II (CRC, Boca Raton, 1989).

Kra 56   K. Kraus and D. Nelson, Paper 837, Geneva Conference, Vol. 7 (1956).

Kra 89   J.V. Kratz et al., Radiochim. Acta **48**,121 (1989).

Mar 79   T.J. Marks and R.D. Fischer, Eds., *Organometallics of the f-Elements*, (Reidel, Dordrecht, 1979).

Mar 82   T.J. Marks, Science **217**, 989 (1982).

Mar 86   T.J. Marks in *Handbook on the Physics and Chemistry of the Actinide Elements*, Vol. 4, A.J. Freeman and C. Keller, Eds. (North-Holland, Amsterdam, 1986) p. 491.

Mar 89   T.J. Marks, M.R. Gagne, S.P. Nolan, L.E. Schock, A.M. Seyam, and D. Stern, Pure Appl. Chem. **61**, 1665 (1989).

Mey 80   R.A. Meyer and E.A. Henry, *Proc. Workshop Nucl. Spectrosc. Fission Products, Grenoble*, 1979, (Bristol, London, 1979) pp. 59–103.

McK 83   D.R. McKelvey, J. Chem. Ed. **60**, 112 (1983).

Mik 73   N.B. Mikheev, V.I. Spitsyn, A.N. Kamenskaya, I.A. Rumer, B.A. Grozdez, N.A. Rosenkevich, and L.N. Auerman, Dokl. Akad. Nauk SSSR **208**, 1146 (1973).

Nug 73   L.J. Nugent, R.D. Baybarz, J.L. Burnett and J.L. Ryan, J. Phys. Chem. **77**, 1528 (1973).

Pit 79   K.S. Pitzer, Acc. Chem. Res. **12**, 271 (1979).

Pow 68   R.E. Powell, J. Chem. Ed. **45**, 558 (1968).

Pyy 79   P. Pyyko and J.P. Desclaux, Acc. Chem. Res. **12**, 276 (1979).

Rey 56   L.T. Reynolds and G. Wilkinson, J. Inorg. Nucl. Chem. **2**, 246 (1956).

Sch 78   M. Schädel, W. Brüchle, B. Haefner, J.V. Kratz, W. Schorstein, N. Trautmann and G. Herrmann, Radiochim. Acta **25**, 111 (1978).

Sch 88a   M. Schädel, W. Brüchle, and B. Haefner, Nucl. Instrum. Method Phys. Res. **A264**, 308 (1988).

Sch 88b   U. Scherer, J.V. Kratz, M. Schädel, W. Brüchle, K.E. Gregorich, R.A. Henderson, D. Lee, M. Nurmia and D.C. Hoffman, Inorg. Chim. Acta **146**, 249 (1988).

Sea 89	G.T. Seaborg, J. Nucl. Mat. **166**, 22 (1989).
Sho 80	R.R. Shoun and W.J. McDowell in *Actinide Separations*, J.D. Navratil and W.W. Schulz, Eds. (American Chemical Society, Washington, 1980) p. 71.
Sil 70	R. Silva, J. Harris, M. Nurmia, K. Eskola and A. Ghiorso, Inorg. Nucl. Chem. Lett. **6**, 871 (1970).
Ska 80	G. Skarnemark, P.O. Aronsson, K. Broden, J. Rydberg, T. Bjornstad, N. Kaffrell, E. Stender, and N. Trautmann, Nucl. Instrum. Methods **171**, 323 (1980); see also G. Skarnemark, M. Skalberg, J. Alstad, and T. Bjornstad, Phys. Scr. **34**, (1987).
Str 68	A. Streitwieser, Jr. and U. Muller-Westhoff, J. Am. Chem. Soc. **90**, 7364 (1968).
Sza 69	A. Szabo, J. Chem. Ed. **46**, 678 (1969).
Whi 31	H.E. White, Phys. Rev. **38**, 513 (1931).
Zhu 89	B.L. Zhuikor, Y.T. Chulurkov, S.N. Timokin, K.U. Jin, and I Zvara, Radiochim. Acta **46**, 113 (1989).
Zva 72	I. Zvara, V.Z. Belov, V.P. Domanov, Yu.S. Korotkin, L.P. Chelnokov, M.R. Shalavskii, V.A. Shchegolev, and M. Hussonis, Radiokhimiya, 119 (1972); Sov. Radiochem. **14**, 115 (1972).
Zva 76	I. Zvara, V.Z. Belov, V.P. Domanov, and M.R. Shalzevskii, Sov. Radiochem. **18**, 328 (1976).

4

NUCLEAR STRUCTURE AND RADIOACTIVE DECAY PROPERTIES

In this chapter we shall discuss the nuclear structure and radioactive decay properties of the elements beyond uranium with special emphasis being placed upon those aspects of nuclear behavior that are unique in these nuclei. Our level of coverage of this complex subject will be such as to introduce readers to the exciting physics involved. Occasionally we discuss more advanced, recent developments. Readers seeking a more detailed discussion of the topics covered in this chapter may wish to consult the references cited in the chapter and the bibliography at the end of the chapter.

4.1 NUCLEAR SHAPES AND SIZES

Heavy nuclei are characterized by an approximately constant density, ρ_0, of ~0.17 nucleons/fm^3. This density is constant through the nuclear volume until near the "edge" of the nucleus where the density decreases to zero in a distance of ~2–3 fm. This can be represented as a Fermi distribution, where the dependence of the density upon the radius r is given by

$$\rho(r) = \frac{\rho_0}{1 + \exp[(r - R)/a]} \qquad (4.1)$$

where a is a surface diffuseness constant. This "skin region" of heavy nuclei of thickness, $t = 2a \ln 9$, is considerably smaller than the total nuclear size which can be represented as a sphere of radius $R = r_0 A^{1/3}$ fm where A is the nuclear mass number and $r_0 \sim 1.1$–1.2.

More accurate studies of transuranium nuclei show they have slightly prolate (i.e., "football") shapes which can be represented by

$$R(\theta) = c[1 + \beta_2 Y_{20}(\theta) + \beta_4 Y_{40}(\theta)] \tag{4.2}$$

where Y_{l0} are the normalized spherical harmonic functions

$$Y_{20}(\theta) = \left(\frac{5}{4\pi}\right)^{1/2} (3\cos^2\theta - 1)/2 \tag{4.3}$$

$$Y_{40}(\theta) = \left(\frac{9}{4\pi}\right)^{1/2} (35\cos^4\theta - 30\cos^2\theta + 3)/8 \tag{4.4}$$

and c is a parameter determined by fitting experimental data. The deformation parameters β_2 and β_4 range from 0.23 to 0.29 and +0.1 to −0.1 for typical transuranium nuclei (Bem 73, Zum 84). The shapes of typical heavy nuclei are shown in Figure 4.1. The nuclei are prolate spheriods with a typical axis ratio of 1.3:1. Evidence also exists that in addition to the prolate deformations, very small stable octupole (pear-shaped) and hexadecapole deformations can be found in certain actinide nuclei.

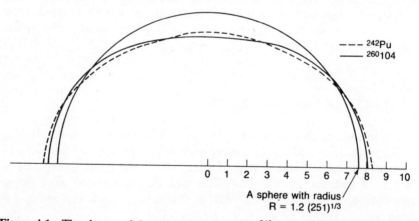

Figure 4.1 The shapes of the transuranium nuclei ^{242}Pu and ^{260}Rf (data from Ras 75). Also shown for comparison is the shape expected for a hypothetical spherical nucleus with $A = 251$.

4.2 NUCLEAR MASSES

Having considered the shapes of the elements beyond uranium, it is appropriate to consider another macroscopic property of these nuclei, their masses. The masses, α-decay energies and β-decay energies for most of the more stable transuranium nuclides (with $A \leq 250$) have been measured and are tabulated in standard references (Wap 88). In the absence of measured values of the masses, two approaches can be taken to obtain values for the masses of the transuranium nuclei. The first of these is a semiempirical approach. This approach is based upon two ideas: (a) the energy released in α- and β-decay of the heavy elements exhibits a smooth variation with changes in the Z, A of the emitting nucleus (see later Figure 4.16) that allows extrapolation and interpolation to unknown systems; and (b) the concept of closed decay-energy cycles. This latter idea is shown in Figure 4.2. Consider the α- and β-decays connecting $^{237}_{93}$Np, $^{241}_{95}$Am, $^{241}_{94}$Pu, and $^{237}_{92}$U. By conservation of energy, one can state that the sum of the decay energies around the cycle connecting these nuclei must be zero (within experimental uncertainty). In those cases where experimental data or reliable estimates are available for three branches of the cycle, the fourth can be calculated by difference. An impressive testimonial to

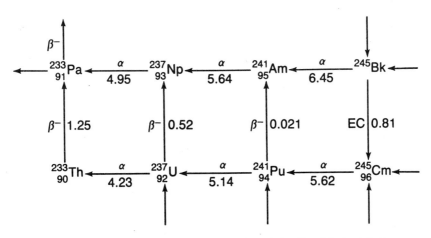

Figure 4.2 Decay cycles for part of the 4n + 1 family. Modes of decay are indicated over the arrows; the numbers indicate total decay energies in MeV.

the power of these semiempirical estimation techniques can be obtained by considering such a compilation made in 1958 (Sea 58). Using modern data to compare to the 1958 estimates of masses and decay energies, one finds an average *percentage* deviation between the predicted values (1958) and known values (1987) of the α- and β-decay energies of 0.80 ± 0.85, -0.56 ± 0.62, -0.16 ± 0.33, and -0.93 ± 0.41 for the 4n, 4n + 1, 4n + 2 and 4n + 3 decay series, respectively. An example of this semiempirical approach to estimating the masses of unknown heavier nuclei is the work of Viola et al. (Vio 74) who calculated the masses of nuclei with $50 \leq Z \leq 118$, $130 \leq A \leq 311$.

An alternative approach to estimating the masses of unknown nuclei, particularly those nuclei that are far away from the regions of known nuclei is that of calculating the masses from a semiempirical, but physically reasonable basis. All such calculations of nuclear masses start with the liquid drop model of nuclei. In this model, the nucleus is likened to a uniformly charged drop of liquid. (Nuclei which have thin skins like the heavy nuclei are especially well described by this model.) The mass of the nucleus containing Z protons and N neutrons is given as:

$$M(Z, A) = ZM_H + (A - Z)M_n - \overset{(1)}{C_1 A} + \overset{(2)}{C_2 A^{2/3}} + C_3 \overset{(3)}{\frac{Z^2}{A^{1/3}}} - C_4 \frac{Z^2}{A} - \delta \quad (4.5)$$

where M_H and M_n are the masses of a hydrogen atom and a neutron, respectively. The coefficients C_1 and C_2 are given as

$$C_1 = a_1 \left[1 - K\left(\frac{N-Z}{A}\right)^2 \right] \quad (4.6)$$

$$C_2 = a_2 \left[1 - K\left(\frac{N-Z}{A}\right)^2 \right] \quad (4.7)$$

The values of the constants a_1, a_2, K, C_3, and C_4 are determined to be 15.677 MeV, 18.56 MeV, 1.79, 0.717 MeV and 1.211 MeV, respectively by fitting data on the masses of known nuclei (Mye 66). The terms labeled 1, 2, and 3 in equation 4.5 represent the volume, surface and coulomb energy terms, respectively. The pairing term, δ, is $+11/A^{1/2}$ for e-e nuclei, 0 for odd A nuclei and $-11/A^{1/2}$ for o-o nuclei.

The volume term ($\propto A$) represents the decrease in mass due to the increased binding by virtue of having more nucleons in the nucleus. Some of these nucleons will be located on the nuclear surface and will not be as tightly bound as nucleons in the interior of the nucleus, hence a correction to the volume energy term which is proportional to the nuclear surface area ($\propto A^{2/3}$) is necessary, that is, the surface energy term. The Coulomb energy term represents the decreased nuclear binding (and increased mass) due to the repulsion between the protons in the nucleus. In Figure 4.3, we compare the predictions of equation 4.5 with the masses of known nuclei. We observe that there is extra stability for nucleon numbers 28, 50, 82, and 126, the so-called magic numbers. These magic numbers are naturally associated with configurations of special nuclear stability as given by the nuclear shell model.

Strutinsky (Str 67) devised a method to calculate these "shell corrections" (and also corrections for nuclear pairing) to the liquid drop model. In this method, the total energy E of a nucleus is taken as the sum of a

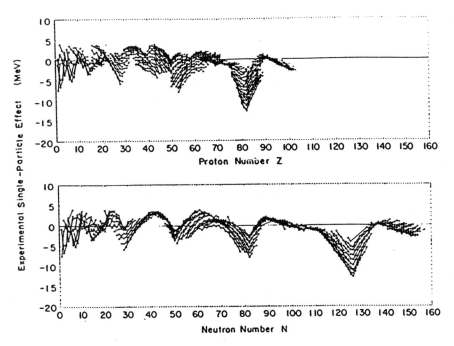

Figure 4.3 Nuclear shell effects on the ground state masses of nuclei (from Mye 77). Plotted is the difference (MeV) between the experimental ground state mass and the mass given by a liquid drop semiempirical mass equation, uncorrected for shells. Note the extra special stabilization energy at nucleon numbers 28, 50, 82, and 126. The extra stability can exceed 10 MeV near 82 protons and 126 neutrons.

liquid drop model energy E_{LDM} and the shell (δS) and pairing (δP) corrections to this energy, that is,

$$E = E_{\text{LDM}} + \sum_{p,n} (\delta S + \delta P) \tag{4.8}$$

The shell corrections (as well as the liquid drop model energies) are functions of the nuclear deformation. One finds that the shell corrections tend to lower the ground state mass of spherical nuclei with near-magic or magic numbers of neutrons and protons (thus reducing the discrepancy between calculated and experimental values of the masses in Figure 4.3). They also tend to lower the ground state mass of midshell nuclei at some finite deformation ($\beta_2 \simeq 0.3$) thus accounting for deformed nuclei like the actinides.

In general, calculations of this type are able to reproduce the masses of known nuclei to ± 1 MeV and are used to estimate the masses of nuclei very far from stability, in particular the very heavy nuclei. The best current estimates of the masses of heavy nuclei are made using the semiempirical prescription of Liran and Zeldes (Lir 76). (It should be noted that the Liran–Zeldes prescription, which has 178 parameters that are fitted to the known masses, is actually based on the nuclear shell model rather than the liquid drop model.)

4.3 FISSION BARRIERS

4.3.1 Introductory Comments

The ability to undergo nuclear fission is one of the most important properties of the transuranium nuclei. The description of this property is closely related to the preceding discussion of the calculation of the masses of transuranium nuclei far from stability. Figure 4.4 shows a schematic view of the fission process. A nucleus with some equilibrium ground state deformation (as specified by equation 4.2) absorbs energy, becoming excited and deforms to a "transition state" or "saddle point" configuration. The formation and decay of this transition state is the rate-determining step for the fission reaction and corresponds to the passage over an activation energy barrier to the reaction. In a very short time, the neck between the nascent fragments in the transition state configuration disappears and the nucleus divides into two fragments at the "scission point." At this scission point one has two deformed, highly charged fragments in contact with each other. The large Coulomb repulsion between the fragments quickly accelerates them to 90% of their final kinetic energy within a time $\sim 10^{-20}$ s. As the accelerated fragments move away from one another, they contract to more spherical shapes, converting the potential energy of deformation into internal excitation energy, that is, they get "hot". This excitation energy is removed by the emission of neutrons (from the fully accelerated fragments) and then, by γ-ray emission followed finally by β^- decay of the n-rich fragments to stability. The overall probability of fission occurring is governed by the rate of passage of the system over the activation energy barrier, that is, the fission barrier.

4.3.2 The Fissionability of Nuclei

The estimation of the height of the fission barrier in heavy nuclei begins with the liquid drop model (equation 4.5). In a volume conserving deformation, the nuclear surface and Coulomb energy act in opposite ways upon the mass (stability) of the deformed system. (As the nucleus initially deforms, the surface energy increases and the Coulomb energy decreases.) If the deformation is not too large, the decrease in Coulomb repulsion allows the increase in surface energy to return the drop to its initial shape. We can parameterize the small *nonequilibrium* deformations of the nuclear surface

$$R(\theta) = R_0(1 + \alpha_2 P_2(\cos \theta)) \qquad (4.9)$$

where α_2 is the quadrupole distortion parameter ($= (5/4\pi)^{1/2}\beta_2$) and P_2 is the second-order Legendre polynomial. For small distortions, the surface, E_s, and Coulomb, E_c, energies are given by

$$E_s = E_s^o(1 + \tfrac{2}{5}\alpha_2^2) \qquad (4.10)$$

$$E_c = E_c^o(1 - \tfrac{1}{5}\alpha_2^2) \qquad (4.11)$$

where E_s^o and E_c^o are the surface and Coulomb energies of undistorted spherical drops. When the change in Coulomb and surface energies ($\Delta E_c = E_c^o - E_c, \Delta E_s = E_s - E_s^o$) are equal, the nucleus becomes spontaneously unstable with respect to fission. At this time

$$\frac{E_c^o}{2E_s^o} = 1 \qquad (4.12)$$

Thus it is natural to express the fissionability of nuclei in terms of a parameter x called the *fissionability parameter* which is given as

(Z_2, A_2)

Scission configuration, two nuclear potentials, fragments highly deformed with the degree of deformation depending upon fragment stiffness; kinetic energies of fragments small (?); occasional small particle emitted although neutron yield may be relatively large.

10^{-20} sec

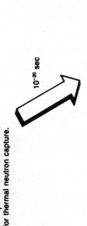

(Z_0, A_0)

Transition state nucleus with saddle deformation and $(E - E_f)$ MeV of excitation. Lifetime depends on $E - E_f$ and is about 10^{-15} sec for thermal neutron capture.

10^{-20} sec

(Z_0, A_0)

Excited nucleus with equilibrium deformation of target and E MeV of excitation energy

Nuclear Reaction

Target nucleus in equilibrium deformation

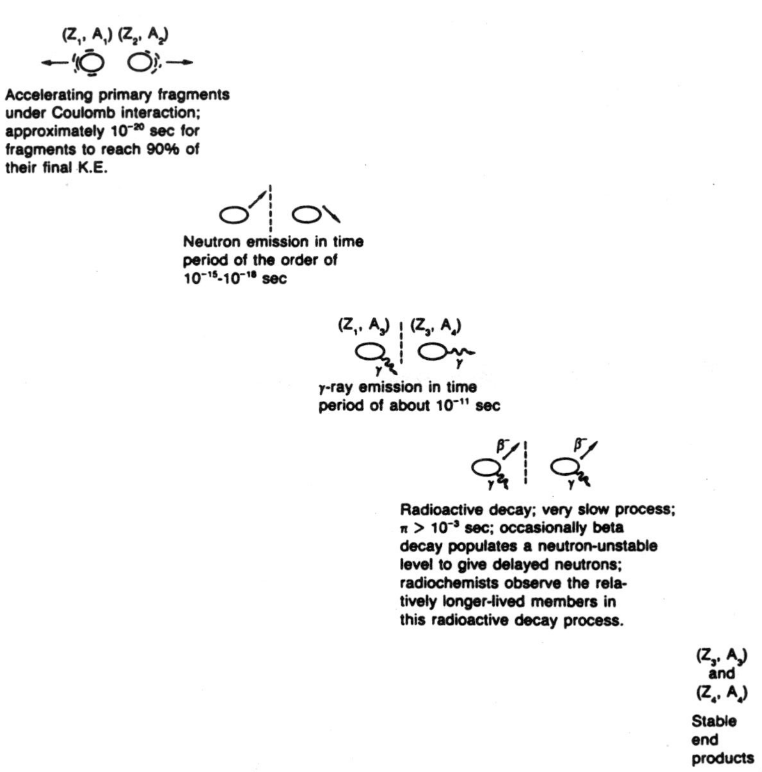

Figure 4.4 A schematic diagram of the fission process (after Gin 68).

$$x \equiv \frac{E_c^\circ}{2E_s^\circ} = \frac{1}{2}\left(\frac{\text{Coulomb energy of a charged sphere}}{\text{Surface energy of the sphere}}\right) \quad (4.13)$$

We can approximate the Coulomb and surface energies of a sphere by the following expressions

$$E_c^\circ = \frac{3}{5}\frac{Z^2 e^2}{R_0 A^{1/3}} = \left(a_c \frac{Z^2}{A^{1/3}}\right) \quad (4.14)$$

where $a_c = 3e^2/5R_0$,

$$E_s = 4\pi R_0^2 S A^{2/3} = a_s A^{2/3} \quad (4.15)$$

where S is the surface tension per unit area and $a_s = 4\pi R_0^2 S$. Then equation 4.13 becomes

$$x = \left(\frac{a_c}{2a_s}\right)\left(\frac{Z^2}{A}\right) = \left(\frac{Z^2}{A}\right) \bigg/ \left(\frac{Z^2}{A}\right)_{\text{critical}} \quad (4.16)$$

where $(a_c/2a_s)^{-1}$ is referred to as $(Z^2/A)_{\text{critical}}$.

Thus the parameter Z^2/A gives a measure of the relative fissionability of nuclei. The greater the value of this parameter, the more "fissionable" the nucleus is. The reader would do well to remember that the Z^2/A factor is simply proportional to the ratio of the disruptive Coulomb energy ($\propto Z^2/A^{1/3}$) to the cohesive surface energy ($\propto A^{2/3}$). More sophisticated treatments of the fissionability of nuclei show that $(Z^2/A)_{\text{critical}}$ varies slightly from nucleus to nucleus and is given by

$$(Z^2/A)_{\text{critical}} = 50.883\left[1 - 1.7826\left(\frac{(N-Z)}{A}\right)^2\right] \quad (4.17)$$

4.3.3 The Calculation of the Fission Barrier Height

The above discussion was intended to show the importance of the parameter Z^2/A in evaluating the relative fissionability of nuclei. It was based upon calculations of the deformation energy of a charged drop for very small distortions from sphericity. The actual calculation of the potential energy of deformation for the liquid drop model involves many deformation coordinates (not just the α_2 of equation 4.9) and is a major computational task. Figure 4.5 shows the qualitative features of the liquid drop

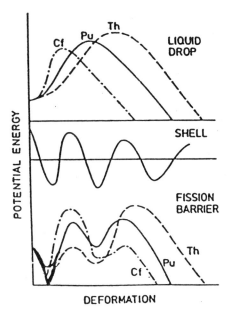

Figure 4.5 Qualitative features of the fission barriers for actinide nuclei (from Bri 82).

model fission barriers. The fission barrier height decreases and the maximum moves to smaller deformations as Z^2/A increases. If we add to this "liquid drop barrier," the same sort of "shell corrections" as we use for the ground state masses of nuclei, significant alterations in the barrier shapes occur (Figure 4.5). Large minima in the realistic shell correction energies occur for ratios of the major/minor nuclear axes of 3/2 and 2/1 (where level bunchings occur in a simple harmonic oscillator potential). The bottom portion of Figure 4.5 shows the results of combining the liquid drop model barriers with the deformation dependent shell corrections. The stable ground state now has some finite deformation ($\beta_2 \sim 0.25$) rather than no deformation (a sphere) and a secondary minimum in the barrier is created at $\beta_2 \sim 0.6$ (axes ratio of 2/1). For the heaviest nuclei where the liquid drop fission barriers are small or nonexistent, the fission barrier heights are enhanced relative to the liquid drop model barriers due primarily to a lowering of the ground state mass by shell corrections.

For nuclei in the uranium-plutonium region, a double-peaked fission barrier with equal barrier heights and a deep secondary minimum is predicted. For heavier nuclei like californium, the first barrier is predicted to be much larger than the second barrier and passage over the first barrier should be rate determining. For lighter nuclei such as thorium, the second barrier should be dominant.

Nuclei can be trapped in the secondary minimum of the fission barrier and will experience considerable hindrance of their γ-ray decay back to the ground state and considerable enhancement of their decay by spontaneous fission. Such nuclei are called fission isomers and they were first observed in 1962 (Pol 62). For the heavy actinides, where the second barrier is lower these isomers should decay predominately by spontaneous fission while in the lighter nuclei, these isomers should decay primarily by γ-ray decay. In addition, experimental results showing subbarrier resonances in fission probability distributions and intermediate structure in neutron induced fission can be explained in terms of the double-humped fission barrier.

Since the initial development of the calculational method by Strutinsky, there have been many calculations and measurements made of the fission barriers of the heavy nuclei. These measurements and calculations have been the subject of encyclopedic review articles by Bjørnholm and Lynn (Bjø 80) and Britt (Bri 80).

There is now ample experimental and theoretical evidence that the lowest energy path in the fission process corresponds to having the nucleus, initially in an axially symmetric and mass (reflection) symmetric shape, pass over the first maximum in the fission barrier with an axially asymmetric, but mass symmetric shape, and then pass over the second maximum in the barrier with an axially symmetric, mass (reflection) asymmetric shape. The fission barriers for the lighter nuclei (radium, thorium) may actually involve a triple-humped fission barrier in which the second or outer maximum in the barrier is split in two. The suggestion has also been made that for nuclei with $N = 154$ the first or inner maximum is split in two.

Both Britt (Bri 80) and Bjørnholm and Lynn (Bjø 80) have independently surveyed all of the available data on fission barriers and have attempted to derive a self-consistent set of barrier systematics. Representing the fission barrier in a one-dimensional manner as shown schematically in Figure 4.6, the combined barrier systematics are given in Table 4.1 and Figure 4.7. The compilations of Britt (Bri 80) and Bjørnholm and Lynn (Bjø 80) disagree slightly in the actual numerical values of the barrier heights E_A and E_B because of different numerical assumptions when calculating fission barrier penetrabilities. By convention, the shape of one maximum of a fission barrier near its top is taken to be that of an inverted harmonic oscillator potential (a parabola) and the transmission coefficient is given by the Hill–Wheeler formula

$$T = (1 + \exp[2\pi(E_{\text{barrier}} - E)/\hbar\omega])^{-1} \qquad (4.18)$$

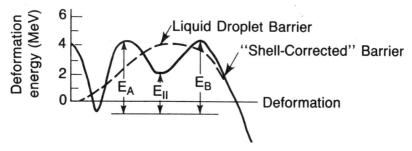

Figure 4.6 A schematic view of the fission barrier showing the barrier as calculated using (a) the liquid drop model (dashed curve) and (b) the shell-corrected liquid drop model (solid curve).

where $\hbar\omega$ is the barrier "curvature" (spacing between the levels in a regular harmonic oscillator potential). Large values of $\hbar\omega$ imply tall, thin barriers with high penetrability; low values of $\hbar\omega$ imply short, fat barriers with low penetrability. Britt (Bri 80) allowed $\hbar\omega$ to vary freely for both the inner and outer barrier while Bjørnholm and Lynn constrained $\hbar\omega$ as follows: e-e nuclei, $\hbar\omega_A = 1.04$ MeV, $\hbar\omega_B = 0.6$ MeV; odd A nuclei, $\hbar\omega_B = 0.45$ MeV. When these differences are taken into account, the derived values of the barrier parameters are in good agreement. In general, these experimental barrier heights should be accurate to 0.3 MeV.

Upon surveying the data in Figure 4.7, one can observe the general trend that the inner barrier height, E_A, is roughly constant (at ~6 MeV) over a wide range of nuclei while the outer barrier height, E_B decreases steadily from about 6 MeV in the thorium isotopes to about 4 MeV in the curium isotopes. In Figure 4.8 we show a comparison between measured (Bri 80) and theoretical values (How 80) for these fission barrier heights. The calculations agree with the experimental barrier heights for the first and second barrier heights within ±1 MeV (or *less* in most cases, see also Kup 84) which is the estimated limit of the accuracy of this method of calculation. There is less accuracy in predicting the height of the second minimum E_{II}, particularly for heavy actinides, as that is a more stringent test of the single particle potentials involved.

Calculations of the barriers for elements with $Z \geq 100$ have been made (Cwi 83, Cwi 85, Ran 76, Möl 86, Bön 86, Sob 87, Möl 87). The calculations generally predict the existence of a single-humped barrier at the approximate position of the first barrier. Experimental barrier curvature energies support the idea of "thin", single humped fission barriers for these nuclei (Hes 86). The height of the fission barrier remains roughly constant or decreases only modestly in going from $Z = 92$ to $Z = 110$. The fission barrier heights deduced from experimental data on the ground state

TABLE 4.1 Measured Fission Barrier Systematics[a]

Isotope	E_A	E_{II} (Relative to Ground State)	E_B
^{238}U	5.7 ± 0.2	2.6 ± 0.1	5.7 ± 0.2
^{239}U	6.3 ± 0.2	1.9 ± 0.3	6.1 ± 0.2
^{240}U	5.7 ± 0.2		5.5 ± 0.2
^{233}Np	5.4		4.7
^{234}Np	5.5 ± 0.2		5.1 ± 0.2
^{235}Np	5.5 ± 0.2		5.2 ± 0.2
^{236}Np	5.8 ± 0.2		5.6 ± 0.2
^{237}Np	5.7 ± 0.2	2.8 ± 0.3	5.4 ± 0.2
^{238}Np	6.1 ± 0.2	2.3 ± 0.3	6.0 ± 0.2
^{239}Np	5.9 ± 0.2		5.4 ± 0.2
^{232}Pu	5.3 ± 0.4		
^{234}Pu	5.8 ± 0.7		
^{235}Pu		2.6 ± 0.4	5.1 ± 0.4
^{236}Pu			4.5 ± 0.4
^{237}Pu	5.9	2.8 ± 0.2	5.2
^{238}Pu	5.5 ± 0.2	2.7 ± 0.2	5.0 ± 0.2
^{239}Pu	6.2 ± 0.2	2.6 ± 0.2	5.5 ± 0.2
^{240}Pu	5.6 ± 0.2	2.4 ± 0.3	5.1 ± 0.2
^{241}Pu	6.1 ± 0.2	1.9 ± 0.3	5.4 ± 0.2
^{242}Pu	5.6 ± 0.2		5.1 ± 0.2
^{243}Pu	5.9 ± 0.2	1.7 ± 0.3	5.2 ± 0.2
^{244}Pu	5.4 ± 0.2		5.0 ± 0.2
^{245}Pu	5.6 ± 0.2		5.0 ± 0.2
^{237}Am		2.4 ± 0.2	
^{238}Am		2.6 ± 0.2	
^{239}Am	6.2 ± 0.3	2.4 ± 0.2	
^{240}Am	6.5 ± 0.2	3.0 ± 0.2	5.2 ± 0.3
^{241}Am	6.0 ± 0.2	2.2 ± 0.2	5.1 ± 0.3
^{242}Am	6.5 ± 0.2	2.9 ± 0.2	5.4 ± 0.3
^{243}Am	5.9 ± 0.2	2.3 ± 0.2	5.4 ± 0.3
^{244}Am	6.3 ± 0.2	2.8 ± 0.4	5.4 ± 0.3
^{245}Am	5.9 ± 0.2		5.2 ± 0.3
^{247}Am	5.5 ± 0.2		
^{241}Cm	6.3 ± 0.3	2.1 ± 0.3	4.3 ± 0.5
^{242}Cm	5.8 ± 0.4		4.0 ± 0.5
^{243}Cm	6.4 ± 0.3	1.9 ± 0.3	
^{244}Cm	5.8 ± 0.2		4.3 ± 0.3
^{245}Cm	6.2 ± 0.2	2.1 ± 0.3	4.4 ± 0.2
^{246}Cm	5.7 ± 0.2		4.2 ± 0.3
^{247}Cm	6.0 ± 0.2		
^{248}Cm	5.7 ± 0.2		
^{249}Cm	5.6 ± 0.2		

TABLE 4.1 (*Continued*)

Isotope	E_A	E_{II} (Relative to Ground State)	E_B
^{250}Cm	5.3 ± 0.2		3.9 ± 0.3
^{245}Bk	6.60^d		
^{246}Bk	6.40^e		
^{247}Bk	6.50^d		
^{248}Bk	6.50^e		
^{249}Bk	6.1 ± 0.2		
^{250}Bk	6.1 ± 0.2		4.1 ± 0.3
^{250}Cf	5.6 ± 0.3		
^{251}Cf	6.15^b		
^{252}Cf	5.30^c		4.80
^{253}Cf	5.4 ± 0.3		
^{249}Es	6.80^h		
^{250}Es	6.70^f		
^{251}Es	6.60^h		
^{255}Es	5.40^g		
^{256}Es	4.80^g ± 0.3		
^{255}Fm	5.70^g ± 0.3		

^{a}Unless otherwise indicated, the above values of the fission barrier parameters are taken from the compilation of Bjørnholm and Lynn (Bjø 80). For these parameters $\hbar\omega$ values are as follows: e-e nuclei $\hbar\omega_A = 1.04$ MeV, $\hbar\omega_B = 0.6$ MeV; odd A nuclei, $\hbar\omega_A = 0.8$ MeV; $\hbar\omega_B = 0.52$ MeV; o-o nuclei, $\hbar\omega_A = 0.65$ MeV; $\hbar\omega_B = 0.45$ MeV.
^{b}From Bri 80; $\hbar\omega_A = 0.75$.
^{c}From Bri 80; $\hbar\omega_A = 1.10$.
^{d}From Bri 80; $\hbar\omega_A = 0.85$.
^{e}From Bri 80; $\hbar\omega_A = 0.45$.
^{f}From Bri 80; $\hbar\omega_A = 0.40$.
^{g}From Britt et al., Phys. Rev. **C21**, 761 (1980).
^{h}From Dah 82.

masses (Arm 85) confirm these predictions (Figure 4.9b). This finding is quite surprising and is due to the large shell corrections to the macroscopic (liquid drop) fission barriers (which become vanishingly small through this region) (Figure 4.9a).

In the region of the heavy fermium isotopes, the structure of the potential surface for fissioning nuclei becomes more complex (Möl 87). The possibility of forming two "magic" fission fragments ($N = 82$) and the resultant stability afforded by these fragment shells causes the appearance of a "new path" to fission leading to a very compact scission configuration

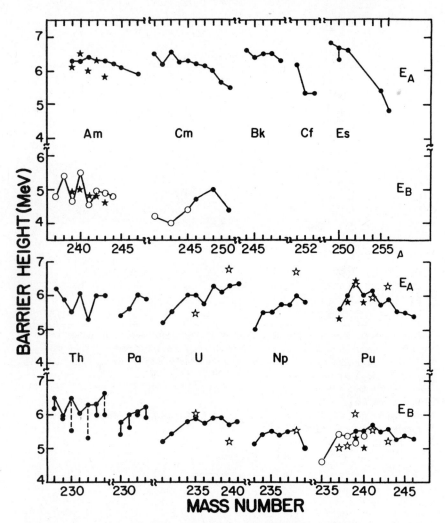

Figure 4.7 Experimental measurements of fission barrier heights from Back and Britt (Bri 82). The symbols E_A and E_B are defined in Figure 4.6.

(which may explain some unusual features of fission in the heavy fermium isotopes—see Section 4.4). Some controversy exists as to whether the "new-path" affects the shape of the fission barrier seen in ground state spontaneous fission and consequently, estimates of the spontaneous fission half-life.

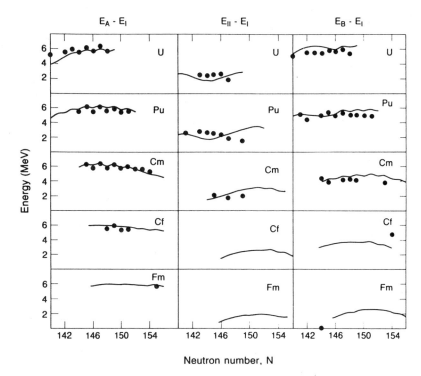

Figure 4.8 A comparison of the experimental fission barrier parameters and those calculated by Howard and Möller (How 80). Left: calculated and experimental (dots) first barrier heights; Middle: calculated and experimental (dots) second minimum heights; Right: calculated (with inclusions of mass asymmetry) and experimental (dots) second barrier heights.

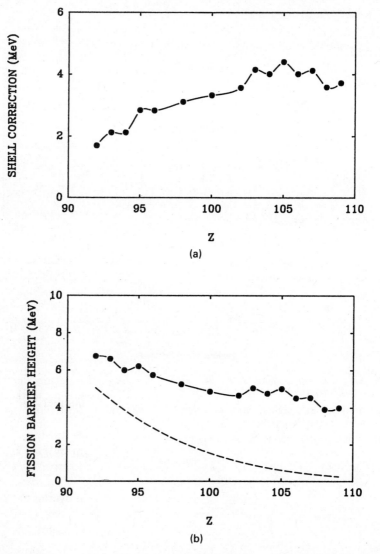

Figure 4.9 (a) A plot of the "experimental" ground state shell corrections deduced from experimental values of the masses of $N - Z = 48$ nuclei. "Experimental" shell corrections were determined by comparing the experimental masses with the macroscopic masses (Möl 88). (b) A plot of the fission barrier heights for $N - Z = 48$ nuclei. The barrier heights were deduced by adding the experimental shell corrections to the macroscopic fission barrier (Dah 82). The dashed line shows the pure liquid drop fission barrier heights (Dah 82) with no shell corrections.

4.4 SPONTANEOUS FISSION

4.4.1 Ground State Spontaneous Fission Systematics

In 1940, Petrzhak and Flerov (Pet 40) found that the ^{238}U nucleus could decay by spontaneously fissioning into two large fragments (with a partial half-life of $\sim 10^{16}$ years). This remarkable finding uncovered what has turned out to be a unique and very important mode of decay for the heavy nuclei, that is, spontaneous fission from the nuclear ground state. Since the pioneering work of Petrzhak and Flerov, many other examples of this decay mode have been found. Spontaneous fission is a rare mode of decay in the light actinides like uranium and increases in importance with increasing atomic number until it is an important (often stability limiting) decay mode for nuclei with $Z \geq 98$. The known, well-established spontaneous fission half-lives change a factor of 10^{28} in going from the longest-lived uranium isotopes to the shortest-lived fermium isotopes.

While the measurement of a spontaneous fission half-life can be a relatively straightforward matter, the assignment of the Z and A of the fissioning system is not. It is not possible, with current technology to simultaneously measure the Z and A of both fragments and the number of emitted neutrons accurately enough to allow deduction of the Z and A of the fissioning system. Therefore other modes of decay (such as α-particle emission) by the fissioning system must be characterized to allow establishment of the Z and A of the fissioning system. Where spontaneous fission is an important mode of decay as in the very heavy nuclei, it is very difficult to reliably assign the atomic and mass numbers of spontaneous fission activities. In Table 4.2, we tabulate the values of all the known, well-established ground state spontaneous fission half-lives.

To understand the data of Table 4.2, we might begin by plotting the measured spontaneous fission half-lives for a number of species versus x, the liquid drop model fissionability parameter as defined in equations 4.16 and 4.17. We show such a plot in Figure 4.10. There is an overall decrease in spontaneous fission half-life with increasing fissionability (increasing Z^2/A of the nucleus) but clearly the spontaneous fission half-life does not depend upon the fissionability parameter alone. One observes that the odd A nuclei have abnormally long half-lives relative to the e-e nuclei. Also the spontaneous fission half-lives of the heaviest nuclei ($Z \geq 104$) are roughly similar with values of the order of milliseconds.

A semiquantitative understanding of some of the main features of these data begins with the idea that spontaneous fission is a barrier penetration phenomenon with the nucleus "tunneling" from its ground state through the fission barrier to the scission point. We expect that the half-life for

TABLE 4.2 Ground State Spontaneous Fission Activities

Nuclide	$J\pi$	$t_{1/2}^{SF}$
^{232}U	0+	8×10^{13} years
^{233}U	5/2+	1.2×10^{17} years
^{234}U	0+	1.42×10^{16} years
^{235}U	7/2−	9.80×10^{18} years
^{236}U	0+	2.43×10^{16} years
^{236}Pu	0+	3.5×10^{9} years
^{237}Np	5/2+	$>1.00 \times 10^{18}$ years
^{238}U	0+	8.08×10^{15} years
^{238}Pu	0+	4.77×10^{10} years
^{239}Pu	1/2+	5.5×10^{15} years
^{240}Pu	0+	1.2×10^{11} years
^{240}Cm	0+	1.9×10^{6} years[a]
^{241}Pu	5/2+	$\sim 6 \times 10^{16}$ years[a]
^{241}Am	5/2−	1.147×10^{14} years
^{242}Pu	0+	6.84×10^{10} years
^{242}Cm	0+	2.6×10^{9} years
^{242}Fm	0+	800 μs
^{243}Am	5/2+	3.4×10^{13} years
^{244}Am	0+	6.61×10^{10} years
^{244}Cm	0+	1.344×10^{7} years
^{244}Fm	0+	3.7 ms
^{245}Cm	7/2+	1.4×10^{12} years[a]
^{246}Cm	0+	1.809×10^{7} years
^{246}Cf	0+	2.0×10^{3} years
^{246}Fm	0+	14 s
^{248}Cm	0+	4.12×10^{6} years
^{248}Cf	0+	3.2×10^{4} years
^{248}Fm	0+	10 h
^{249}Bk	7/2+	1.9×10^{9} years
^{249}Cf	9/2−	6.7×10^{10} years
^{250}Cm	0+	$<1.13 \times 10^{4}$ years
^{250}Cf	0+	1.7×10^{4} years
^{250}Fm	0+	~ 10 years[a]
^{250}No	0+	250 μs
^{252}Cf	0+	85.5 years
^{252}Fm	0+	125 years
^{252}No	0+	8.55 s
^{253}Es	7/2+	6.4×10^{5} years
^{253}Rf	−	~ 3.6 s
^{254}Cf	0+	60.7 days
^{254}Fm	0+	228 days
^{254}Rf	0+	500 μs

TABLE 4.2 (*Continued*)

Nuclide	$J\pi$	$t_{1/2}^{SF}$
^{255}Es	(7/2+)	2700 years
^{255}Fm	7/2+	9500 years
^{255}Rf	–	2.7 s[a]
^{255}Ha	–	~7.5 s
^{256}Cf	0+	12.3 min
^{256}Fm	0+	2.86 h
^{256}No	0+	18 min
^{256}Rf	0+	6.9 ms[a]
^{257}Fm	(9/2+)	131 years
^{257}Rf	–	27 s
^{257}Ha	–	11 s
^{258}Fm	0+	0.380 ms
^{258}No	0+	1.2 ms
^{258}Rf	0+	13 ms
^{259}Fm	–	1.6 s[a]
^{259}Md	(7/2–)	1.6 h
^{259}Rf	–	38 s
^{260}Md	–	33 days[a]
^{260}No	–	106 ms[a]
^{260}Rf	0+	21 ms
^{260}Ha		16 s
^{260}106	0+	7.2 ms[a]
^{261}Ha		7.2 s
^{261}Lr		39 min
^{262}No	0+	5 ms
^{262}Lr		216 min
^{262}Rf	0+	47 ms
^{262}Ha		47 s
^{263}106		1.1 s

[a]D.C. Hoffman and L.P. Somerville in *Charged Particle Emission from Nuclei*, Vol. II (CRC Press, Boca Raton, 1988). Primary data source for this table was Bro 86b.

spontaneous fission (in seconds) will be given as

$$T_{1/2}^{SF} = \frac{\ln 2}{fp} \quad (4.19)$$

where f is the frequency of assaults on the fission barrier ($\sim 10^{20}$/s) and p is the barrier penetrability. Clearly the most important factor is the barrier

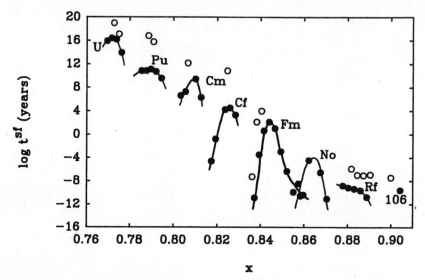

Figure 4.10 The spontaneous fission half-life of e-e (●) and e-o (○) transuranium nuclei as a function of the fissionability parameter x, calculated using the parameters suggested by Kra 79.

penetrability p. If one ignores the changes which take place in the nascent fragments during barrier penetration the penetrability is given as

$$p = \exp(-(2/\hbar) \int_{\epsilon_1}^{\epsilon_2} \sqrt{2M(V-E)}\, d\epsilon) = e^{-k} \qquad (4.20)$$

where M is the reduced mass of the separating fragments and $V - E$ is their "negative kinetic energy" at distortion ϵ. The range of integration extends over the thickness of the barrier. This situation is depicted in Figure 4.11. The penetrability of the barrier depends exponentially upon "the area underneath the barrier." As indicated in Section 4.3.3, this area depends roughly on Z^2/A (the fissionability parameter x). Because the penetrability depends *exponentially* on $V - E$, small changes in V or E can have large effects on the penetrability.

Many years ago, Swiatecki realized this (Swi 55) and pointed out that the deviations of the spontaneous fission half-life from a smooth dependence on Z^2/A might be due to small fluctuations in E, in particular, deviations, δm, of the ground state mass from a smooth liquid drop model mass surface (these deviations are exactly the shell and pairing corrections

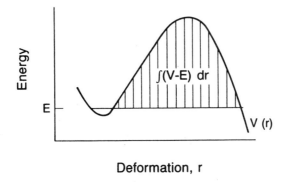

Figure 4.11 Schematic diagram of the penetration of the fission barrier by a nucleus initially in a state E.

calculated in Section 4.2). A change of 1 MeV in the ground state mass was found to change the spontaneous fission half-life by a factor of 10^5.

In Figure 4.12, we show a comparison of the measured spontaneous fission half-life of various transuranium nuclei and that which one would calculate using equations 4.19 and 4.20, assuming the potential energy V was given by a smooth, droplet-like representation (Pat 89). The measured half-lives are a few to fifteen orders of magnitude longer than those one would expect based upon a simple smooth droplet-like potential energy

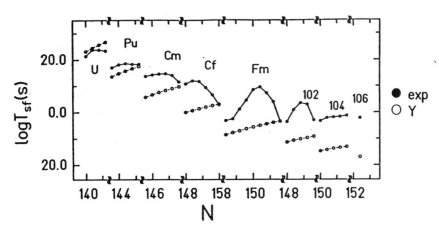

Figure 4.12 Comparison of the measured spontaneous fission half lives (●) for e-e nuclei with those expected in a simple droplet model approach (○). (Pat 89)

surface which reflects essentially the dependence on Z^2/A. The difference between the measured and calculated values of $t_{1/2}^{SF}$ are thought to be due to "shell effects," primarily in the value of the ground state mass. Without this lengthening of the spontaneous fission half-lives by shell effects, the heaviest nuclei would have half-lives that are so short ($t \sim 10^{-18}$–10^{-16} s) as to render them unobservable. Thus the existence of the heaviest transuranium nuclei can be said to represent a dramatic demonstration of the existence of shell effects in nuclei.

The "retardation" of spontaneous fission for odd A and o-o nuclei relative to e-e nuclei is thought to be due primarily to two factors. They are (a) changes in the inertial mass M of equation 4.20 due to a reduction of pairing among the nucleons and (b) due to the need to conserve angular momentum and parity of the odd nucleon(s) during deformation, the nucleus cannot follow the lowest energy path to fission, resulting in an increased barrier height. This increase in barrier height due to the presence of odd particle(s) is referred to as the "specialization energy."

This retardation of spontaneous fission for odd A nuclei can be parameterized in the form of "hindrance factors" (Hof 88) as

even Z, odd N

$$\log(HF) = \log[t_{1/2}^{SF}(^A Z)] - \tfrac{1}{2}\{\log[t_{1/2}^{SF}(^{A-1} Z)] + \log[t_{1/2}^{SF}(^{A+1} Z)]\} \quad (4.21)$$

odd Z, even N

$$\log(HF) = \log[t_{1/2}^{SF}(^A Z)] - \tfrac{1}{2}\{\log[t_{1/2}^{SF}(^{A-1} Z - 1)] + \log[t_{1/2}^{SF}(^{A+1} Z + 1)]\} \quad (4.22)$$

Typical values of the hindrance factor, HF, are $\sim 10^5$ although values for individual nuclei can range from ~ 1 to 10^{10}.

Calculation of spontaneous fission half-lives is difficult. Previously we have noted that calculations of fission barrier heights are generally accurate to ± 1 MeV; yet as observed above, a 1-MeV change in the fission barrier height changes the spontaneous fission half-lives by 10^5. Furthermore, since one is dealing with "the area underneath the barrier" in spontaneous fission, one is sensitive to the barrier shape as well as its height along with any changes in the inertial mass M with deformation (equation 4.20). Nonetheless, numerous predictions of the spontaneous fission half-life of unknown heavy transuranium nuclei have been made (Vio 66b, Ran 76, Loj 85, Bön 86, Möl 87, Sob 87, Loj 88, Möl 89). In Figure 4.13, we show an example of such predictions for $Z = 104$–110. The crucial point for those who would synthesize new nuclei is that spon-

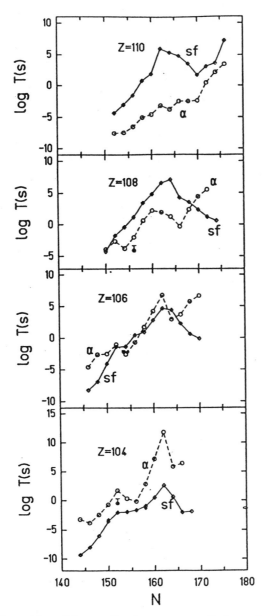

Figure 4.13 Logarithm of the predicted spontaneous fission and alpha decay half lives for e-e nuclei (Loj 88). Measured values are indicated by filled diamonds and circles.

taneous fission is not the stability-limiting mode of decay for the heaviest elements and the spontaneous fission half-lives are not predicted to fall precipitously with increasing Z. One also sees evidence in Figure 4.13 for a special stability associated with $N = 162–164$.

4.4.2 Spontaneously Fissioning Isomers (Van 77, Poe 88)

As previously mentioned, the first spontaneously fissioning isomer was discovered in 1962 by Polikanov et al. (Pol 62). During the late 1960s and early 1970s, a considerable number of additional isomers were found and characterized. A systematics of the isomer half-lives was developed. A good deal of attention has been given to investigating the nuclear spectroscopy of these unusual nuclear states. From these investigations, a reasonably complete description of this phenomenon has emerged. The central feature of this description, as mentioned earlier, is the fact that spontaneously fissioning isomers are stationary states in the second minimum in the fission barrier.

A list of the known spontaneously fissioning isomers and their half-lives is given in Table 4.3 and illustrated in Figure 4.14. Note that the isomeric half-lives are generally in the range of 10^{-9}–10^{-3} s whereas the ground state half-lives are $\sim 10^{25}$–10^{30} times longer. As seen in Figure 4.14, these isomers ranging from thorium to berkelium form an island about a point of maximum stability at ^{242}Am. Gamma-ray decay limits the stability of isomers with lower Z and N than those in the island while high rates of spontaneous fission limit the stability of nuclei with higher Z and N. The isomer half-lives show a strong dependence upon neutron number and a strong odd-even effect. Metag (Met 75) has made a phenomenological analysis of these effects by fitting the spontaneous fission half-lives of 33 isomers with the equation

$$t_{1/2}^{SF} = (\ln 2)(4 \times 10^{-21})\left[1 + \exp\frac{2\pi}{\hbar\omega}[ax + b(N - N_0)^2 + d + \delta]\right] \quad (4.23)$$

with

$$\hbar\omega = \hbar\omega_0 \begin{matrix} 1/\delta & \text{odd-odd} \\ 1 & \text{odd-even} \\ \delta & \text{odd-even} \end{matrix} \quad \text{and} \quad s = \begin{matrix} 2\delta_0 & \text{odd-odd} \\ \delta_0 & \text{odd-even} \\ 0 & \text{even-even} \end{matrix} \quad (4.24)$$

where fissionability parameter x was assumed to be

$$x = \frac{Z^2/A}{2(a_s/a_c)\{1 - [k(N - Z)^2/A^2]\}} \quad (4.25)$$

TABLE 4.3 Occurrence and Half-Life of Spontaneously Fissioning Isomers[a]

Nucleus	$t_{1/2}$	Nucleus	$t_{1/2}$
^{233}Th	17 ns		
^{236}U	115 ns	^{240}Am	910 μs
^{238}U	240 ns	^{241}Am	1.5 μs
^{237}Np	45 ns	^{242}Am	14 ms
^{235}Pu	30 ns		
^{236}Pu	34 ns	^{243}Am	5.5 μs
	37 ps	^{244}Am	6.5 μs
			900 μs
^{237}Pu	110 ± 9 ns	^{245}Am	640 ± 60 μs
	1100 ± 80 ns	^{246}Am	73 ± 10 μs
^{238}Pu	600 ps	^{240}Cm	10 ps
	6 ns		55 ns
^{239}Pu	8 μs	^{241}Cm	15 ns
	3 ns		
^{240}Pu	3.8 ns	^{242}Cm	180 ns
^{241}Pu	24 μs		50 ps
	30 ns		
^{242}Pu	3.6 ns	^{243}Cm	42 ns
	50 ns	^{244}Cm	>100 ns
			≥5 ps
^{243}Pu	60 ns	^{245}Cm	13 ± 2 ns
^{244}Pu	400 ps	^{242}Bk	600 ns
^{245}Pu	90 ± 30 ns[b]		9.5 ns
^{237}Am	5 ns	^{244}Bk	820 ns
^{238}Am	35 μs		
^{239}Am	163 ns	^{245}Bk	2 ± 1 ns

[a] Data from Bro 86b unless otherwise indicated. Half-life values are the total half-life for the isomeric state; the spontaneous fission partial half-life may be different. From Figure 4.15 (or similar data compilations) one can determine the energies of the α-particles.
[b] Met 80.

with $k = 1.87$, $a_s = 17.64$ MeV, $a_c = 0.72$ MeV, and $N_0 = 146$. Least square fitting gave

$$\frac{a}{\hbar\omega_0} = -49.4 \pm 5.2 \tag{4.26}$$

$$\frac{b}{\hbar\omega_0} = (3.9 \pm 0.4) \times 10^{-2} \tag{4.27}$$

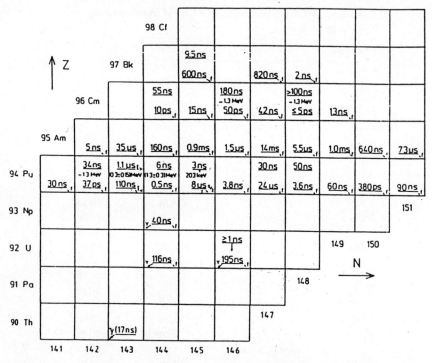

Figure 4.14 Part of the nuclide chart giving the half-lives of all fission isomers known at present. Two values for the same nucleus indicate spin-isomeric states in the second minimum (Hab 89).

$$\frac{d}{\hbar\omega_0} = 46.0 \pm 4.4 \qquad (4.28)$$

$$\frac{\delta_0}{\hbar\omega_0} = 0.43 \pm 0.26 \qquad (4.29)$$

$$\delta_0 = 1.16 \pm 0.08 \qquad (4.30)$$

The quality of the fits to the half-lives is sufficiently good to allow use of the above relations to predict $t_{1/2}$ for unknown isomers.

Spontaneously fissioning isomers have become of interest to nuclear spectroscopists because of their large deformation and special symmetry (Met 80). The first major development in isomer spectroscopy was the measurement of the conversion electrons (Spe 72) from the rotational transitions for the ground state band and for the band built on the spontaneously fissioning isomeric state of ^{240}Pu. It was found that the moment of inertia for the isomeric state is more than twice as large as the moment of inertia for the ground state band providing unequivocal evidence for the large deformation of the isomeric state. The measurement of the electric quadrupole moment of the isomeric state confirms this measurement and provides additional information about the exact deformation of the isomeric state. The measurements all indicate that these isomeric states have quadrupole moments typical of prolate ellipsoids with a ratio of axes of 2:1.

4.5 ALPHA-PARTICLE DECAY

Among the most important nuclear properties of the transuranium nuclei are their radioactive decay properties. These properties of the known transuranium nuclei are summarized in Figure 4.15 and Appendix I. While it may not be obvious in Figure 4.15, all of the elements beyond uranium are unstable with respect to α-particle emission. (For this reason, there are no stable transuranium nuclides.) While there are some transuranium nuclei whose primary decay mode involves spontaneous fission or β-decay, the most common mode of decay for these heavy nuclei is α-particle emission.

From Figure 4.15 (or similar data compilations) one can determine the energies of the alpha particles which characterize the decay of a particular radionuclide. When coupled with a measurement of the half-life of a given species, this information can be used to identify the nuclide. The α-decay energies for unknown nuclei can be predicted with some confidence from predicted values of the nuclear masses (see Section 4.2). In Figure 4.16 we show for e-e transuranium nuclei the energy released in α decay, Q_α, as a function of the neutron number N of the emitting nuclide for each atomic number, Z. The energy released in alpha decay, Q_α, differs from the energy of the emitted alpha particle by two small correction factors: (a) the energy of the recoiling daughter nucleus $(M_\alpha/M_{\text{daughter}})E_\alpha$; and (b) the change in total electronic binding energies in the decay ($\sim 65.3Z^{7/5} - 80Z^{2/5}$ eV.) For most nuclides, the -decay energies for the isotopes of a given element generally decrease with increasing mass number and for a series of isobars they increase with increasing Z. These trends are simple consequences of the shape of the nuclear mass surface. Superimposed on these general trends are very sharp discontinuities at the neutron magic numbers of 126 (and 184) due to shell effects on

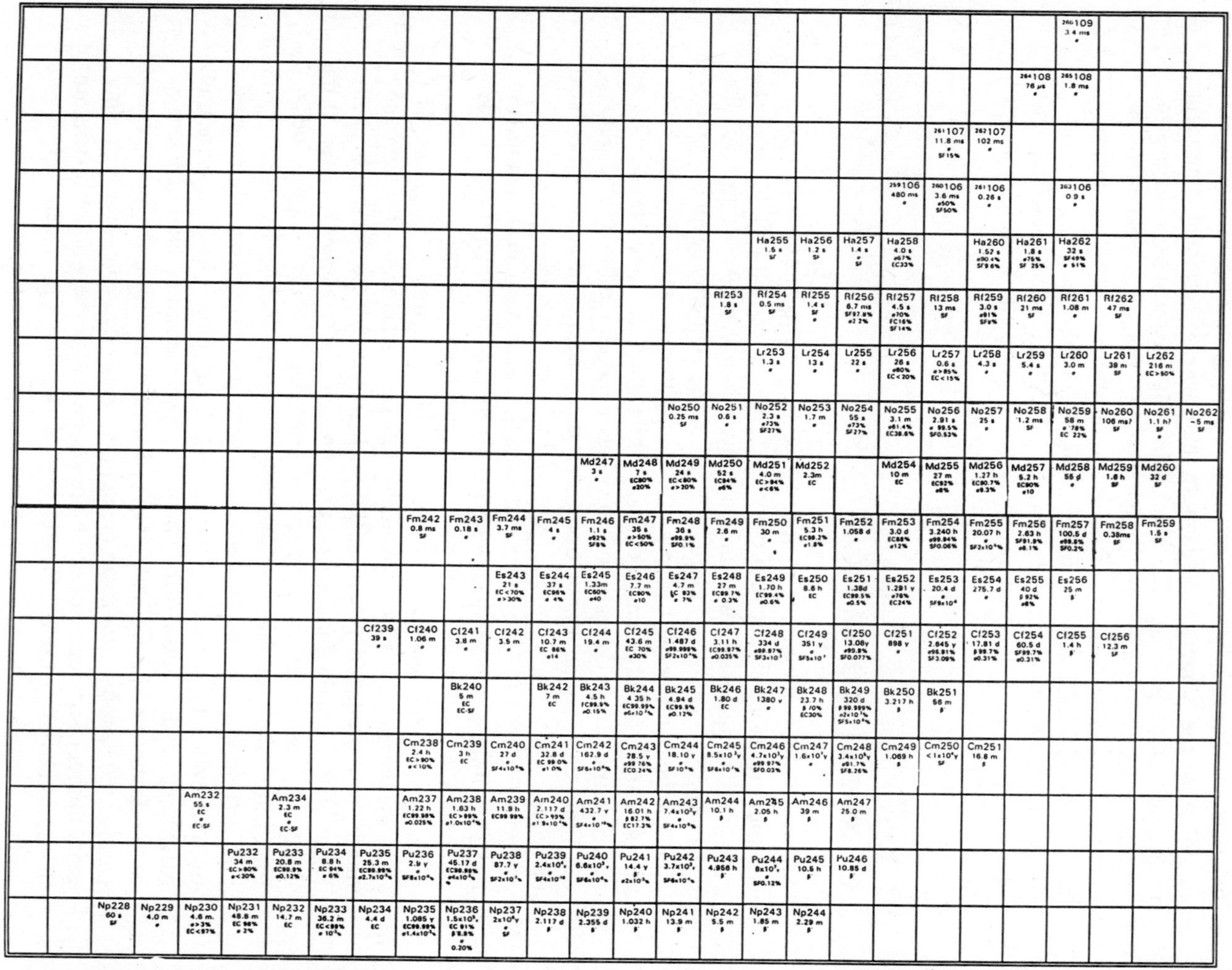

Figure 4.15 A portion of the nuclide chart showing the decay properties of the

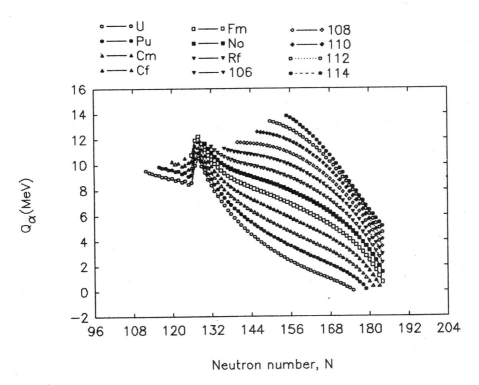

Figure 4.16 The energy release in alpha decay, Q_α, as a function of neutron number N for e-e nuclei. Values of Q_α were calculated using the Liran–Zeldes mass formula (Lir 76).

the mass surface. A small decrease in Q_α near $N = 152$ occurs and is due to a gap in the neutron level structure at $N = 152$ (see Figure 4.17).

The extreme sensitivity of the decay rate to the value of Q_α is responsible for the fact that alpha decay only populates excited states within a few hundred kilovolts of the ground state. The exact correlation is that log $t_{1/2}$ is a linear function of $1/\sqrt{Q_\alpha}$ for isotopes of a given element. This dependence which may seem peculiar at first glance can be shown to be an approximate consequence of the quantum mechanical treatment of alpha decay which involves the calculation of barrier penetration probabilities. Taagepera and Nurmia (Taa 61) have used this correlation to construct a semiempirical relationship between the alpha half-life and the daughter atomic number Z and the α-particle energy E_α for e-e nuclei, that is,

$$\log_{10} t_{1/2}^{\alpha} \text{ (years)} = 1.61 \, (ZE_{\alpha}^{-1/2} - Z^{2/3}) - 28.9 \qquad (4.31)$$

An improved representation due to Keller and Munzel (Kel 70) has the form

$$\log_{10} t_{1/2}^{\alpha}(s) = a(Z/\sqrt{Q_{\alpha}} - Z^{2/3}) + b \qquad (4.32)$$

where a and b are fitted to different groups of nuclei. The best-fit values of a and b are: e-e nuclei $a = 1.61$, $b = -20.261$; e-o nuclei $a = 1.65$, $b = 20.238$; o-e nuclei $a = 1.66$, $b = 20.726$; o-o nuclei, $a = 1.77$, $b = 20.657$.

The actual quantum mechanical WKB treatment of the penetration of a one-dimensional barrier for $\ell = 0$ alpha particles gives the following equation for the penetration factor P

$$P \simeq \exp\left(\frac{-2}{\hbar} \int 2M(V(r) - E) \, dr\right)^{1/2} \qquad (4.33)$$

where the symbols have similar meaning as for equation 4.20 and Figure 4.11. Using a simple Coulomb potential, $V(r) = 2Ze^2/r$, equation 4.33 reduces to

$$P \simeq \exp\left\{\frac{-2}{\hbar}(2MB)^{1/2}R[X^{-1/2} \arccos X^{1/2} - (1-X)^{1/2}]\right\} \qquad (4.34)$$

where $X = Q_{\alpha}/2Ze^2$, $B = 2Ze^2/R$ and R is the nuclear radius of the daughter. With a proper choice of constants and approximations, equation 4.34 reduces to the Taagepera–Nurmia or Keller–Munzel relation, equation 4.31 or 4.32. Using either equation 4.32 or 4.34, the α-decay properties of e-e nuclei decaying to the ground states of their daughters are reasonably well described. For decay to excited states of the daughter ($\ell \neq 0$), a small correction to the potential energy $V(r)$ is made in the form of a centrifugal potential

$$V_{\text{cent}} = \frac{\hbar^2}{2mr^2} \ell(\ell + 1) \qquad (4.35)$$

where ℓ is the orbital angular momentum of the emitted alpha particle.

There are, of course, some strict selection rules for alpha decay based on conservation of total angular momentum and parity. Thus we have $I_i - I_f \leq L \leq I_i + I_f$ where I_i and I_f are the nuclear spins (total angular momenta) of the initial and final nuclear states and L is the orbital angular momentum of the emitted alpha particle. Only even values of L are allowed if the parity (π) of the initial and final nuclear states is the same, and only odd values of L are allowed if there is a parity change. If either I_i or I_f is zero, it is possible to have cases where alpha decay is absolutely forbidden (to the extent that parity is conserved in the decay). For example, alpha decay of an even-even nucleus (0+) cannot populate odd spin states of even parity, or even spin states of odd parity.

Up to now, we have only considered the decay rates of e-e alpha emitters. The odd nucleon alpha emitters, especially in ground state transitions, decay at a slower rate than that suggested by the simple one-body formulation as applied to e-e nuclei. This means that the decay data for such nuclides would lie above the lines shown in Figure 4.16, or for odd Z nuclides the points would lie above the hypothetical interpolated lines. These are referred to as "hindered" decays and the "hindrance factor" may be defined as the ratio of the measured partial half-life for a given α-transition to the half-life which would be calculated from simple, one-body, α-decay theory as applied to e-e nuclei. These hindrance factors for odd A nuclei may be divided into five classes:

a. If the hindrance factor is between 1 and 4, the transition is called a "favored" transition. In such decays, the emitted alpha particle is assembled from two low lying pairs of nucleons in the parent nucleus, leaving the odd nucleon in its initial orbital.
b. A hindrance factor of 4–10 indicates a mixing or overlap between the initial and final nuclear states involved in the transition.
c. Factors of 10–100 indicate that spin projections of the initial and final states are parallel.
d. Factors of 100–1000 indicate transitions with a change in parity but with projections of initial and final states being parallel.
e. Hindrance factors of >1000 indicate that the transition involves a parity change and a spin flip, that is, the spin projections of the initial and final states are antiparallel.

4.6 HEAVY PARTICLE RADIOACTIVITY

One of the favorite pedagogical exercises of the authors when discussing α-particle decay in introductory nuclear chemistry classes was to pose the question "Why don't heavy nuclei spontaneously decay by emitting carbon ions?" The student was supposed to realize that although the Q value for such decays was a large, positive number, the barrier penetrability was so small along with the small probability of "preforming" a carbon ion in the emitting nucleus as to preclude observation of such a decay. The industrious student would realize though the probability of such a decay was not zero, it was very small compared to the probability of α-particle decay. In a beautiful experiment reported in 1984 (Ros 84), Rose and Jones found the first documented case of such heavy particle radioactivity in the decay ^{223}Ra $\rightarrow$ ^{14}C + ^{209}Pb. The probability for the decay was $\sim 10^{-9}$ of the α-decay probability. In six months of counting a 3.3 μCi source of ^{223}Ra (in secular equilibrium with 21 y ^{227}Ac), eleven events in which 29.8 MeV ^{14}C nuclei were emitted ($Q = 31.9$ MeV) were observed. Since that pioneering experiment, a number of other examples of heavy particle radioactivity have been observed, such as ^{24}Ne and ^{28}Mg emission by nuclei with $Z \geq 88$ (allowing the daughter nucleus to be stabilized by the $Z = 82$ or $N = 126$ shell closures).

4.7 NUCLEAR STRUCTURE

We have come as far as we can in our discussion of radioactive decay without involving any details of the nuclear structure of the transuranium nuclei. However, to progress further in our discussion of radioactive decay, and to get a proper picture of one of the most important aspects of the transuranium nuclei, we need to discuss the main features of the nuclear structure of these nuclei. (A detailed discussion of this topic is beyond the scope of this book. There are a number of excellent elementary, intermediate and advanced textbooks dealing with the general features of nuclear structure [Fri 81, Coh 71, Mar 69, DeS 74, Kra 88, Boh 69]. Especially recommended discussions of the nuclear structure of the heaviest elements are found in Ras 75, Sol 76, and Cha 77.)

4.7.1 Intrinsic States

Discussions of nuclear structure begin with the simple nuclear shell model which considers the motions of the unpaired nucleons in an average field generated by the other nucleons in the nucleus. This average nuclear field is assumed to be isotropic and is mathematically approximated by a square well potential

$$V(r) = -V_0, \quad r < R \qquad (4.36)$$

$$V(r) = 0, \quad r \geq R$$

(where $V(r)$ is the potential at distance r from the center of the nucleus and R is the nuclear radius), or by a harmonic oscillator potential

$$V(r) = -V_0\left[1 - \left(\frac{r^2}{R}\right)\right] \qquad (4.37)$$

or by a special form such as the Woods–Saxon potential

$$V(r) = \frac{-V_0}{1 + \exp[(r-R)/a]} \qquad (4.38)$$

where a is a constant ($\simeq 0.6$ fm).

In this simple shell model, the spectroscopic state of all odd A nuclei is determined by the odd nucleon. The spin, s ($=1/2$), and orbital angular momentum, ℓ, of this odd nucleon combine to give j, the total angular momentum of the particle (and the nucleus). Due to spin-orbit coupling, states of higher j ($=\ell + 1/2$) are more stable than states of lower j ($=\ell - 1/2$). The parity (π) of the nucleus is positive or negative according to whether the value of π of the odd nucleon is even or odd, respectively. All of the nucleons are paired in the ground states of e-e nuclei, leading to zero spin and positive parity, and excited states result from unpairing and excitation of nucleons to higher levels. More complicated considerations are involved in determining the spectroscopic states of odd-odd nuclides (containing both an odd number of neutrons and protons) and of nuclides in general where the interactions of two or more unpaired nucleons are important. Extra stability associated with filling a major shell is found for spherical nuclei containing 2, 8, 20, 28, 50, and 82 protons; 2, 8, 20, 28, 50, 82, and 126 neutrons; and predicted for 114 protons and 184 neutrons—the so-called magic numbers.

As discussed previously, the elements beyond uranium assume nonspherical shapes. The theory of single particle states in a nonspherical potential was first investigated by S.G. Nilsson and these states have become known as Nilsson states. To help us to understand the nature of the Nilsson states, we need to remember that as one goes from a spherical potential to a spheroidal (nonspherical) potential, the angular momentum of the single particle state, j, is not a constant, but its projection on the symmetry axis, Ω, is. Thus a $d_{5/2}$ state (which is 6-fold degenerate in a spherical potential because all $2j + 1$ projections of the angular momentum lie at the same energy) splits into three doubly degenerate states ($\Omega = \pm 5/2, \pm 3/2, \pm 1/2$) in a spheroidal nucleus (because the states Ω and $-\Omega$ are degenerate while those of different Ω are not). For prolate deformations that occur in transuranium nuclei, states of low Ω are more stable than those of high Ω.

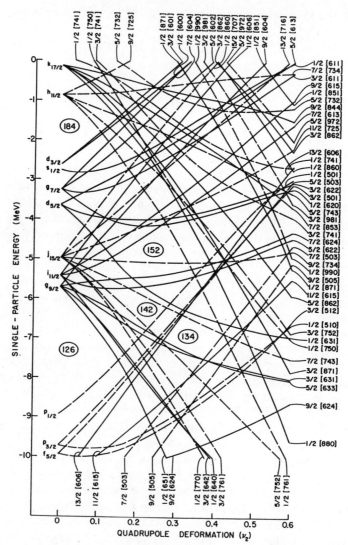

Figure 4.17 Actinide neutron single particle levels obtained from a Woods–Saxon potential as a function of quadrupole deformation, v_2 (Cha 77). As customary, levels of odd parity are indicated by dashed lines, while levels of even parity are indicated by solid lines. See Cha 77 for exact definition of v_2. The numbers in brackets are the asymptotic quantum numbers N, n_z, and Λ.

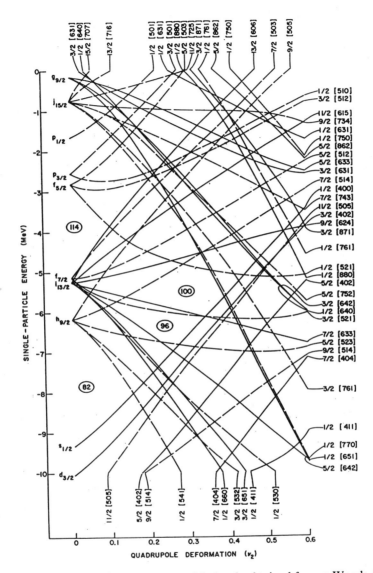

Figure 4.18 Actinide proton single particle levels obtained from a Woods–Saxon potential as a function of quadrupole deformation, v_2 (Cha 77). The numbers in brackets are the asymptotic quantum numbers N, n_z, and Λ.

The actual splitting of single particle states in spheriodal nuclei was investigated by Nilsson using an anisotropic three-dimensional harmonic oscillator potential. Each state in the deformed nucleus is characterized by the quantities $\Omega^\pi [N, n_z, \Lambda]$; Ω and π are good quantum numbers while the terms in the bracket are termed the asymptotic quantum numbers which describe the state at large deformation. N is the principal oscillator quantum number, $\pi = (-1)^N$ is the parity of the state, n_z, the oscillator quantum number along the nuclear symmetry axis (the number of planar nodes perpendicular to the symmetry axis), and Λ is the projection of the orbital angular momentum on the symmetry axis. Ω must, of course, be equal to $\Lambda \pm 1/2$. Other quantum numbers may be deduced from this set. The intrinsic spin projection must equal the difference of total and orbital projections; thus $\Sigma = \Omega - \Lambda$. Typical sets of Nilsson states relevant to transuranium nuclei are shown in Figures 4.17 and 4.18. (Both these diagrams were calculated assuming a Woods–Saxon potential instead of a harmonic oscillator). Note the gaps in the number of levels at the spherical "magic" numbers and the emergence of a gap at $N = 152$ at finite deformations. In transuranium nuclei, neutrons (protons) fill each orbital above the closed shell of 126 (82) pairwise and the ground state spin and parity of an odd A nucleus is simply the Ω^π of the state occupied by the last unpaired nucleon.

4.7.2 Collective States

Since the nuclei beyond uranium have stable nonspherical shapes, they have distinguishable orientations in space and are expected to show rotational levels in addition to the single particle or intrinsic states discussed above. Furthermore since addition of energy to these nuclei can cause them to deform away from their equilibrium states, one expects these nuclei to be able to oscillate about their equilibrium states and thus show vibrational levels.

4.7.2.1 *Rotational Levels.*
The rotation of an axially symmetric deformed nucleus is described using the vector quantities I, K, and M (see Figure 4.19). I is the total nuclear angular momentum, M is the projection of I upon a space fixed axis z, and K is the projection of I upon the nuclear symmetry axis. In the simplest mode of rotational motion (Figure 4.19b), the angular momentum of an odd nucleon, j, lies along the symmetry axis and $K = \Omega$. The collective rotation of the nucleus gives rise to a component of the angular momentum R that is perpendicular to the nuclear symmetry axis.

For this simplest mode of rotation, the rotational energy levels are

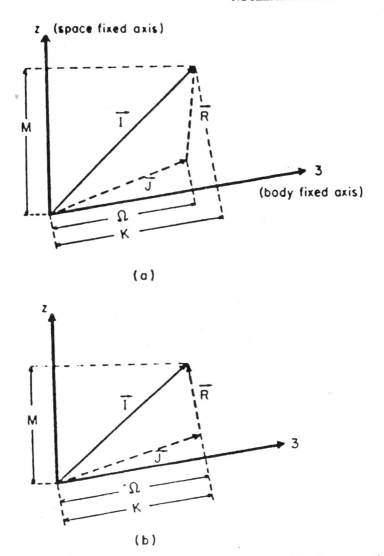

Figure 4.19 Coupling diagram for odd A deformed nuclei (under assumption of strong coupling). The total angular momentum I has projection M on the space fixed axis and projection K on the body fixed symmetry axis. I is resolved into its components, j (the angular momentum of the last unpaired nucleon (with its projection on the symmetry axis L)) and R, the collective rotational angular momentum. (a) represents the general case, K while (b) represents the special case, $K = \Omega$.

given by

$$E_{ROT} = \frac{\hbar^2}{2\mathfrak{I}}\langle \vec{R}^2 \rangle \qquad (4.40)$$

where $\mathfrak{I}$ is the effective moment of inertia of the nucleus (less than $\mathfrak{I}_{rigid}$). But $\vec{R} = \vec{I} - \vec{j}$ and therefore we have, by the cosine law, and quantum mechanics

$$\langle \vec{R}^2 \rangle = [I(I+1) + j(j+1) - 2I_3j_3 - 2I_yj_y - 2I_xj_x] \qquad (4.41)$$

The quantity I_3j_3 is K^2 while the last two terms can be neglected except when $K = 1/2$.

For e-e nuclei in their ground state $\Omega = 0$, thus $K = 0$. The rotational band built on the ground state has the form

$$E_1 = \frac{\hbar}{2\mathfrak{I}} I(I+1) \qquad (4.42)$$

where I has the allowed values 0, 2, 4, 6, 8 ... and $\pi = +$. For odd A nuclei or in excited states of e-e nuclei with $K \neq 0$, the base of each rotational band has

$$I_0 = K = \Omega$$

and the rotational sequence is

$$I = I_0, I_0 + 1, I_0 + 2, \ldots$$

the energies of the levels are

$$E_{I,K} = \epsilon_K + \frac{\hbar^2}{2\mathfrak{I}}[[I(I+1) - 2K^2 + \delta_{K,1/2}a(-1)^{I+1/2}(I+1/2)] \qquad (4.43)$$

where ϵ_K is the energy level of any odd nucleon, a is a "decoupling constant" for $K = 1/2$ bands and $\delta_{K,1/2}$ is the Kronecker delta. The effect of the last term is to spoil the nice characteristic rotational spacing pattern for $K = 1/2$ levels. This term is due to a Coriolis or rotation-particle interaction associated with the last two terms in equation 4.41. The term $\hbar^2/2\mathfrak{I}$ is ~7 keV for e-e nuclei, 6 keV for odd A nuclei, and ~5.7 for o-o nuclei unless there is significant Coriolis mixing, which can cause substantial deviations from these values. When $K \neq 1/2$, the excitation of each level above the ground state of the band is

$$\Delta E_I = \frac{\hbar^2}{2\mathcal{J}}[I(I+1) - K(K+1)]$$

and this relation can be used to determine K from the level spacings.

4.7.2.2 Vibrational Levels. The simplest modes of vibrational motion of prolate nuclei such as the transuranium nuclei involve two types of vibration: β vibrations in which the length of the major axis expands and contracts periodically with the "cigar" becoming thinner or fatter and γ-vibrations which keep the length of the major axis constant but squeeze the nucleons periodically to make its cross section elliptical in the plane perpendicular to the symmetry axis. The quantum numbers n_β and n_γ specify the number of phonons* associated with each vibrational mode. Rotational bands are built on each vibrational state. An idealized sequence of I, K, π values for members of the bands built on low lying vibrational states is shown in Figure 4.20. Another type of simple vibrational motion

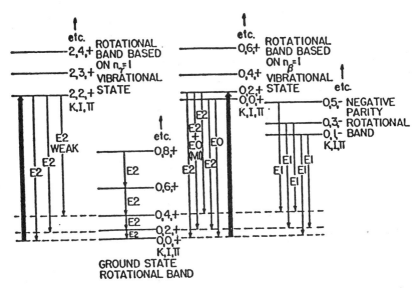

Figure 4.20 Pattern of rotational and vibrational levels for an idealized e-e deformed nucleus. A typical gamma ray deexcitation sequence is also shown.

---

*In analogy with the quantum theory of electromagnetic radiation in which a unit of radiation is called a *phonon*.

found in transuranium nuclei is the octupole vibration. This vibrational mode can carry 0–3 units of angular momentum. The lowest mode is a vibration mode with axial symmetry of the form $P_3(\cos\theta)$, that is, a pear-shaped deformation. The corresponding nuclear levels in an e-e nucleus have $I\pi = 1^-, 3^-, 5-, \ldots$ with $K = 0$.

4.7.3 Pairing Effects

There is a last general aspect of nuclear structure that should be discussed before we make a detailed examination of the nuclear structure of the transuranium nuclei. It is the subject of nuclear pairing. If we would examine the sequence of single particle levels shown in Figures 4.17 and 4.18, we would find the level spacings are roughly 300 keV and on this basis, we would expect to find excited nonrotational states in even nuclei to appear at this energy. But, in fact, the lowest lying single particle states in such nuclei appear at an energy of about 1 MeV. This observation is accounted for by the difficulty of breaking a particle pair in the ground state due to strength of the pairing force.

The pairing force is a short-range attractive force between identical nucleons who form a pair with zero total angular momentum. (We say that the pair have the same quantum numbers except for opposite signs of the projection of the angular momentum.) Realistic calculations of the effect of the pairing force are quite complex. To show the essential features of the problem, we shall describe the solutions obtained using the relatively simple BCS method originally developed to explain superconductivity due to electron pairing in metals.

We denote the average occupation number of state k as V_k^2 where $0 \leq V_k^2 \leq 1$. Figure 4.21 shows the occupation probability of various single particle states as a function of energy for two situations: (a) no pairing force acting, states are filled up to the Fermi energy and all states above this are empty; and (b) pairing force acting, the occupation number does *not* go to zero at the Fermi energy but does so over an energy interval centered about the Fermi energy.

Formally the average occupation numbers are given by

$$V_k^2 = \frac{1}{2} - \left(1 - \frac{\epsilon_k - \lambda}{[(\epsilon_k - \lambda)^2 + \Delta^2]^{1/2}}\right) \qquad (4.44)$$

where ϵ_k is the energy of the single particle state, λ is essentially the Fermi energy and Δ is called the gap parameter and is proportional to the strength of the pairing interaction.

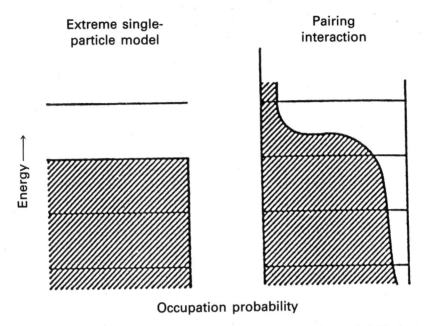

Figure 4.21 Schematic diagram of the ground state occupation probability in an e-e nucleus. The extreme single particle model prediction is shown on the left while the pairing model result is shown on the right (from Fri 81).

In e-e nuclei, the lowest single particle excited state corresponds to breaking a nucleonic pair and moving a nucleon to a higher lying state and thus leaving a "hole" in the original state. The term "quasiparticle" is used to refer collectively to these holes and unpaired particles. Their energy above the ground state is

$$\epsilon_k = [(\epsilon_k - \lambda) + \Delta^2]^{1/2} \qquad (4.45)$$

In an e-e nucleus the lowest intrinsic state is thus a two-quasiparticle state with energy

$$E_{k_1} + E_{k_2} + [(\epsilon_{k_1} - \lambda)^2 + \Delta^2]^{1/2} + [(\epsilon_{k_2} - \lambda)^2 + \Delta^2]^{1/2} \qquad (4.46)$$

For the lowest state, $\epsilon_{k_1} - \lambda$ and $\epsilon_{k_2} - \lambda$ are small compared to Δ and thus the minimum excitation is 2Δ. Configuration mixing may also be treated by considering the pairing interaction (Coh 71).

Detailed calculations of the single particle level spacings and the level occupation probabilities have been made for the actinides (Cha 77) and compared to experimental data. The pairing force is found to be the most important residual interaction for the actinide nuclei. Its effect is to modify the occupation probabilities of the nucleon orbitals and to change the level spacings relative to those of the simple single particle model. One nucleon transfer reactions such as the (d, p) or (d, t) reactions can be used to check these calculations and the data agree well with the predictions using the pairing interaction (Cha 77).

4.7.4 Energy Levels and Illustrations of Nuclear Structure Models

Data concerning the ground and excited states of transuranium nuclei will be considered now in light of our elementary discussions of nuclear structure. Such data are obtained from nuclear spectroscopic studies of radioactive decay and nuclear reactions. Table 4.4 shows the experimentally assigned ground state configurations for some odd mass transuranium nuclei. We can use the Nilsson diagrams in Figures 4.17 (for neutrons) and 4.18 (for protons) to compare with these assignments. If one knows the deformation of the nucleus, ν_2 ($\sim \beta_2 \sim 0.25$ for most actinides), one

TABLE 4.4 Ground State Configuration of Some Odd A Transuranium Nuclei[a]

	Configuration	Examples
N		
143	7/2⁻ [743]	^{237}Pu
145	1/2⁺ [631]	^{239}Pu
147	5/2⁺ [622]	^{241}Pu, ^{243}Cm
149	7/2⁺ [624]	^{243}Pu, ^{245}Cm, ^{247}Cf
151	9/2⁻ [734]	^{247}Cm, ^{249}Cf, ^{251}Fm
153	1/2⁺ [620]	^{249}Cm, (^{251}Cm), ^{251}Cf, ^{253}Fm
155	7/2⁺ [613]	^{253}Cf, ^{255}Fm
Z		
93	5/2⁺ [642]	^{235}Np, ^{237}Np, ^{239}Np
95	5/2⁻ [523]	^{239}Am, ^{241}Am, ^{243}Am, ^{245}Am
97	3/2⁻ [521]	^{247}Bk, (^{251}Bk)
	7/2⁺ [633]	^{249}Bk
99	3/2⁻ [521]	^{251}Es
	7/2⁺ [633]	^{253}Es, (^{255}Es)

[a] The numbers in the brackets are the asymptotic quantum numbers N, n_z, and Λ. Parentheses around a nuclide indicate an uncertain assignment.

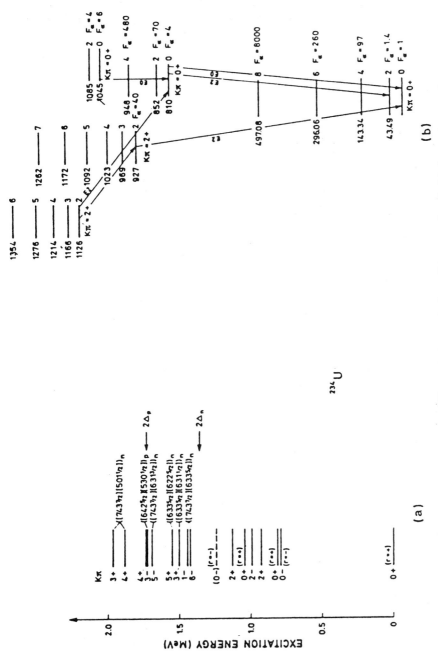

Figure 4.22 Spectrum of low-lying states in ^{234}U. (a) General view, (b) more detailed information on the rotational structure of the low-lying even parity levels (from Boh 69). F_α is the α-decay hindrance factor. The numbers in brackets are the quantum numbers N, n_z, Λ, and Ω.

merely moves to the correct place on the abscissa corresponding to this deformation, then starts adding particles to the levels (putting 2 particles/level) beginning with the first labeled gap in the level diagram until the state corresponding to the last particle is reached. This state should correspond to the observed ground state configuration. If one or more Nilsson states happen to lie close together, it may be that the ground state will correspond to one of the other states (due to approximations in general diagrams such as Figures 4.17 and 4.18). If one does this exercise, one finds the agreement between measured and predicted configurations to be quite good.

Over 150 excited Nilsson state configurations are known for actinide nuclei (Cha 77, Led 78) and most can be reasonably understood in terms of the levels of Figures 4.17 and 4.18 and pairing considerations.

The low energy excited states of a typical heavy e-e nucleus, ^{234}U, are shown in Figure 4.22. One can easily identify a ground state rotational

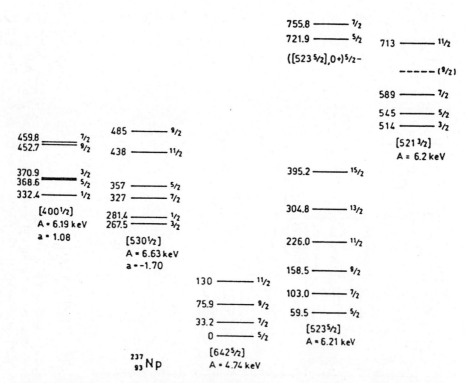

Figure 4.23 Level scheme of ^{237}Np (from Boh 69.). The numbers in brackets are the quantum numbers N, n_z, Λ, and Ω.

band, low lying β- and γ-vibrational levels with rotational bands built upon them as well as various quasiparticle excitations. Note the gaps in the levels near the two quasiparticle excitations $2\Delta_{proton}$ and $2\Delta_{neutron}$.

The energy levels of odd A nuclei are generally even more complex than those of e-e nuclei. Figure 4.23 shows the low lying excited states of a typical odd A nucleus, ^{237}Np. The observed single particle states can be identified with Nilsson states (in Figures 4.17 and 4.18) with rotational bands built upon them. Note the ordering of the $K = 1/2$ bands is "anomalous" compared to the other bands due to Coriolis effects. The $K\pi = 5/2^-$ state at 722 keV results from the coupling of a particle in the 5/2$^-$ [523] configuration with a β vibration.

4.7.5 Alpha-Decay Hindrance Factors

The strong dependence of α-decay transition rates upon the "gross variables" such as the energy of the emitted alpha particle and the atomic number of the emitting nucleus, and some strict selection rules, were discussed in Section 4.5. There are additional factors, however, related to nuclear structure that can affect α transition rates. It is our purpose in this section to touch briefly on these matters.

As discussed previously, α-decay transition rates involve two factors; a "macroscopic factor" related to barrier penetration and "microscopic factor" related to the probability of preforming an alpha particle in the nucleus. It is this latter factor which we wish to address further here. The probability of preforming the alpha particle in the nucleus is sensitive to neutron and proton pair correlations in the nucleus. Transitions from ground state to ground state in e-e nuclei and transitions in odd A nuclei with hindrance factor of ~ 1 (favored transitions) involve the emission of two pairs of nucleons in paired orbits. In these transitions $\Delta K = 0$, $\pi_{initial} = \pi_{final}$. Other transitions where there is less overlap between the initial and final state are far less probable. A practical demonstration of this latter statement is found in the decay of ^{241}Am to ^{237}Np. ^{241}Am has $K\pi = 5/2^-$ and decays overwhelmingly ($HF \sim 1$) to an excited state of ^{237}Np at 59.5 keV with $K\pi = 5/2^-$. The decay to the ground state of ^{237}Np with $K\pi = 5/2^+$ is inhibited by a factor of 600. The difference between the states is that the ^{241}Am ground state and the 59.5 keV state of ^{237}Np have the configuration 5/2$^-$ [523] while the ground state of ^{237}Np has the configuration 5/2$^+$ [642].

As discussed previously a rotational band corresponds to states having the same intrinsic single particle structure but with the different states representing different frequencies of rotation. The relative population of these states in alpha decay characterized by a definite L value (neglecting the energy dependence) will depend only on structure independent quantities, such as simple Clebsch–Gordan coefficients involving I and K of the initial and final states and dependent on L.

4.8 ELECTROMAGNETIC TRANSITION RATES

Electromagnetic transitions may proceed by the emission of a photon (γ-ray decay) or the radiationless ejection of an orbital electron (internal conversion). These transitions are collectively called gamma transitions and are described in terms of the process in which a photon is actually emitted.

Nuclear gamma transitions can be conveniently classified according to multipolarity. Except in special cases, a transition will proceed by the lowest multipolarity, since the lifetimes of competing higher multipoles will generally be much longer.

By way of review, let us set down the strict selection rules for gamma transitions, selection rules involving the good quantum numbers, I, the total nuclear angular momentum, and π the parity. Radiation of multipolarity L carries off an angular momentum of L; hence, it follows from conservation of angular momentum that the allowed L values comprise the integers ranging between the difference and the sum of the initial and final spins.

$$I_i - I_f \leq L \leq I_i + I_f$$

Electric multipole radiation has the associated parity $(-1)^L$, while magnetic multipole radiation has parity $(-1)^{L+1}$. Conservation of parity imposes an additional restriction, and the complete strict selection rules are implicit in Table 4.5, where the letter E means electric and M means magnetic radiation.

The most powerful experimental method for establishing gamma multipolarities is the determination of conversion coefficients and comparison with theoretical values. Another general method involves angular correlation studies.

Gamma lifetimes (emission partial lifetimes) are often compared with the Weisskopf single particle estimates of transition probabilities (Bla 52). E1 transitions are retarded by factors of 10^3–10^7 relative to the single proton estimates while retardation factors for M1 and E2 transitions are 10^1–10^3. The transitions best obeying the formula are the M4 isomeric transitions whose measured lifetimes agree well with the formula except

TABLE 4.5 Standard Notation of Gamma Transitions

L Name	0 Monopole[a]	1 Dipole	2 Quadrupole	3 Octupole	4 Hexadecapole	5 2^5-pole
No parity change		M1	E2	M3	E4	M5
Parity change		E1	M2	E3	M4	E5

[a] A radiative (1 photon) transition is not allowed for $L = 0$, but the E0 transition may occur by pure internal conversion of penetrating electrons ($s_{1/2}$ or $p_{1/2}$).

that the decay is one order of magnitude slower than the calculated rate. The Weisskopf single particle partial γ-ray half-lives are given in Table 4.6. In these formulae E_γ is the transition energy in MeV, A the mass number, and S a statistical factor dependent on the initial and final spins and L values, and on the multipolarity of the transition. The factor S is often assumed to be unity for purposes of comparisons of the calculated with the experimental values.

In another method of comparison of gamma lifetimes the so-called reduced transition probability, a quantity proportional to the nuclear matrix element squared, is calculated. A relevant formula, given by A. Bohr and B. Mottelson (Boh 69), for both electric and magnetic multipoles is

TABLE 4.6. Partial γ-ray Half-lives According to the Weisskopf Estimate for Different Multipole Transitions Using a Nuclear Radius of $1.2A^{1/3}$ fm and a Statistical Factor $S = 1$ (A = mass number and $E\gamma$ = transition energy in MeV)[a]

$t_{(1/2)\gamma w}(E1) = 6.762 A^{-2/3} \quad E_\gamma^{-3} \times 10^{-10}$ s
$t_{(1/2)\gamma w}(E2) = 9.523 A^{-4/3} \quad E_\gamma^{-5} \times 10^{-9}$ s
$t_{(1/2)\gamma w}(E3) = 2.044 A^{-2} \quad E_\gamma^{-7} \times 10^{-2}$ s
$t_{(1/2)\gamma w}(E4) = 6.499 A^{-8/3} \quad E_\gamma^{-9} \times 10^{4}$ s
$t_{(1/2)\gamma w}(E5) = 2.893 A^{-10/3} \quad E_\gamma^{-11} \times 10^{11}$ s
$t_{(1/2)\gamma w}(M1) = 2.202 \quad E_\gamma^{-3} \times 10^{-14}$ s
$t_{(1/2)\gamma w}(M2) = 3.100 A^{-2/3} \quad E_\gamma^{-5} \times 10^{-8}$ s
$t_{(1/2)\gamma w}(M3) = 6.655 A^{-4/3} \quad E_\gamma^{-7} \times 10^{-2}$ s
$t_{(1/2)\gamma w}(M4) = 2.116 A^{-2} \quad E_\gamma^{-9} \times 10^{5}$ s
$t_{(1/2)\gamma w}(M5) = 9.419 A^{-8/3} \quad E_\gamma^{-11} \times 10^{11}$ s

[a] Note that $t_{1/2\gamma}$ is the *partial* half-life for gamma emission only; the occurrence of internal conversion will always shorter the measured half-life.

$$B_L = \frac{L[(2L+1)!!]^2}{8\pi(L+1)} \hbar \left(\frac{\hbar c}{E_\gamma}\right)^{2L+1} P_\gamma \qquad (4.46)$$

where E_γ is the gamma energy, c is the speed of light, P_γ is the partial γ-ray transition probability, and B_L is reduced transition probability. Numerical values of B_L are shown in Table 4.7.

There are selection rules involving the other quantum numbers. Of especial importance is the selection rule (Ala 55) involving the quantum number, K, which states that the multipolarity of the transition must equal or exceed the change in K (i.e., $L \geq \Delta K$). The quantity $\Delta K - L = \nu$ expresses the degree of K-forbiddenness of a given transition. For every unit of ν, the transition rate is retarded by 10^2–10^3. The weaker selection rules in the asymptotic quantum numbers N, n_z, Λ, and Σ will be discussed later and are summarized in Table 4.8.

4.8.1 Transitions in Even-Even Nuclei

The even-even nuclei in the region of spheroidal distortion exhibit several general spectral features. As mentioned earlier, based on the 0^+ ground state is a closely spaced rotational band with the 2^+, 4^+, 6^+, 8^+, et cetera, level sequence (~42–100 keV first excited states). The gamma transitions connecting these rotational states are pure E2 transitions. The reduced transition probabilities between the ground and the first excited state are one to three orders of magnitude faster than those estimated from the single proton formula. These fast rates result from a rotational transition of the deformed nucleus, which for a large intrinsic quadrupole moment Q_0, is equivalent to the cooperative motion of many nucleons.

The next rotational band often found above the ground state band in the heavy even-even nuclei is a band with a 1^- base state (octupole vibrational states). The 1^- state decays to the ground (0+) and first excited (2^+) states by electric dipole transitions. The reduced transition probability to the first excited state in the numerous cases studied is about twice that to the ground state. Such ratios of reduced transition rates to two or more states of a common rotational band are treated as simply involving geometrical considerations of vector addition. The ratios should be functions of the multipolarity, the nuclear spins, and the associated K quantum number of the initial and final states. (To the extent that K is not an absolutely good quantum number, the theoretical ratios may be inexact.) The experimental ratios for transition from 1^- states yield the important result that the K quantum number of the lowest 1^- state is zero.

TABLE 4.7 Relations Between the Reduced Transition Probabilities $B(XL; I_i \rightarrow I_f)$ and the Partial γ-ray Transition Probabilities $P\gamma(XL; I_i \rightarrow I_f)$ for Different Multipole Transitions[a]

$B(E1; I_i \rightarrow I_f) = 6.288 \times 10^{-16}$	$E_\gamma^{-3} P_\gamma(E1; I_i \rightarrow I_f)$
$B(E2; I_i \rightarrow I_f) = 8.161 \times 10^{-10}$	$E_\gamma^{-5} P_\gamma(E2; I_i \rightarrow I_f)$
$B(E3; I_i \rightarrow I_f) = 1.752 \times 10^{-3}$	$E_\gamma^{-7} P_\gamma(E3; I_i \rightarrow I_f)$
$B(E4; I_i \rightarrow I_f) = 5.893 \times 10^{3}$	$E_\gamma^{-9} P_\gamma(E4; I_i \rightarrow I_f)$
$B(E5; I_i \rightarrow I_f) = 2.892 \times 10^{10}$	$E_\gamma^{-11} P_\gamma(E5; I_i \rightarrow I_f)$
$B(M1; I_i \rightarrow I_f) = 5.687 \times 10^{-14}$	$E_\gamma^{-3} P_\gamma(M1; I_i \rightarrow I_f)$
$B(M2; I_i \rightarrow I_f) = 7.381 \times 10^{-8}$	$E_\gamma^{-5} P_\gamma(M2; I_i \rightarrow I_f)$
$B(M3; I_i \rightarrow I_f) = 1.584 \times 10^{-1}$	$E_\gamma^{-7} P_\gamma(M3; I_i \rightarrow I_f)$
$B(M4; I_i \rightarrow I_f) = 5.329 \times 10^{5}$	$E_\gamma^{-9} P_\gamma(M4; I_i \rightarrow I_f)$
$B(M5; I_i \rightarrow I_f) = 2.615 \times 10^{12}$	$E_\gamma^{-11} P_\gamma(M5; I_i \rightarrow I_f)$

[a] The transition energy $E\gamma$ given in MeV, $B(EL; I_i \rightarrow I_f)$ in units of $e^2(\text{fm})^{2L}$, $B(ML; I_i \rightarrow I_f)$ in units of $(e\hbar/2M_p c)^2 (\text{fm})^{2L-2}$, and the decay rate $P_\gamma(XL; I_i \rightarrow I_f)$ is given in s^{-1}.

TABLE 4.8 Selection Rules for Asymptotic Quantum Number; Transitions in Deformed Nuclei

Multipole	ΔK	ΔN	Δn_z	$\Delta\Lambda$
E1	1	±1	0	0
	0	±1	±1	0
	2	0, ±2	0	2
E2	1	0, ±2	±1	1
	0	0, ±2	0, ±2	0
	3	±1, ±3	0	3
E3	2	±1, ±3	±1	2
	1	±1, ±3	0, ±2	1
	0	±1, ±3	±1, ±3	0
M1	1	0, ±2	0, ±1	0, 1
	0	0	0	0
	2	±1, ±3	0, ±1	1, 2
M2	1	±1, ±3	0, ±1, ±2	0, 1
	0	±1, ±3	0, ±1	0, ±1
	3	0, ±2, ±4	0, ±1	2, 3
M3	2	0, ±2	0, ±1, ±2	1, 2
	1	0, ±2, ±4	0, ±1, ±2	0, ±1
	0	0, ±2, ±4	0, ±1, ±2	0, ±1

In the vicinity of just below 1 MeV in several even-even nuclei a 0^+ state is observed which is the first excited, β-vibrational level. This state decays to the ground state by pure electric monopole radiationless internal conversion at a rate roughly comparable to its decay by E2 radiation to the first excited state. Slightly above the β-vibrational band, in several cases one finds a band with spin sequence 2^+, 3^+, et cetera. This 2^+ state decays characteristically by E2 radiation to the 0^+, 2^+, and 4^+ states of the ground band although the decay to the 4^+ state is weak.

The vector addition relations neatly explain the relative gamma intensity pattern if this 2^+ state is assigned a K quantum number of 2. The lack of detectable M1 radiation in the $2^+ \rightarrow 2^+$ transition to the first excited state may be attributed to operation of the approximate K selection rule, that the multipolarity L must be greater than or equal to the change in K, $|K_i - K_f|$. This $K = 2$ band is the γ-vibrational band.

4.8.2 Transitions in Odd-Mass Nuclei

The greater level density of the odd-mass nuclei over even-even nuclei makes for greater experimental difficulties, and a greater diversity of gamma transition types.

As in even-even nuclei, we find rotational bands, but with spin sequences I_0, $I_0 + 1$, $I_0 + 2$, etc. The characteristic intraband radiation pattern involves fast pure E2 transitions connecting alternate levels and M1-E2 mixed transitions connecting adjacent levels.

For interpretation of the radiations connecting different rotational bands the K selection rules and additional weaker selection rules in the other approximate quantum numbers have proved most useful. As in even-even nuclei, the odd nuclei provide numerous examples of the validity of the K selection rules.

An interesting class of interrotational-band transitions is that of the numerous low energy E1 transitions. Many of them have measured lifetimes and are retarded from the single particle estimate by factors of 10^4–10^6 even though they do not violate the K selection rules. Examples of even greater retardation are known for K forbidden E1 transitions—as, for example, in ^{239}Pu.

The extraordinary slowness of the low energy E1 transitions has been associated with the fact that all of them violate approximate selection rules which are strict in the limits of no spheroidal deformation and of very large spheroidal deformation. The approximate selection rules in the spherical limit will be rules involving ℓ and j, the odd-nucleon orbital angular momentum and total angular momentum, respectively. For E1 transitions the only allowed cases are $\Delta \ell = \pm 1$, $\Delta j = 0$, ± 1 and the j

selection rule would be violated for E1 transitions within a major shell. Of greater utility for deformed nuclei are the selection rules in the large deformation limit. As this limit is approached, the asymptotic quantum numbers N, n_z, Λ, and Σ discussed previously will signify properties of the odd-nucleon wave function which are nearly constants of the motion. The first three of these quantum numbers are given beside the various orbitals in the Nilsson energy level diagrams (Figures 4.17 and 4.18). The asymptotic selection rules are given in Table 4.8. These rules are only approximately valid. If a transition is forbidden by these rules, it probably is greatly reduced but it is not completely forbidden.

4.9 BETA-DECAY RATES AND ENERGETICS

Whereas all nuclides above bismuth are unstable with respect to α-particle emission, β decay is not unique in this region of the periodic table and not all heavy nuclides are beta unstable. For example, the plutonium isotopes with $A = 236$, 238, 239, 240, 242, and 244 are all beta stable. If these isotopes were not alpha-unstable with half-lives less than 10^8 years, they would all be found in nature.

Beta decay is a nuclear process in which a nuclear proton is converted to a neutron or vice versa, changing the atomic number by one unit with no change in mass number. Nuclei with a neutron excess emit an electron and an antielectron neutrino (β^- decay); heavy neutron-deficient nuclei absorb an orbital electron and emit an electron neutrino (EC decay).

In Figure 4.24, the energy released in β decay of the transuranium nuclei is shown as a function of mass number. Solid points designate measured values, open circles designate values derived from systematics, while the lines indicate the predictions of the Möller–Nix mass formula (Möl 81). From nuclear mass tables or mass formula predictions, it is possible to determine which transuranium nuclides are β-stable. This important information is summarized in Table 4.9.

From a knowledge of the energetics of β decay, one can make estimates of the lifetimes for β decay. Two approaches seem appropriate for the heavy elements. The first of these is a semiempirical approach which is useful in predicting the half-lives of species near the known nuclei. It is based upon the observation of Viola and Seaborg (Vio 66b) that the average $\log(ft)$ values for β^- decay of heavy nuclei are 6.0 ± 0.7, 6.4 ± 0.6, 6.5 ± 0.7 and 7.2 ± 0.7 for e-e, e-o, o-e, and o-o nuclei, respectively, with the corresponding values for electron capture are 6.2 ± 1.3, 6.2 ± 1.3, 6.1 ± 0.8, and 6.9 ± 0.7. From a knowledge of Q_β, and using the above values of $\log(ft)$ a value of t is obtained (Led 78).

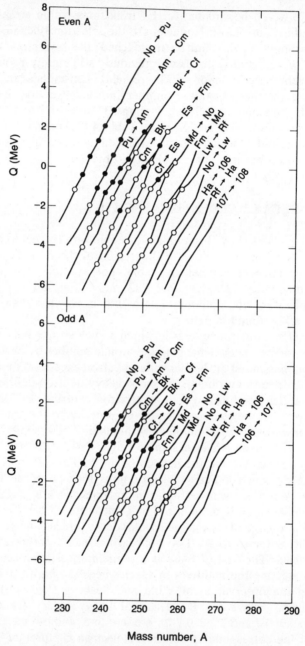

Figure 4.24 Q_β for even A and odd A transuranium nuclei, as calculated using the Möl 81 mass formula. Solid points designate measured values, open circles designate values derived from systematics. Negative values of Q_β indicate decay in the reverse direction by electron capture.

TABLE 4.9 Beta-Stable Transuranium Nuclides[a]

Np	^{237}Np
Pu	^{236}Pu, ^{238}Pu, ^{239}Pu, ^{240}Pu, ^{242}Pu, ^{244}Pu
Am	^{241}Am, ^{243}Am
Cm	^{242}Cm, ^{244}Cm, ^{245}Cm, ^{246}Cm, ^{248}Cm,
Bk	^{247}Bk
Cf	^{248}Cf, ^{249}Cf, ^{250}Cf, ^{251}Cf, ^{252}Cf, ^{254}Cf, ^{256}Cf
Es	^{253}Es
Fm	^{252}Fm, ^{254}Fm, ^{255}Fm, ^{256}Fm, ^{257}Fm, ^{258}Fm, (^{260}Fm)
Md	^{259}Md
No	^{258}No, (^{260}No), (^{261}No), (^{262}No), (^{263}No), (^{264}No)
Lr	(^{265}Lr)
Rf	(^{264}Rf), (^{266}Rf), (^{267}Rf), (^{268}Rf)
Ha	(^{269}Ha)
106	(^{270}106), (^{271}106), (^{272}106)
107	(^{273}107)

[a]Unknown nuclei in parentheses. Q_β for these nuclei from Moo 89.

A second approach which is probably more appropriate for large extrapolations from the region of known nuclei is the use of the "gross theory" of β decay (Tak 73). Using this theory, values of the β decay lifetimes for isotopes of elements with $93 \leq Z \leq 100$ have been calculated (Tak 73) for species with $210 \leq A \leq 290$. An improved version of these calculations which concentrates on the n-rich nuclei of importance in astrophysics and the decay heat of nuclear reactors has been made (Kla 81). This latter approach was used to predict the β-decay lifetimes used in constructing Figure 1.2.

In the actinides the log(ft) values for "allowed" transitions (no parity change: $\Delta I = 0, \pm 1$) and those for "first-forbidden" transitions (a parity change; $\Delta I = 0, \pm 1, \pm 2$) cluster between 5.5 and 8.0. For other transitions, the log(ft) value is greater than 8.0. Thus observation of log(ft) < 8.0 implies that $\Delta I = 0, \pm 1$, a finding that is useful in understanding the nuclear spectroscopy of the actinides. It is even possible to systematize the log(ft) information further to reveal the Nilsson states in the initial and final nuclei (Cha 77).

In β decay, the same K selection rules apply as for electromagnetic decay. In addition there are approximate selection rules for changes in the asymptotic quantum numbers N, n_z, Λ. Thus for allowed transitions ($\Delta I = , \pm 1, \Delta \pi = $ no), the other selection rules give $\Delta K = 0, =1, \Delta N = 0$, $\Delta n_z = 0$, $\Delta \Lambda = 0$. For first forbidden transitions ($\Delta I = 0, \pm 1, \pm 2, \Delta \pi = $ yes), the other rules give $\Delta K = 0, \pm 1, \Delta N = \pm 1, \Delta n_z = 0, \Delta \Lambda = 0, \pm 1$.

4.10 GROUND STATE AND LOW ENERGY FISSION PROPERTIES OF THE TRANSURANIUM NUCLEI

It can be argued that the most important decay mode of transuranium nuclei in their ground and excited states is that of nuclear fission. In this section, we describe further the general properties of this decay mode for transuranium nuclei with emphasis on their ground states or with excitation energies less than 10 MeV. In particular, we shall focus our attention on what is termed "scission point" or "post-fission" phenomena, that is, the observed fission product isobaric and isotopic distributions, the kinetic energy release and the distribution of neutrons and photons emitted at or beyond the point at which the nucleus divides into two fragments. Other aspects of the fission process such as the fragment angular distributions or the probability of fission are treated in later chapters or not at all. Our treatment of fission phenomena is, to some extent, superficial and the reader who is interested in a more authoritative account is referred to the excellent monograph of Vandenbosch and Huizenga (Van 73).

4.10.1 Fission Product Mass Distribution

The fission fragment mass distribution is one of the most studied, most important and least understood aspects of the fission process. Early investigations of the thermal-neutron-induced fission of uranium and plutonium isotopes and the spontaneous fission of ^{252}Cf found that the fissioning nuclei divided asymmetrically, that is, the ratio of the abundant heavy fragment mass number to that of the light fragment M_H/M_L was 1.3–1.5. (The liquid drop model, discussed earlier in this chapter, would predict that the most probable fission mass split would be a symmetric one ($M_H/M_L = 1.0$).) This is illustrated in Figure 4.25 where the mass distributions for the thermal neutron induced fission of the "big three nuclides," ^{233}U, ^{235}U, and ^{239}Pu, are shown. Symmetric fission is suppressed by at least two orders of magnitude relative to asymmetric fission.

The key to understanding the nature of this preference for asymmetric fission amongst the lighter actinides is found in the data of Figures 4.25 and 4.26. In these figures, we show that as the mass of the fissioning system increases, the position of the heavy fragment peak in the mass distribution remains constant while the light fragment peak position increases with increasing fissioning system mass. This observation along with the observation that the lower edge of the heavy fragment peak is anchored at $A = 132$ has suggested to many that the special stability associated with having one fragment with $Z = 50$, $N = 82$ (i.e., a doubly magic nucleus) is causing this preference for asymmetric fission. Qualita-

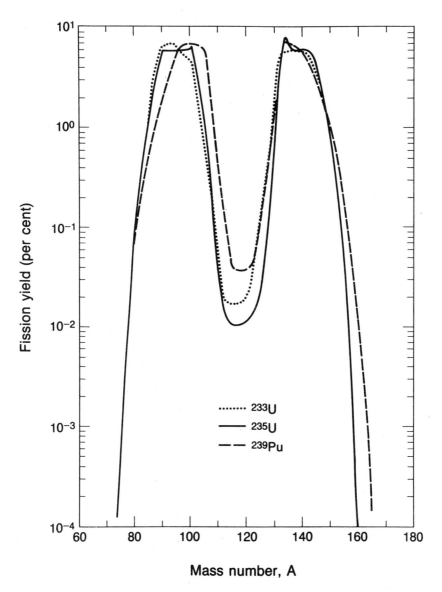

Figure 4.25 Smoothed fragment mass distributions for the thermal neutron induced fission of ^{235}U, ^{233}U, and ^{239}Pu.

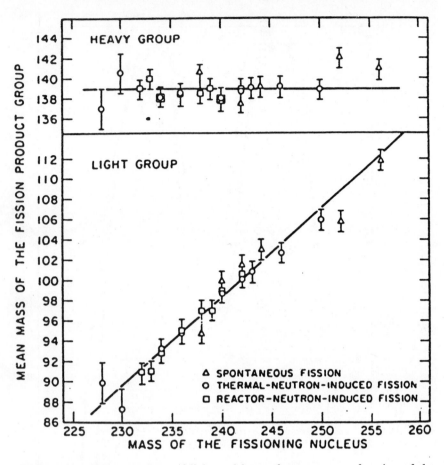

Figure 4.26 Average masses of light and heavy fragments as a function of the mass of the fissioning nucleus (from Fly 72).

tively, if this were the case, one would expect that as one raised the excitation energy of the fissioning system, the influence of the ground state shell structure of the nascent fragments would decrease and the fission mass distribution would show a greater amount of symmetric fission. In fact, this is exactly what happens and at high enough excitation energies, all heavy nuclei fission symmetrically.

Further dramatic evidence for the influence of fragment shell structure is given in Figure 4.27 where the fragment mass distributions from spontaneous fission of the heaviest nuclei are shown (Hof 88). One's attention is specifically directed to the yield distributions for the fermium isotopes

FISSION PROPERTIES OF THE TRANSURANIUM NUCLEI 179

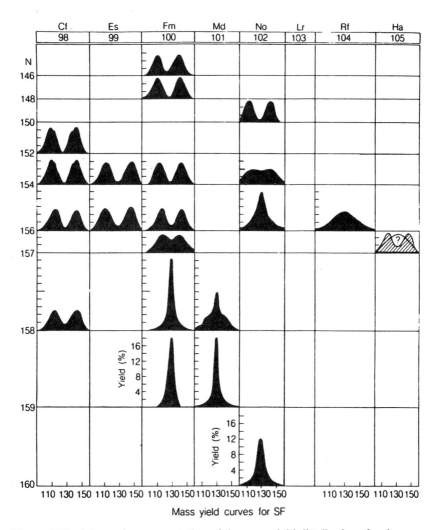

Figure 4.27 Schematic representation of the mass-yield distributions for the spontaneous fission of the trans-berkelium nuclides (from Hof 88).

shown in greater detail in Figure 4.28. A sharp transition between asymmetric and symmetric fission is seen in going from ^{257}Fm to ^{258}Fm. Also the very narrow symmetric mass distribution seen for the spontaneous fission of ^{258}Fm broadens considerably when this nucleus is produced with 26 MeV excitation energy by the ^{257}Fm (n, f) reaction. For fermium, mendelevium, and nobelium one observes a change from asymmetric to

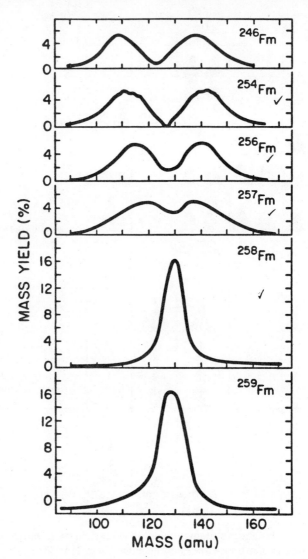

Figure 4.28 Mass distribution for the spontaneous fission of the fermium isotopes (from Hof 84).

symmetric mass division as the fissioning system mass number equals or exceeds 258.

These observations on the transuranium element fission mass distributions can be understood in terms of a general model of the scission point introduced by Wilkins, Steinberg, and Chasman (Wil 76). To understand this model, one needs to describe the sequence of events leading to scission

(Figure 4.4). There is considerable evidence that the transition of the nucleus from the saddle point to the scission point is primarily nonadiabatic. Nuclei behave as viscous fluids and the decrease in nuclear potential energy in going from saddle to scission point which appears as collective motion is transferred into the form of internal excitation energy, that is, "heat" in the nascent fragments. If there is sufficient nonadiabatic mixing of the energy among the single particle states of these fragments, statistical equilibrium may be established. The Wilkins, Steinberg, and Chasman model assumes some nuclear viscosity leading to a quasi equilibrium between collective degrees of freedom (characterized by a temperature T_{coll}) and some viscous heating leading to population of single particle levels with a characteristic temperature, T_{int}. The relative probability P of forming any fragment pair is given by

$$P(N, Z, T_{\text{int}}, d) = \int_{\beta_1=0}^{\beta_{\max}} \int_{\beta_2=0}^{\beta_{\max}} \exp[-V(N, Z, T_{\text{int}}, d)/T_{\text{coll}}] \, d\beta_1 \, d\beta_2$$

where β_1 and β_2 are the deformation parameters describing fragment 1 and 2 separated by distance d. Note that small changes in the interaction potential V leads to large changes in fragment yields. Thus, this model, like all statistical models, has a natural "amplification" mechanism built into it. The potential energy of the two nearly touching fragments V is calculated (using the Strutinsky procedure described earlier in Section 4.2) as the sum of the liquid drop and shell and pairing correction terms for each fragment, and the Coulomb and nuclear potential terms describing the interaction between them. A single choice of values of d, T_{int}, T_{coll} is used to describe all nuclei. Figure 4.29 shows the calculated proton and neutron shell corrections as a function of fragment deformation. At zero deformation, one sees the strong spherical shells at $N = 50$, 82 and $Z = 50$ along with other shell stabilization regions at other deformations denoted by various letters. The fragment mass distributions calculated by this model are shown in Figure 4.30. Considering the choice of a single set of parameters for all calculations, the agreement between predictions and observations is quite good. Similar successes also exist in predicting the general features of the total kinetic energy and fragment excitation energy distributions. (The model is less successful in describing Z_p, the most probable fragment charge, as a function of mass split, particularly in the region near the $Z = 50$ closed shell, possibly due to a neglect of the octupole degree of freedom.)

To understand qualitatively how this model works, consider the typical mass asymmetric fragment distributions for the actinides. The heavy fragment peak is accounted for by the influence of the strong deformed

182 NUCLEAR STRUCTURE AND RADIOACTIVE DECAY PROPERTIES

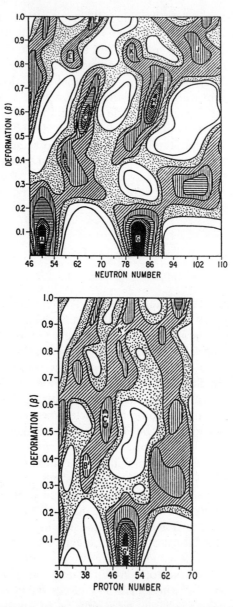

Figure 4.29 (a) Neutron shell corrections calculated as a function of deformation parameter (β) and neutron number. The contours are plotted at 1 MeV intervals with the black regions (representing the largest shell corrections) containing all values less than -4 MeV and the inner white region (representing the smallest shell corrections) containing all values greater than $+2$ MeV. (b) Proton shell corrections as a function of β and atomic number (from Wil 76).

FISSION PROPERTIES OF THE TRANSURANIUM NUCLEI 183

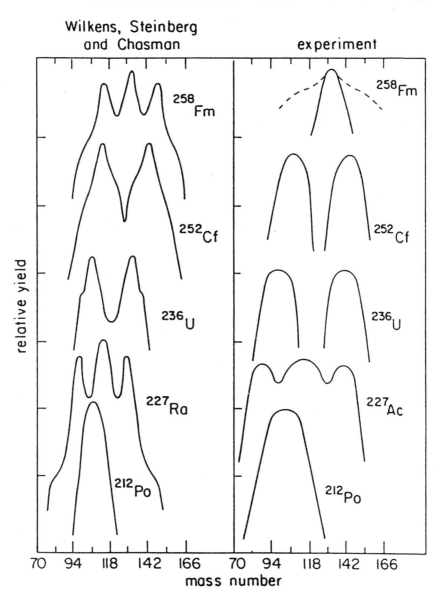

Figure 4.30 The left panel shows the mass distributions for various fissioning systems as calculated using the Wilkins, Steinberg and Chasman model with a single set of parameters for all systems ($T_{coll} = 1.0$ MeV, $T_{int} = 0.75$ MeV, and $d = 1.4$ fm) (from Wil 76). The right panel shows experimental data (Van 89).

neutron shell at point H (Figure 4.29) with $N \simeq 88$ and $\beta = 0.65$. The complement of this fragment will involve the areas designated B-C-D. Thus the simplest arguments invoking the influence of the spherical shell $N = 82$ upon the heavy fragment are really not right. It is rather a matter of the "magic numbers" changing with nuclear deformation. It has been shown (Wah 88) that the maximum yields occur at the neutron and proton numbers suggested by this model. For the heavy fermium isotopes, fragments are thought to populate the points G and K leading to symmetric fission, with one spherical and one highly deformed fragment. This split is enhanced by presence of the strong, spherical proton shell at G' and the proton shell near K' (Figure 4.29). One also predicts that as the fissioning system gets heavier ($N \sim 165$–176) the fission mass distribution will become asymmetric again with symmetry dominating beyond $N \sim 176$.

4.10.2 Total Kinetic Energy Release in Fission

To a first approximation, one can understand that the kinetic energies of the fragments produced in the fission process are the result of the Coulomb repulsion between the fragments following scission. Thus one might expect that the total kinetic energy release (TKE) in fission might vary as $Z^2/A^{1/3}$ of the fissioning system and that a measurement of fission fragment kinetic energies could serve as a rough identification of the (Z, A) of the fissioning system. Viola (Vio 66a), in fact showed that a fairly large amount of data on fission TKE could be summarized in a semiempirical equation of the form

$$\text{TKE (MeV)} = 0.1071 \frac{Z^2}{A^{1/3}} + 22.2 \qquad (4.48)$$

or more recently (Vio 85)

$$\text{TKE (MeV)} = 0.1189 \frac{Z^2}{A^{1/3}} + 7.3 \qquad (4.49)$$

where Z, A refer to the atomic and mass numbers, respectively, of the fissioning system (Figure 4.31). Unik et al. (Uni 74) considering only the nuclei with $230 \leq A \leq 250$ arrived at a correlation equation of the form

$$\text{TKE (MeV)} = 0.13323 \frac{Z^2}{A^{1/3}} + 11.64 \qquad (4.50)$$

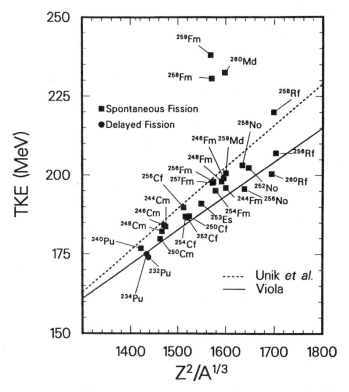

Figure 4.31 Average total kinetic energy, TKE, as a function of $Z^2/A^{1/3}$ for heavy nuclides. Solid line is prediction of Vio 66a, dashed line is prediction of Uni 74. Data for ^{258}Fm, ^{259}Fm, ^{259}Md, and ^{260}Rf are most probable TKEs (from Hof 84).

Most TKE data (for $A < 250$) can be described by one of these equations. These equations are consistent with the idea that the fragment kinetic energies are due to their mutual Coulomb repulsion following scission provided the fragments are ~1.6 times more elongated than a spherical fragment. In the region of the heavy fermium isotopes (^{258}Fm, ^{259}Fm) and for ^{260}Md one sees (Hul 80, Hof 80) unusually high values of the fragment TKE (~230–240 MeV) as compared to predictions of the aforementioned semiempirical systematics (TKE = 190–200 MeV) (Figure 4.31). This observation of an anomalously high TKE release in the spontaneous fission of ^{258,259}Fm and ^{260}Md suggests an obvious interpretation in terms of the correlation between the total kinetic energy release, the fragment mass distribution and the influence of nuclear shell structure upon the fission process. The nuclides showing an anomalously high TKE release also exhibit anomalous symmetric mass distributions which in turn appear to be related to the special stabilization offered by the spherical nuclear shell at $N = 82$. Thus one might suggest that because of the presence of a spherical $N = 82$ shell in one of the fragments (point G in Figure 4.29) the scission configuration is unusually compact, leading to a high TKE release. It is considered to be one of the most important successes of the Wilkins, Steinberg and Chasman model (WSC model) that it predicted the occurrence of high TKE release in the heavy fermium isotopes.

Another important triumph of the WSC model is to explain the data shown in Figure 4.32a where the average TKE release ($\overline{\text{TKE}}$) for various fissioning systems is shown. For plutonium and uranium isotopes, $\overline{\text{TKE}}$ is independent of the mass of the fissioning system, A_F, while for curium and californium, it depends on A_F. For nuclei in this region, the heavy fragment configuration remains constant (at point H in Figure 4.29a) as the mass of the fissioning system changes. Thus $\overline{\text{TKE}}$ is dictated by the deformation of the light fragment. Figure 4.32b shows the predicted deformation of the light fragment for various fissioning systems. For $A = 230$–244, the deformation of the light fragment remains constant (leading to constant $\overline{\text{TKE}}$). But at $A \sim 244$, an abrupt transition occurs and the fragment deformation increases markedly with further increases in fissioning system mass (leading to a decreasing $\overline{\text{TKE}}$).

The observation of the unusually high values of the fragment TKE has motivated additional, revealing measurements of the *distribution* of TKE values for a given fissioning system (Figure 4.33). The TKE distributions for the fermium-rutherfordium nuclei are skewed and perhaps bimodal, with two different TKE values, a low, regular value (~ 200 MeV) and high value (~ 230–240 MeV) being evident (Hul 86, Hul 89). The mass distributions associated with the high $\overline{\text{TKE}}$ events are symmetric and very narrow while those associated with "normal" $\overline{\text{TKE}}$ events are symmetric, but with widths that are three to four times larger.

These complex distributions have motivated several studies of the effects of nuclear shell structure upon the potential energy surfaces describing the latter stages of the fission process (Bro 86a, Möl 87, Möl 89, Cwi 88, Cwi 89). In these studies, two competing paths to fission for ^{258}Fm were found leading to elongated and compact scission shapes, respectively, corresponding to the low and high $\overline{\text{TKE}}$ events, respectively. This view is given further credence by the recent observations of Wild et al. (Wil 90) who show the high TKE events have a low neutron multiplicity befitting a compact shape while the low TKE events have a high neutron multiplicity reflecting a deformed shape. Dramatic effects of nuclear shells and the fission potential energy surface are not restricted to the heaviest actinides

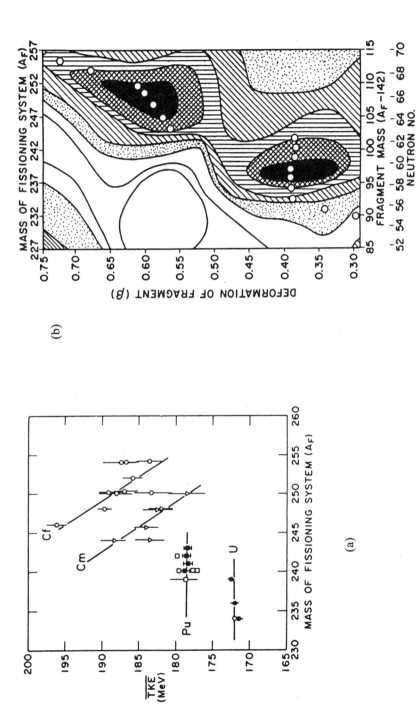

Figure 4.32 (a) The average total kinetic energy release, TKE, as a function of the mass of the fissioning system, A_F, for the isotopes of U, Pu, Cm, and Cf (from Wil 76). (b) The sum of the neutron and proton shell corrections at different deformations, β, for the light fragment complementary to a heavy fragment of $A = 142$ at $\beta_H = 0.65$ (point H in Figure 4.29). Contours have same meaning as in Figure 4.29. The open circles represent the deformation of light fragment ($A_F = 142$) for various fissioning systems (from Wil 76).

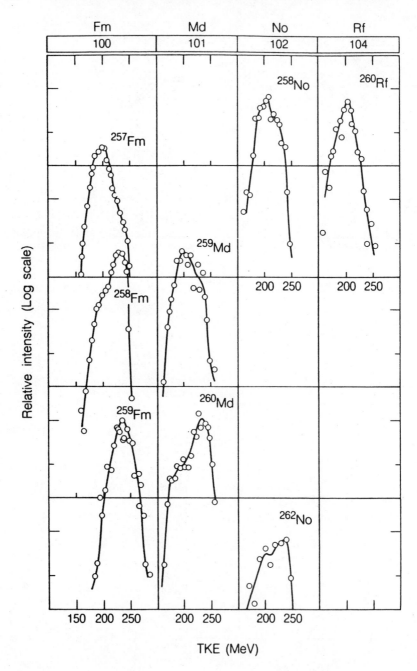

Figure 4.33 TKE distributions for the spontaneous fission of some trans-einsteinium nuclides (from Hof 88).

FISSION PROPERTIES OF THE TRANSURANIUM NUCLEI

but are also seen in careful measurements of the fragment mass and energy distributions in the plutonium isotopes (Wag 89).

4.10.3 Charge Distributions in Low Energy and Spontaneous Fission (Wah 87, Wah 88)

As Halpern (Hal 59) has pointed out, if one were to plot the yield of a fragment of given atomic number versus its atomic number for a fissioning system (Figure 4.34), the results would be similar to a plot of the fragment mass distribution. The point is that nuclear matter is not very polarizable and to first order, the protons in the fissioning nucleus divide like the neutrons. The primary fission products, labeled "IN" in Figure 4.34, are on the neutron-rich side of β-stability. The enhanced yields for even Z nuclides relative to odd Z nuclides are associated with the stabilization from proton pairing.

A more detailed treatment would show the independent yield of isobars

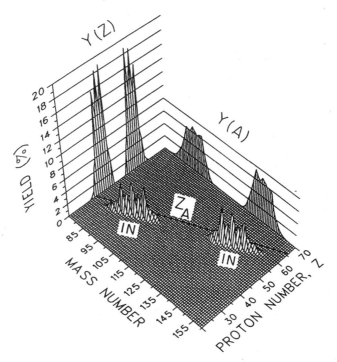

Figure 4.34 Yields of products from thermal-neutron-induced fission of ^{235}U. The Z_A line indicates the approximate location of the line of β-stability. IN designates the independent or primary fission product yields (from Wah 87).

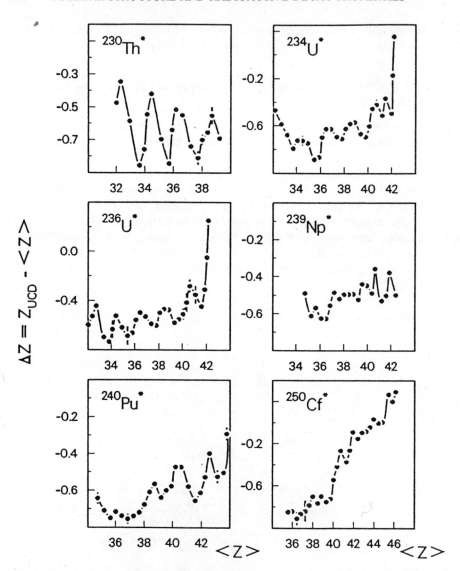

Figure 4.35 The measured values of $\Delta Z(Z_{UCD} - Z_p^L)$ for light products from thermal-neutron-induced fission of various nuclei (Boc 89).

has a Gaussian form given as

$$P(Z) = \frac{1}{\sqrt{c\pi}} \exp[-(Z - Z_p)^2/c] \quad (4.51)$$

where c has an average value of 0.80 ± 0.14 and Z_p is the most probable primary fragment atomic number (non-integer) for that isobaric series. (The more familiar Gaussian width parameter σ can be related to c by $c = 2[\sigma^2 + 1/12]$.) The dependence of Z_p upon fragment mass split is shown in Figure 4.35 where $Z_{ucd} - Z_p$ is plotted versus $\langle Z \rangle$ (Z_{ucd} refers to the idea of an "unchanged charge distribution," that is, same Z/A as that of the fissioning system). Clearly there is a deviation from the idea of an unchanged charge distribution ($\Delta Z = 0$) with the light fragment receiving more than its "share" of the protons (and the heavy fragment receiving less). These observations can be explained qualitatively in terms of the shape of the isobaric mass parabolas and the Coulomb interaction energy (Van 73) or quantitatively by including the effects of fragment shell structure (Wil 76). The large oscillations with a two charge unit periodicity are due to odd-even effects.

The other notable feature of the fragment charge distributions is their narrow width. For a given A, only a few isobars have significant yields. Two effects tend to favor narrow charge dispersions: (a) the high energetic cost of unfavorable charge splits; and (b) the existence of known ground state correlations between neutrons and protons in the fragments.

4.10.4 The Excitation Energies of the Fission Fragments

By subtracting the value of the average total kinetic energy release in fission from the total energy release, one can determine the excitation energies of the fission fragments. For the thermal neutron induced fission of ^{235}U, the average total energy release in forming the primary fragments is ~ 200 MeV while the total kinetic energy release is ~ 172 MeV. This implies that the total excitation energy of the fragments should be ~ 28 MeV. The average number of neutrons emitted is about 2.4 with each neutron having a kinetic energy of ~ 2 MeV and a binding energy of ~ 5.5 MeV. Thus about 18 MeV of the fragment excitation energy ends up in the form of emitted neutrons. Prompt photon emission carries away ~ 7.5 MeV which leaves us only ~ 2.5 MeV short of balancing our energy accounts in these crude estimates.

The prompt neutrons are emitted from the fully accelerated fragments after scission. The average number of such neutrons ν_T, as a function of the mass of the fissioning system is shown in Figure 4.36. The general increase in ν_T with fissioning system mass is due to the increase in the available fragment excitation energy as the mass of the fissioning system increases. For very heavy systems ($Z \sim 114$) ν_T is predicted to be ~ 10, allowing the critical mass for a self-sustaining fission reaction to be quite small for such nuclei.

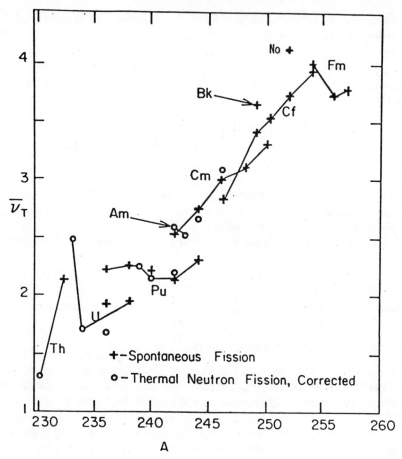

Figure 4.36 The average total number of neutrons $\bar{\nu}_T$, as function of the mass number of the fissioning system, A. Measurements of $\bar{\nu}_T$ for thermal neutron induced fission have been corrected to zero excitation energy, assuming $d\bar{\nu}_T/dE^* = 0.11 \text{ MeV}^{-1}$ (from Hof 84).

The average neutron kinetic energy is ~2 MeV. The distribution of energies in the frame of the moving fission fragment is Maxwellian in character, $P(E_n) = E_n e^{-E_n/T}$. Transforming this spectrum into the laboratory frame, the neutron energy spectrum is predicted to have the Watt form, that is,

$$P(E_n) = e^{-E_n/T} \sinh(4E_n E_f/T^2)^{1/2}$$

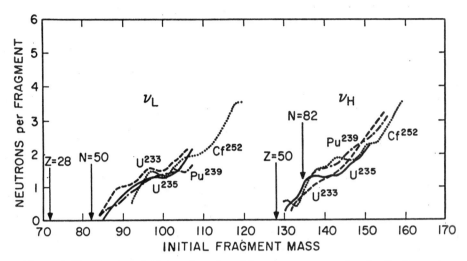

Figure 4.37 Neutron yields as a function of fragment mass number for the thermal neutron induced fission of ^{233}U, ^{235}U, and ^{239}Pu and the spontaneous fission of ^{252}Cf (from Ter 65).

where E_n and E_f are the laboratory system energies of the neutron and fission fragment (in MeV/nucleon) and T is the nuclear temperature (typically $T = 0.965$ MeV, $E_f = 0.54$ MeV/nucleon).

Another important aspect of the neutron emission process is the variation of the neutron yield as a function of fragment mass number, $\nu(A)$ (Figure 4.37). The striking feature of Figure 4.37 is the nearly universal character of $\nu(A)$, that is, the weak dependence of the shape of $\nu(A)$ upon fissioning system mass and the strong dependence upon fragment mass. These data represent one of the more compelling arguments as to the importance of fragment nuclear structure upon post-fission phenomena. As indicated by the energy balance considerations presented earlier, if one understands the dependence of the TKE upon fragment mass (discussed earlier), then the sawtooth $\nu(A)$ curve is also understood.

Prompt γ emission follows or competes with the last stages of fragment deexcitation by neutron emission. These photons are generally emitted on a time scale of 10^{-9} s or less. One observes a γ-ray multiplicity of 7–10 photons/fission which carry away a total energy of 7.5 MeV. This γ-ray yield is considerably larger than one would predict from simple statistical arguments assuming γ-ray emission to follow neutron emission and not to compete with it. However, because of the significant angular momentum of the fission fragments ($\sim$7–10$\hbar$) even in spontaneous fission, photon emission can compete with neutron emission. The emitted γ-rays are mostly dipole radiation with some significant quadrupole radiation due to "stretched" transitions ($J_i \rightarrow J_f = J_i - L$) in the fission fragments.

REFERENCES

Ala 55   G. Alaga, K. Alder, A. Bohr, and B.R. Mottelson, Kgl. Danske Videnskab. Selskab., Mat-Fys. Medd. **29**, No. 9 (1955).

Arm 85   P. Armbruster, Annu. Rev. Nucl. Part. Sci. **35**, 135 (1985).

Bem 73   C.E. Bemis, Jr., F.K. McGowan, J.L.C. Ford, Jr., W.T. Milner, P.H. Stelson, and R.L. Robinson, Phys. Rev. **C8**, 1466 (1973).

Ben 84   R. Bengtsson, P. Möller, J.R. Nix, and Jing-ye Zhang, Phys. Scr. **29**, 402 (1984).

Bjø 80   S. Bjørnholm and J.E. Lynn, Rev. Mod. Phys. **52**, 725 (1980).

Bla 52   J.M. Blatt and V.F. Weisskopf, *Theoretical Nuclear Physics* (Wiley, New York, 1952) p. 583.

Boc 89   J.P. Bocquet and R. Brisset, Nucl. Phys. **A502**, 213c (1989).

Boh 69   A. Bohr and B.R. Mottelson, *Nuclear Structure*, Vols. I and II (Benjamin, Reading, 1975).

Bön 86   K. Böning, Z. Patyk, A. Sobiczewski, and S. Cwiok, Z. Phys. **A325**, 479 (1986).

Bri 80   H.C. Britt, *Physics and Chemistry of Fission*, 1979, *Vol.I*. (IAEA,Vienna, 1980) p. 31.

Bri 82   H.C. Britt in *Actinides in Perspective*, N.M. Edelstein, Ed. (Pergamon, Oxford, 1982) p. 245.

Bro 86a   U. Brosa, S. Grossman, and A. Muller, Z. Phys. **A325**, 241 (1986).

Bro 86b   E. Browne, R.B. Firestone, and V.S. Shirley, *Table of Radioactive Isotopes* (Wiley, New York, 1986) The spontaneous fission branching ratios for ^{230}Th, ^{232}Th, ^{231}Pa, ^{234}U, ^{235}U, ^{236}U, ^{238}U, ^{237}Np, ^{236}Pu, ^{240}Pu, and ^{242m}Am were inadvertently omitted from this reference.

Cha 77   R.R. Chasman, I. Ahmad, A.M. Friedman, and J.R. Erskine, Rev. Mod. Phys. **49**, 833 (1977).

Coh 71   B.L. Cohen, *Concepts of Nuclear Physics* (McGraw-Hill, New York, 1971).

Cwi 83   S. Cwiok et al., Nucl. Phys. **A410**, 254 (1983).

Cwi 85   S. Cwiok, Z. Lojewski, and V.V. Paskevich, Nucl. Phys. **A444**, 1 (1985).

Cwi 88   S. Cwiok, P. Rozmej, and A. Sobiczewski, *Proc. 5th Int. Conf. on Nuclei Far From Stability*, 1987, *Rosseau Lake, Ontario, Canada*, I.S. Towner, Ed. (AIP Conf. Proc. 164, New York, 1988) GSI-87-61.

Cwi 89   S. Cwiok, P. Rozmej, A. Sobiczewski, and Z. Patyk, Nucl. Phys. **A491**, 281 (1989).

Dah 82	M. Dahlinger, D. Vermeulen, and K.-H. Schmidt, Nucl. Phys. **A376**, 94 (1982).
DeS 74	A. De Shalit and H. Feshbach, *Theoretical Nuclear Physics, Volume I: Nuclear Structure* (Wiley, New York, 1974).
Fly 72	K.F. Flynn, E.P. Horwitz, C.A.A. Bloomquist, R.F. Barnes, R.K. Sjoblom, P.R. Fields, and L.E. Glendenin, Phys. Rev. **C5**, 1725 (1972).
Fri 81	G. Friedlander, J.W. Kennedy, E.S. Macias, and J.M. Miller, *Nuclear and Radiochemistry*, 3rd Edn. (Wiley, New York, 1981).
Gin 68	J.E. Gindler and J.R. Huizenga, Nuclear Fission in *Nuclear Chemistry*, Vol. II (Academic Press, New York, 1968) p. 1.
Hab 89	D. Habs, Nucl. Phys. **A502**, 105c (1989).
Hal 59	I. Halpern, Annu. Rev. Nucl. Sci. **9**, 245 (1959).
Hes 86	F.P. Hessberger et al., J. Less-Common Met. **122**, 445 (1986).
Hof 80	D.C. Hoffman et al., Phys. Rev. **C21**, 972 (1980).
Hof 84	D.C. Hoffman, Acc. Chem. Res. **17**, 235 (1984).
Hof 88	D.C. Hoffman and L.P. Somerville in *Particle Emission from Nuclei*, Vol. III, D. Poenaru and M.S. Ivascua, Eds. (CRC Press, Boca Raton, 1988) pp. 1–40.
How 80	W.M. Howard and P. Möller, Atom. Data Nucl. Data Tables **25**, 219 (1980).
Hul 80	E.K. Hulet et al., Phys. Rev. **C21**, 966 (1980).
Hul 86	E.K. Hulet, et al., Phys. Rev. Lett. **56**, 313 (1986).
Hul 89	E.K. Hulet et al., Phys. Rev. **C40**, 770 (1989).
Kel 70	K.A. Keller and H. Munzel, Nucl. Phys. **A148**, 615 (1970); Z. Phys. **255**, 419 (1972).
Kla 81	H.V. Klapdor, J. Metzinger, and H. Oda, Max Plank Institute Preprint MPI-4-1981-V47.
Kra 79	H.J. Krappe, J.R. Nix, and A.J. Sierk, Phys. Rev. **C20**, 992 (1979).
Kra 88	K.S. Krane, *Introductory Nuclear Physics* (Wiley, New York, 1988).
Kup 84	V.M. Kupriyanov, G.N. Smirenkin, and B.I. Fursov, Yad. Fiz **39**, 281 (1984); Sov. J. Nucl. Phys. **39**, 176 (1984).
Led 78	C.M. Lederer and V.S. Shirley, *Table of Isotopes*, 7th Edn. (Wiley, New York, 1978)
Lir 76	S. Liran and N. Zeldes, Atom. Data and Nucl. Data Tables **17**, 431 (1976).
Loj 85	Z. Lojewski and S. Baran, Z. Phys. **A322**, 695 (1985).
Loj 88	Z. Lojewski and A. Baran, Z. Phys. **A329**, 161 (1988).
Mar 69	P. Marmier and E. Sheldon, *Physics of Nuclei and Particles* (Academic Press New York, 1969).
Met 75	V. Metag, Nukleonika **20**, 789 (1975).
Met 80	V. Metag, D. Habs, and H.J. Specht, Phys. Rep. **65**, 1 (1980).

Möl 81   P. Möller and J.R. Nix, Atom. Data Nucl. Data Tables **26**, 165 (1981).
Möl 86   P. Möller, G.A. Leander, and J.R. Nix, Z. Phys. **A323**, 41 (1986).
Möl 87   P. Möller, J.R. Nix, and W.J. Swiatecki, Nucl. Phys. **A469**, 1(1987).
Möl 88   P. Möller, W.D. Myers, W.J. Swiatecki, and J. Treiner, At. Data and Nucl. Data Tables **39**, 225 (1988)
Möl 89   P. Möller, J.R. Nix, and W.J. Swiatecki, Nucl. Phys. **A492**, 349 (1989).
Moo 89   K.J. Moody, Private communication.
Mus 78   M.G. Mustafa and R.L. Ferguson, Phys. Rev. **C18**, 301 (1978).
Mye 66   W.D. Myers and W.J. Swiatecki, Nucl. Phys. **81**, 1 (1966).
Mye 77   W.D. Myers, *Droplet Model of Atomic Nuclei* (Plenum Press, New York, 1977).
Pat 89   Z. Patyk, A. Sobiczewski, P. Armbruster, and K.H. Schmidt, Nucl. Phys. **A491**, 267 (1989).
Pet 40   K. A. Petrzhak and G.N. Flerov, C.R. Acad. Sci. USSR **28**, 500 (1940).
Poe 88   D.N. Poenaru, M.S. Ivascu, and D. Mazilu, in *Particle Emission from Nuclei*, Vol. III, D. Poenaru and M.S. Ivascu, Eds. (CRC Press, Boca Raton, 1988) pp. 41–61.
Pol 62   S.M. Polikanov, V. Druin, N. Karnaukov, V. Mikheev, A. Pleve, N. Skobelev, N. Subbotin, G. ter'Akopian, and B. Fomichev, Sov. Phys. JETP **15**, 1016 (1962).
Ran 76   J. Randrup et al., Phys. Rev. **C13**, 229 (1976).
Ras 75   J.O. Rasmussen in *Nuclear Reactions and Spectroscopy*, Part D, J. Cerny, Ed. (Academic Press, New York, 1975) p. 97.
Ros 84   H.J. Rose and G.A. Jones, Nature **307**, 245 (1984).
Sea 58   G.T. Seaborg, *The Transuranium Elements* (Yale University Press, New Haven and Addison-Wesley, Reading, 1958).
Sob 87   A. Sobiczewski, Z. Patyk, and S. Cwiok, Phys. Lett. **B186**, 6 (1987).
Sol 76   V.G. Soloviev, *Theory of Complex Nuclei*, P. Vogel, Translator (Pergamon, Oxford, 1976).
Spe 72   H.J. Specht, J. Weber, E. Konecny, and D. Heunemann, Phys. Lett. **B41**, 43 (1972).
Str 67   V.M. Strutinsky, Nucl. Phys. **A95**, 420 (1967).
Swi 55   W.J. Swiatecki, Phys. Rev. **100** 937 (1955).
Taa 61   R. Taagepera and M. Nurmia, Ann. Acad. Sci. Fennicae, Ser A. Vi, No. 78 (1961) p. 1.
Tak 73   K. Takahashi, Y. Yamada, and T. Kondoh, Atom. Data Nucl. Data Tables **12**, 101 (1973).
Ter 65   J. Terrell, *Proc. IAEA Symp. Phys. Chem. Fission, Salzburg*, 1965, Vol. **2** (IAEA, Vienna, 1965) p. 3.
Uni 74   J.P. Unik et al., *Phys. and Chem. of Fission*, 1973, *Vol. II* (IAEA, Vienna, 1974) p. 19.

Van 73	R. Vandenbosch and J.R. Huizenga, *Nuclear Fission* (Academic Press, New York, 1973).
Van 77	R. Vandenbosch, Annu. Rev. Nucl. Sci. **27**, 1 (1977).
Van 89	R. Vandenbosch, Nucl. Phys. **A502**, 1c (1989).
Vio 66a	V.E. Viola, Jr., Nucl. Data **A1**, 391 (1966).
Vio 66b	V.E. Viola, Jr. and G.T. Seaborg, J. Inorg. Nucl. Chem. **28**, 741 (1966).
Vio 74	V.E. Viola, J.A. Swant, and J. Graber, Atom. Data and Nucl. Data Tables **13**, 35 (1974).
Vio 85	V.E. Viola, Jr., K. Kwiatkowski, and M. Walker, Phys. Rev. **C31**, 1550 (1985).
Wag 89	C. Wagemans, P. Schillebeeckx, and A. Deruytter, Nucl. Phys. **A502**, 287c (1989).
Wah 62	A.C. Wahl, R.L. Ferguson, D.R. Nethaway, D.E. Trautner, and K. Wolfsberg, Phys. Rev. **126**, 1112 (1962).
Wah 87	A.C. Wahl, in *New Directions in Physics*, N. Metropolis, D.M. Kerr, and G.C. Rota, Eds. (Academic Press, Boston, 1987) pp. 163–189.
Wah 88	A.C. Wahl, Atom. Data Nucl. Data Tables **39**, 1 (1988).
Wal 81	M. Walker and V.E. Viola, Jr., Private Communication (1981).
Wap 83	A.H. Wapstra and K. Bos, Atom. Data Nucl. Data Tables (1983).
Wap 88	A.H. Wapstra, G. Audi, and R. Hockstra, Atom. Data Nucl. Data Tables **39**, 281 (1988).
Wil 76	B.D. Wilkins, E.P. Steinberg, and R.R. Chasman, Phys. Rev. **C14**, 1832 (1976).
Wil 89	J.F. Wild et al., Phys. Rev. C (submitted for publication); Lawrence Livermore National Laboratory Report UCRL-100887, May 1989.
Zum 84	J.D. Zumbro, E.B. Shera, Y. Tanaka, C.E. Bemis, Jr., R.A. Naumann, M.V. Holhn, W. Reuter, and R.M. Steffen, Phys. Rev. Lett **53**, 1888 (1984).

5

EXPERIMENTAL TECHNIQUES

Transuranium element research requires specialized techniques and facilities. The relatively small quantities of most of these elements that are available, the intense radioactivities, the short lifetimes of many species, et cetera are among the problems facing the prospective transuranium investigator. However, through combinations of ingenious techniques and superior manipulative skills, experimentalists have found ways to study the chemical, physical, and nuclear properties of these elements and their reactions even in cases where only a few short-lived atoms of the species in question were present. In this chapter, we will discuss some of the more important of these techniques.

5.1 AVAILABILITY OF MATERIALS

One important factor in most nuclear and chemical experiments involving the transuranium elements is the availability of a suitable quantity of a given element or isotope. In Table 5.1, we summarize information about the current availability of transuranium materials to qualified individuals for use in research in the United States. Weighable quantities of isotopes of elements with $Z \leq 100$ are available, while studies of the transfermium elements will probably always be limited to tracer techniques. However, the quantity of a given element or isotope that is available may be less important than the associated specific activity with its implied self-heating

TABLE 5.1 Availability of Transuranium Element Materials

Nuclide	$t_{1/2}$	Decay Mode	Amounts Available	Specific Activity (dpm/μg)
^{237}Np	2.14×10^6 years	α,SF(10^{-10}%)	kg	1565.
^{238}Pu	87.7 years	α,SF(10^{-7}%)	kg	3.8×10^7
^{239}Pu	2.41×10^4 years	α,SF(10^{-4}%)	kg	1.38×10^5
^{240}Pu	6.56×10^3 years	α,SF(10^{-6}%)	10–50 g	5.04×10^6
^{241}Pu	14.4 years	β,α,(10^{-3}%)	1–10 g	2.29×10^8
^{242}Pu	3.76×10^5 years	α,SF(10^{-3}%)	100 g	$8.73\ 10^3$
^{244}Pu	8.00×10^7 years	α,SF(0.1%)	10–100 mg	39.1
^{241}Am	433 years	α,SF(10^{-10}%)	kg	7.6×10^6
^{243}Am	7.38×10^3 years	α,SF(10^{-8}%)	10–100 g	4.4×10^5
^{242}Cm	162.9 days	α,SF(10^{-5}%)	100 g	7.4×10^9
^{243}Cm	28.5 years	α,ϵ(0.2%)	10–100 mg	1.15×10^8
^{244}Cm	18.1 years	α,SF(10^{-4}%)	10–100 g	1.80×10^8
^{248}Cm	3.40×10^5 years	α,SF(8.3%)	10–100 mg	9.4×10^3
^{249}Bk	320 days	β,α,(10^{-3}%), SF(10^{-8}%)	10–50 mg	3.6×10^9
^{249}Cf	350.6 years	α,SF(10^{-7}%)	1–10 mg	9.1×10^6
^{250}Cf	13.1 years	α,SF(0.08%)	10 mg	2.4×10^8
^{252}Cf	2.6 years	α,SF(3.1%)	10–1000 mg	1.2×10^9
^{254}Cf	60.5 days	SF,α(0.3%)	μg	1.9×10^{10}
^{253}Es	20.4 days	α,SF(10^{-5}%)	1–10 mg	5.6×10^{10}
^{254}Es	276 days	α	1–5 μg	4.1×10^9
^{257}Fm	100.5 days	α,SF(0.2%)	1p g	1.1×10^{10}

effect, possible radiation exposure, and radiation damage, sample purity, activity of any daughter species, et cetera. For macroscopic transplutonium chemical studies, the nuclides of choice are ^{243}Am, ^{248}Cm, ^{249}Bk, ^{249}Cf and ^{253}Es. For tracer studies of the heaviest elements, the nuclides of choice (1989) are 20.1 h ^{255}Fm ($\sim 10^{11}$ atoms/expt.), 77 min ^{256}Md (10^6 atoms/expt.), 3.1 min ^{255}No (10^3 atoms/expt.), 3 min ^{260}Lr (10 atoms/expt.), 65 s ^{261}Rf (1 atom/expt.), and 35 s ^{262}Ha (0.1 atom/expt.) In the United States, the longer-lived nuclides have been available from the transuranium production program associated with the High Flux Isotope Reactor (HFIR) at the Oak Ridge National Laboratory while the shorter-lived nuclides must be made and studied at heavy ion particle accelerators. One should also note that while large quantities of ^{239}Pu are available, it is classified as a Special Nuclear Material because of its use in weaponry and very strict regulations govern the possession and use of this nuclide.

The purity of a particular transuranium element sample is often of great importance. In this regard, one should note that a high sample specific activity makes that sample more reactive with materials in which it comes in contact, thereby lessening sample purity and possibly altering chemical composition and crystal structure. Frequently such small amounts of a given nuclide are available that little can be spared for analysis; nondestructive testing methods are preferred. All materials used in microscale chemical preparations should be of the highest purity, that is, made of quartz, polycarbonate, et cetera. The surface-to-volume ratio of all samples should be minimized to prevent contamination.

5.2 CHEMICAL MANIPULATIONS (Pet 80, Hai 82)

Although multigram quantities of several transuranium elements are now available, the fundamental chemical studies of these elements were (and the current studies of many of the heaviest elements are) carried out using microchemical techniques. One of the most important of these techniques, *the single ion-exchange resin bead technique* for the concentration and manipulation of actinide ions, was developed by the late B.B. Cunningham at Berkeley (Cun 61). Individual resin beads are loaded to saturation by equilibrating them with a dilute acid solution of the actinide. Excess actinide and surface contaminants are removed by washing the loaded bead in water or dilute acid. The amount of actinide sorbed is determined by the size of the resin bead chosen. A Dowex 50X4 cation exchange resin bead with a 0.15 mm diameter will adsorb 1 μg of a typical trivalent actinide ion. The resulting actinide-loaded bead has a concentration of ~2 M, is easily manipulated and because of its spherical shape, has minimum contact with its container, thus minimizing possibilities for contamination. Using this technique, loaded beads of americium through californium compounds weighing up to ~10 μg have been prepared.

Actinide-loaded resin beads can act as the starting point for compound preparation on a microscale basis. Typically the bead will be converted to an oxide by calcining it in air or oxygen at 1200°C. The resulting oxide can be used to synthesize other transuranium compounds or studied directly by X-ray powder diffraction, absorption spectrophotometry, et cetera.

The heaviest elements ($Z > 102$) are made in quantities of a few atoms at a time (see Chapter 2) and studies of their chemistry must involve still more sensitive techniques (Kel 77, Hul 83, Hof 87). Generally the half-lives of the nuclides are short and thus the procedures must be fast and capable of giving the same results as a macroscopic experiment even though only a few atoms are present. Experiments involving ion exchange, solvent extraction, gas chromatography, or other methods that subject the atoms to many chemical reactions of an identical nature give reliable results. It has been argued that subjecting one atom to many chemical reactions is equivalent to subjecting many atoms to one reaction (Kel 77). Techniques which involve an atom reacting only once, such as coprecipitation may give unreliable results. Because of the short lifetimes of the isotopes of the heaviest elements, it is important to minimize the time between production of the atom in a nuclear reaction and the start of any chemical procedure. Quite often, this is accomplished through the use of *a helium jet*. The atoms produced in a nuclear reaction recoil out of a thin target and are stopped in ~1 atm of helium gas in the target chamber. The gas contains an aerosol, typically an alkali halide, to which charged reaction product recoils attach themselves via van der Waals forces. The helium gas (and the aerosol particles) escape through a small orifice to a vacuum chamber, with the gas achieving sonic velocity. The aerosol particles (and the attached heavy element atoms) are collected by allowing the gas stream to strike a collector surface. The resulting deposit can be counted directly or dissolved for further chemical processing. The aerosol loaded gas stream (jet) can also be used to transport the atoms through a thin capillary a distance of several meters in a few seconds.

The chemical procedures used in such experiments are generally conventional although it is common to use sophisticated computer-controlled apparatus (or armies of graduate students) that allow chemical operations to be carried out rapidly and repeatedly. The first investigation of the aqueous chemistry of hafnium, for example, involved 801 separate chemical experiments in which ~ 24 events due to the decay of ^{262}Ha were detected (Gre 88). Along with ion exchange, extraction chromatography, solvent extraction, and gas chromatography, electrochemical techniques have been very useful. Radiopolarographic techniques have been developed and applied to determine half-wave potentials at a dropping mercury electrode (Sam 76). From these measurements standard reduction potentials (III $\to$ 0, II $\to$ 0) have been deduced. Atomic beam measurements have been undertaken to determine magnetic moments (Goo 71) and it may be possible to utilize ion or neutral atom traps and laser techniques such as resonance ionization spectroscopy to gain information about the heaviest elements.

Preparation of single crystals of actinide metals and compounds for solid state research is a very time consuming and expensive effort, requiring considerable expertise (Spi 89). The equipment needed to prepare such samples is generally available only at large national or international research laboratories.

5.3 PHYSICAL TECHNIQUES

5.3.1 Transuranium Targetry

The preparation of actinide targets for use in accelerators in nuclear reaction studies is similar to that used for lower Z materials except that the target material is radioactive. Frequently used target preparation techniques are electrospraying, electromagnetic isotope separation, electrodeposition, and vacuum deposition (Par 68). Because these elements are α-particle emitters, and because of the large radiological "quality" factors associated with exposure to α-particle emitters, the maximum permissible body burdens, atmospheric concentrations, et cetera of these nuclei are very small (generally < 0.1 μg or 10^{-12} Ci/ml). This means that target preparation should take place within hoods or glove boxes, and evaporators, electroplating cells, et cetera used for the preparation of a specific actinide target should not be used for other targets until they have been thoroughly decontaminated. Once prepared, actinide targets should be stored in dry, inert atmosphere in sealed containers which are periodically monitored. The high specific activities of the heavier actinides may limit the amount of target material that can be used without incurring significant radiation stability problems. When actinide targets are used in very intense heavy-ion beams as in attempts to make new elements or radionuclides, special attention must be given to the problem of the radiation stability of the target. For example, early attempts to make superheavy elements using the ^{238}U + ^{248}Cm reaction were hampered (Sch 82) due to the premature failure of the ^{248}Cm targets after only $\sim 10^{15}$ ^{238}U ions had passed through the target. [This problem was solved later by new methods of target mounting and cooling (Hul 82)]. Molitoris and Nitschke (Mol 81) have studied the properties of thin metal films as targets or backing materials. They found helium to be the best cooling gas and molybdenum or tantalum the most durable target backing material. Gäggeler et al. (Gäg 79) have reported that coating the targets with an 0.03 mg/cm^2 carbon film will improve radiative cooling, while Marx et al. (Mar 79) find that a rotating target wheel will allow the use of higher beam intensities by distributing the deposited heat over a larger area.

Because of the intense radioactivity of many actinide targets, the low maximum permissible airborne concentrations of these materials and the possibility of target rupture, use of these targets in heavy-ion accelerators generally requires special techniques to isolate the targets from the main accelerator vacuum system. Moody et al. (Moo 81) and Schädel et al. (Sch 82) describe typical actinide target handling systems in use at the heavy-ion accelerators at Berkeley and Darmstadt, respectively.

5.3.2 Identification of Transuranium Reaction Products

5.3.2.1 General Considerations. In the study of heavy-ion reactions resulting in transuranium reaction products, it is of paramount importance to be able to isolate and uniquely identify the products as to their Z, A, and formation cross section. Indeed the claim to discovery of a new element must involve identification of Z (Har 76) while the claim of discovery of a new nuclide must involve measurement (and/or deduction) of both Z and A. Nitschke (Nit 77) has classified the commonly used techniques of isolating transuranium reaction products by the half-life of the products and the minimum detectable cross section. His classification scheme is shown in Figure 5.1. Some of the isolation techniques shown in Figure 5.1 such as chemistry, magnetic spectrometers, et cetera can also serve as methods of establishing the Z and/or A of the species involved.

5.3.2.2 Chemical Methods. For reaction products with the longest half-lives, chemical separation techniques offer a convenient method of isolating individual reaction products and establishing their atomic numbers. These techniques offer the greatest sensitivity of all methods because of the large amounts of target material that can be used.

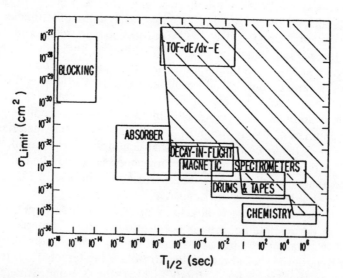

Figure 5.1 Classification of techniques used to isolate transuranium reaction products by minimum detectable half life and production cross section. See text for a discussion of these techniques (from Nit 77).

5.3.2.3 The Helium Jet, Drums, Tapes, and Wheels.

For species with half-lives in the range from $0.1 \leq t_{1/2} \leq 10$ s, the helium jet is a superior method of isolating reaction products, as discussed in Section 5.2. When used in studies of nuclear reaction products, the helium jet will deposit its activity upon a collection device such as a wheel or tape or drum which is moved after a preset collection time in front of semiconductor radiation detectors. Because the helium jet collects all recoils (except inert gases) coming from the nuclear reaction, it does not select or discriminate against any parti-

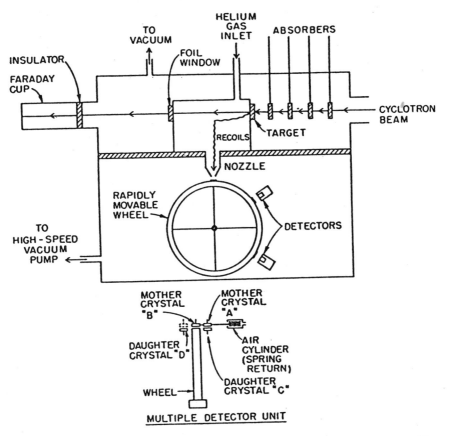

Figure 5.2 Schematic representation of a "gas jet" recoil transport assembly. Thermalized product atoms are transported in the He gas stream and collected on the periphery of a wheel or other suitable collection device. Periodically, the wheel is moved to position the spot in front of the detectors. A "mother-daughter" detector assembly is illustrated in the lower portion of the figure and is used to establish a genetic link between known and unknown activities (from Bem 74).

206   EXPERIMENTAL TECHNIQUES

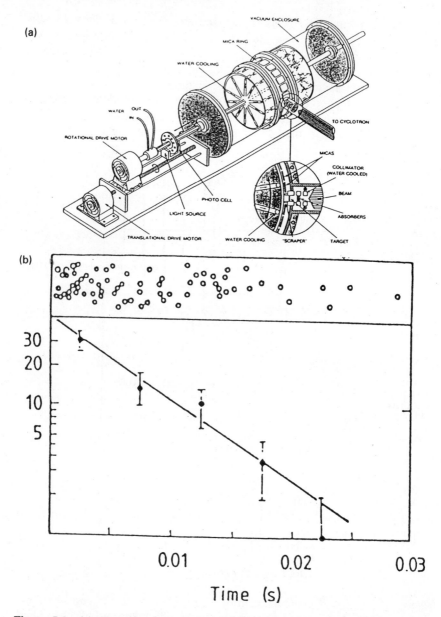

Figure 5.3 (a) Rotating and scanning drum system for the detection of short-lived spontaneously fissioning nuclei (from Nit 80). (b) Typical results from the use of a rotating drum system to study spontaneous fission activities in the reaction ^{208}Pb (^{50}Ti, 2n) ^{256}Rf (Oga 75). Upper part shows the distribution of etched tracks along the mica detectors while the lower part shows the corresponding time distribution of fission decays.

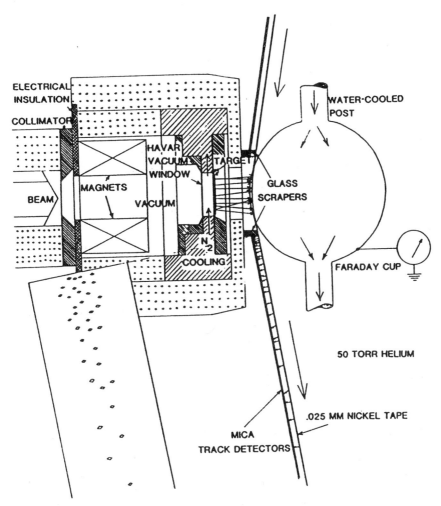

Figure 5.4 Details of a tape system for the collection and detection of short-lived spontaneously fissioning nuclei (from Nit 80).

cular nuclear reaction mechanism. Identification of the collected reaction products can be made with a variety of techniques. Perhaps the most important of these techniques is the "mother-daughter" or "double-recoil" method which establishes a genetic link between the unknown reaction product and known daughter and/or granddaughter activities. In this technique (see Figure 5.2), the heavy recoil atom produced by the -decay of the collected initial reaction product imbeds itself in a "mother crystal."

The mother crystal is then moved in front of a "daughter crystal" which can detect the -decay of the imbedded atom in the mother crystal. If the α-particle decay characteristics of the daughter nucleus are known, then a genetic link is established and the (Z, A) of the parent are established. This technique was used in the discovery of several elements and isotopes (Ghi 69, Ghi 70, Ghi 74, Esk 71).

A technique of exceptional power to identify the Z of collected reaction products is the x-ray method (Dit 71). In this technique one observes the coincidences between the alpha particles emitted by the decay of the collected recoils and the x-rays of the daughter nuclei (produced as a result of internal conversion decay in the daughter). The energies and relative intensities of the x-ray lines serve to identify the Z of the daughter nucleus (Car 69, Lu 71).

For species whose half-lives are in the range $1 \text{ ms} \leq t_{1/2} \leq 100 \text{ ms}$, the product collection device is placed in close proximity to the irradiated target and catches the recoils emerging from the target directly. In such systems, the heavy-ion beam after passing through the target will strike the collection surface (drum, tape, et cetera). Schematic diagrams of two such collection devices are shown in Figures 5.3a and 5.4 (Nit 80). Unfortunately, such devices offer little selectivity as to which reaction products are collected, and it is difficult to detect the radioactive decay of the reaction products amidst a high α-particle background. Therefore these devices are used frequently to detect new spontaneously fissioning nuclides (Figure 5.3b). Since spontaneous fission cannot, in general, be used to identify the Z and A of the fissioning system, experimenters frequently resort to arguments based upon nuclear reaction energetics and cross section systematics to identify the collected products.

5.3.2.4 Magnetic Spectrometers, Velocity Filters.
The principal limitation of the isolation devices discussed previously (tapes, jets, et cetera) is that the reaction product must be stopped and mechanically transported to radiation detectors before product identification can occur. This restricts their use to studies of nuclei whose $t_{1/2} \geq 1$ ms. For detection and identification of species whose $t_{1/2} < 1$ ms, one employs instruments based upon magnetic and/or electrostatic deflection of target recoils. The most spectacularly successful of these devices in recent years is the velocity filter SHIP (Separator for Heavy Ion reaction Products) based at the UNILAC at GSI (Mün 81a). A schematic diagram of this separator is shown in Figure 5.5. Evaporation residues produced in compound nucleus reactions emerge from the target and pass through a thin carbon foil which has the effect of equilibrating the ionic charge distribution of the residues. The ions then pass through two filter stages consisting of electric deflectors,

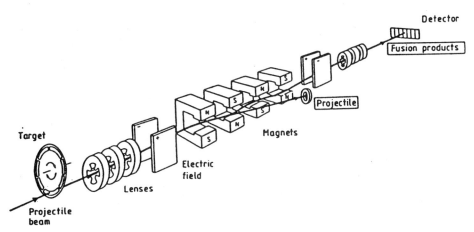

Figure 5.5 A schematic diagram of the velocity filter SHIP at GSI (Mün 81a).

dipole magnets, and a quadruple triplet for focusing. The solid angle of acceptance of the separator is 2.7 msr with a separation time for the reaction products of $\sim 2\ \mu s$ with a total efficiency of collecting evaporation residues of $\sim 20\%$ for $A_{proj} \geq 40$. Since complete fusion evaporation residues have very different velocities and angular distributions than targetlike transfer and deep-inelastic products (a factor of ~ 2 difference in velocity between transfer products and evaporation residues), the separator with its $\pm 5\%$ velocity acceptance range and narrow angular acceptance very effectively separates the evaporation residues from the other reaction products. Following separation, the residues pass through a large area time of flight detector and are stopped in an array of seven position-sensitive detectors (Figure 5.6). From their time of flight and the energy deposited as they stop in the position sensitive detectors, a rough estimate of their mass may be obtained ($\Delta A/A \sim 0.01$). The final genetic identification of the residues is made by recording the time correlations between the original position signals from the detectors and subsequent decay signals from the same location (due to alpha or spontaneous fission decay) and/or signals from γ or x-ray detectors placed next to the position-sensitive detector. Half-lives are determined using the maximum likelihood method because of the generally low count rates. The time intervals between subsequent events are plotted in logarithmic time bins. The distribution has a maximum at the average lifetime of the nuclide (Figure 5.7). SHIP has been used in the discovery of elements 107 (Mün 81b), 108 (Mün 84), and 109 (Mün 82) and the identification of the new nuclides ^{247}Md, ^{243}Fm, and ^{239}Cf (Mün 81a) along with new isotopes of element 106.

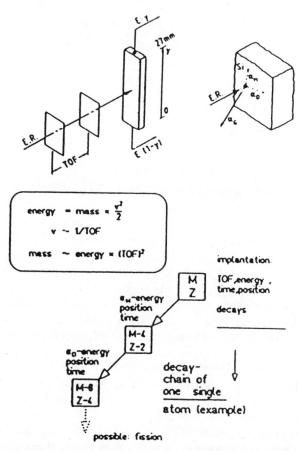

Figure 5.6 The detector system behind SHIP allowing time-of-flight and energy measurements of the recoils and the registration of position, time, and decay energy for correlated members of a decay chain. In the upper left corner the TOF energy measuring system to determine the mass of the implanted recoil is shown. The Si detector in the right corner shows an implanted evaporation residue and its decay by three α-particles in three subsequent generations. The lower part shows a decay chain with the quantities defining the event (from Arm 85). Reproduced, with permission, from the Annual Reviews of Nuclear and Particle Science, Volume 35, Copyright 1985, by Annual Reviews, Inc.

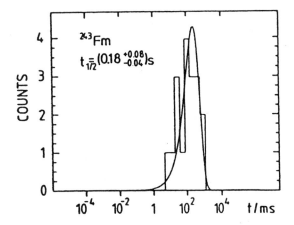

Figure 5.7 The maximum likelihood method applied to the recoil alpha correlations observed in the reaction ^{206}Pb(^{40}Ar,3n)^{243}Fm, in order to determine the halflife of ^{243}Fm (from Arm 85). Reproduced, with permission, from the Annual Reviews of Nuclear and Particle Science, Volume 35, Copyright 1985, by Annual Reviews, Inc.

Separators like SHIP are quite expensive and represent major instrumentation projects. A less sophisticated spectrometer which costs considerably less and is capable of allowing one to measure the formation cross sections, recoil range distributions, and angular distributions of short-lived ($t_{1/2} \leq 1$ ms) alpha emitters formed in heavy-ion reactions has been described by Dufour et al. (Duf 81).

Another type of device used to isolate and identify transuranium nuclei is the gas-filled separator, typified by the Small Angle Separatory System (SASSY) in use at the Lawrence Berkeley Laboratory (Lei 81, Ghi 88), shown in Figure 5.8. In this system, heavy product recoils from a nuclear reaction enter a helium-filled (~ 1 Torr) magnetic separator. The deflection of each ion in the magnetic field depends on its average charge, velocity, and mass. The separator has to be calibrated with well-known heavy-ion beams or with reaction products, whose decay properties are well known. (A similar device is located at the SHIP velocity separator.)

The SASSY system was used to study the ^{208}Pb (^{48}Ca, 2n) ^{254}No reaction (Ghi 88). The suppression of projectile ions was 10^{15}, target recoils 10^3 and the transmission of the separator was measured to be $\sim 40\%$. The time of flight and energy of the recoil nuclei are measured, giving an approximate value of the product mass number. The recoil nuclei which are imbedded in the energy detectors located in the focal plane are identified by their α-particle decay and the decay of their daughters.

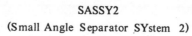
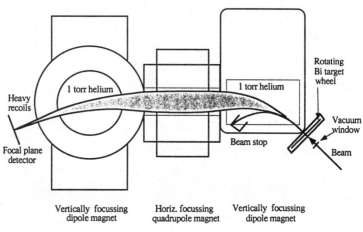

Figure 5.8 The gas-filled magnetic separator system SASSY.

The Dubna laboratories have developed (Yer 89) an electrostatic separator, VASSILISSA, for the study of complete fusion reaction products. It features the ability to look at very short-lived reaction products ($t \sim 10^{-6}$ s).

In Table 5.2, we compare the properties of some techniques for isolating and identifying transuranium reaction product recoils.

5.3.2.5 Time-of-Flight (TOF), Decay-in-Flight (DIF), and Blocking Techniques.
To detect species whose lifetimes are substantially less than 1 s, special techniques must be employed. They include time-of-flight

TABLE 5.2 Comparison of Devices for Isolating Transuranium Reaction Products

Method	He Jet	Rotating Wheel	Velocity Separator	Gas-filled Separator
Separation time (s)	10^{-1}	10^{-3}	10^{-6}	10^{-6}
Total efficiency	0.4	0.6	0.2	0.4
Rejection of transfer products	no	no	10^3	10^2
Decay mode detected	α-chain,	fission	Evaporation residues α-chain, fission	Evaporation residues α-chain, fission

(TOF) techniques which when combined with a measurement of the product energy will give information about the product mass number. For suitable mass resolution, the time of flight must be $\gtrsim 10$ ns. When searching for rare events, some selection process (like SHIP) must be employed to reduce the "background" levels in the apparatus. The decay-in-flight technique (whose use is described in Ghi 77) and the crystal blocking technique (Gib 74) ($10^{-18} \leq t_{1/2} \leq 10^{-14}$ s) give very little information about the identity of a reaction product other than its existence and its approximate lifetime.

5.4 HEALTH AND SAFETY ASPECTS OF TRANSURANIUM ELEMENT USE

As discussed previously, the transuranium nuclei often decay by α-emission or spontaneous fission. Many of them have high specific activities (Table 5.1). For ordinary research use of these elements, the external radiation hazards are not significant. Exceptions to this statement can arise when large quantities of the fissionable nuclei are used, posing possible criticality problems (see Section 3.9 for discussion) or in special cases, such as the use of ^{252}Cf, ^{241}Am, ^{254}Es, et cetera. With ^{252}Cf, the external radiation hazard is due to the neutrons emitted during spontaneous fission of ^{252}Cf which gives neutron fluxes and dose rates of 1.9×10^7 n/cm^2 s at a distance of 1 m from 1 g of ^{252}Cf (~ 2400 rem/h). In the case of ^{241}Am or ^{254}Es or similar nuclei, moderate or high energy photons are emitted in the decay of the parent nuclide or its daughter that may cause concern. The γ-radiation level at 4 cm from 4 μg of ^{254}Es (a typical accelerator target) is ~ 3 rem/h with the beta dose being ~ 50X higher (due to the decay of the ^{250}Bk daughter).

The principal health hazards associated with the transuranium elements are internal radiation hazards posed by inhalation or ingestion of these nuclides. All of the transuranium elements are considered to be bone seekers (where they accumulate in hemosiderin deposits close to the surfaces of the bone in the reticuloendothelial cells of the bone marrow.) When inhaled or injected, they are distributed about equally between the skeleton and the liver. The effective biological half-lives of these materials in the body are several decades. The principal cause of death in experimental animals receiving large doses of the transuranium elements is bone cancer.

The acute toxicity of the transuranium elements is generally great due to the high specific activities. In Table 5.3, we show a selection of the data on the most restrictive ICRP annual limits of intake (ALI) for inha-

TABLE 5.3 Annual Limits of Intake (ALI) for Inhalation and Derived Air Concentrations (DAC) of Some Heavy Elements

	ALI (inhalation)		DAC (40h week)	
	Bq	μCi	Bq/m^3	μCi/m^3
^{226}Ra	2.4×10^4	6.5×10^{-1}	1×10^1	2.7×10^{-10}
^{232}Th	4×10^1	1.1×10^{-3}	2×10^{-2}	5.4×10^{-13}
^{238}U	2×10^3	5.4×10^{-2}	7×10^{-1}	1.9×10^{-11}
^{239}Pu	2×10^2	5.4×10^{-3}	8×10^{-2}	2.2×10^{-12}
^{244}Pu	2×10^2	5.4×10^{-3}	9×10^{-2}	2.4×10^{-12}
^{241}Am	2×10^2	5.4×10^{-3}	8×10^{-2}	2.2×10^{-12}
^{248}Cm	5×10^1	1.4×10^{-3}	2×10^{-2}	5.4×10^{-13}
^{249}Bk	8×10^4	2.2×10^0	3×10^1	8.1×10^{-10}
^{252}Cf	1×10^3	2.7×10^{-2}	4×10^{-1}	1.1×10^{-11}
^{254}Es	4×10^3	1.1×10^{-1}	2×10^0	5.4×10^{-11}

lation and the derived air concentration (DAC) of some transuranium nuclides (Rec 78).

Because of the very low ALIs generally associated with some of these elements, the general strategy for safe use is containment by design. The biggest problem in their use is contamination control. One tries to avoid airborne activity, uses glove boxes for all procedures that could cause atmospheric release of material, continuously monitors (and informs personnel by alarm of) atmospheric activity in work areas, et cetera. A number of excellent guides to the safe use of the transuranium nuclei exist. A highly recommended book by Stewart is among these (Ste 80). All use of these elements should be made in cooperation with and under the guidance of professional health and safety personnel.

REFERENCES

Arm 85	P. Armbruster, Annu. Rev. Nucl. Part. Sci. **35**, 135 (1985).
Bem 74	C.E. Bemis in *Proc. Intl Conf. on Reactions Between Complex Nuclei*, Vol 2., (1974) p. 529.
Car 69	T.A. Carlson, C.W. Nestor, F.B. Malik, and T.C. Tucker, Nucl. Phys. **A135**, 57 (1969).
Cun 61	B.B. Cunningham, Microchemical Techniques, Microchem. J. Symp. Ser. **1**, 69 (1961).
Dit 71	P.F. Dittner, C.E. Bemis, Jr., D.C. Hensley, R.J. Silva, and C.D. Goodman, Phys. Rev. Lett. **26**, 1037 (1971).

Duf 81	J.P. Dufour, R. Del Moral, A. Fleury, F. Hubert, Y. Llabador, M.B. Manhourat, R. Bimbot, D. Gardes, and M.F. Rivet, Proc. of the Intl. Conf. on Nuclei Far From Stability, Helsingor, Denmark, (1981).
Esk 71	K. Eskola, P. Eskola, M. Nurmia, and A. Ghiorso, Phys. Rev. C **4**, 632 (1971).
Gäg 79	H. Gäggeler, A.S. Iljin, G.S. Popeko, W. Seidel, G. M. Ter'Akopian, and S.P. Tretya, Z. Phys. **A289**, 415 (1979).
Ghi 69	A. Ghiorso, M. Nurmia, J. Harris, K. Eskola, and P. Eskola, Phys. Rev. Lett. **22**, 1317 (1969).
Ghi 70	A. Ghiorso, M. Nurmia, K. Eskola, J. Harris, and P. Eskola, Phys. Rev. Lett. **24**, 1498 (1970).
Ghi 74	A. Ghiorso, J.M. Nitschke, J.R. Alonso, C.T. Alonso, M. Nurmia, G.T. Seaborg, E.K. Hulet, and R.W. Lougheed, Phys. Rev. Lett. **33**, 1490 (1974).
Ghi 77	A. Ghiorso, J.M. Nitschke, M.J. Nurmia, R.E. Leber, L.P. Somerville, and S. Yashita, Lawrence Berkeley Laboratory Report LBL-6575 (1977).
Ghi 88	A. Ghiorso, S. Yashita, M Leino, L. Frank, J. Kalnins, P. Armbruster, J.P. Dufour and P.K. Lemmertz, Nucl. Instrum. Methods **A269**, 192 (1988).
Gib 74	W.M. Gibson and N. Maruyama in *Channeling*, D.V. Morgan, Ed., (Wiley, New York, 1974).
Goo 71	L.S. Goodman, H. Diamond, H.E. Stanton, and M.S. Fried, Phys. Rev. **A4**, 473 (1971).
Gre 88	K.E. Gregorich et al., Radiochim. Acta **43**, 233 (1988).
Hai 82	R.G. Haire in *Actinides in Perspective*, N.M. Edelstein, Ed. (Pergamon, Oxford, 1982) p. 309.
Har 76	B.G. Harvey, G. Herrmann, R.W. Hoff, D.C. Hoffman, E.K. Hyde, J.J. Katz, O.L. Keller, Jr., M. Lefort, and G.T. Seaborg, Science **193**, 1271 (1976).
Hof 87	D.C. Hoffman et al., Lawrence Berkeley Laboratory Report LBL-23367, April, 1987; J. Radioanal. Chem. **124**, (1988).
Hul 82	E.K. Hulet and R.W. Lougheed, Lawrence Livermore National Laboratory Report UCAR-10062-81/1, January 1982.
Hul 83	E.K. Hulet, Radiochim. Acta **32**, 7 (1983).
Kel 77	O.L. Keller, Jr. and G.T. Seaborg, Annu. Rev. Nucl. Sci. **27**, 139 (1977).
Lei 81	M.E. Leino, S. Yashita, and A. Ghiorso, Phys. Rev. C **24**, 2370 (1981).
Lu 71	C.C. Lu, F.B. Malik, and T.A. Carlson, Nucl. Phys. **A175**, 289 (1971).

Mar 79	D. Marx, F. Nickel, G. Munzenberg, K. Guttner, H. Ewald, W. Faust, S. Hofmann, and H.J. Schott, Nucl. Instrum. Methods **163**, 15 (1979).
Mol 81	J.D. Molitoris and J.M. Nitschke, Lawrence Berkeley Laboratory Report LBL-9725, January 1981; Nucl. Instrum. Methods **186**, 659 (1981).
Moo 81	K.J. Moody, M.J. Nurmia, and G.T. Seaborg, Lawrence Berkeley Laboratory Report LBL-11588, March 1981, p. 188.
MüGn 79	G. Münzenberg, W. Faust, S. Hofmann, P. Armbruster, K. Guttner, and H. Ewald, Nucl. Instrum. Methods **161**, 65 (1979).
Mün 81a	G. Münzenberg et al., Z. Phys. **A302**, 7 (1981).
Mün 81b	G. Münzenberg, S. Hofmann, F.P. Hessberger, W. Reisdorf, K.H. Schmidt, J.H.R. Schneider, P. Armbruster, C.C. Sahm, and B. Thuma, Z. Phys. **A300**, 107 (1981).
Mün 82	G. Münzenberg et al., Z. Phys. **A309**, 89(1982).
Mün 84	G. Münzenberg et al., Z. Phys. **A317**, 235 (1984).
Nit 77	J.M. Nitschke in *Superheavy Elements*, M.A.K. Lodhi, Ed. (Pergamon, New York, 1978).
Nit 80	J.M. Nitschke, Lawrence Berkeley Laboratory Report LBL-11712, September 1980.
Oga 75	Y.T. Oganessian, A.G. Demin, A.S. Ilj, S.P. Tretya, A.A. Pleve, Y.E. Penionzhkevich, M.P. Ivanov, and Y.P. Tretyakov, Nucl. Phys. **A239**, 157 (1975).
Par 68	W.C. Parker and H. Slatis, in *Alpha, Beta, and Gamma-Ray Spectroscopy*, Vol. I., K. Siegbahn, Ed., (North-Holland, Amsterdam, 1968) p. 379.
Pet 80	J.R. Peterson, in *Lanthanide and Actinide Chemistry and Spectroscopy*, N.M. Edelstein, Ed. (ACS, Washington, D.C., 1980) p. 221.
Rec 78	Recommendations of the International Commission on Radiological Protection, ICRP Publ. 30 (Pergamon, London, 1978).
Sam 76	K. Samhoun and F. David, *Transplutonium Elements*, W. Muller and R. Lindner, Eds. (North-Holland, Amsterdam, 1976) p. 297.
Sch 82	M. Schädel et al., Phys. Rev. Lett. **43**, 852 (1982).
Spi 89	J.C. Spirlet, J. Nucl. Mat. **166**, 41 (1989).
Ste 80	D.C. Stewart, *Handling Radioactivity*, (Wiley, New York, 1980).
Yer 89	A.V. Yeremin et al., Nucl. Instrum. Methods **A274**, 528 (1989).

6

NUCLEAR SYNTHETIC TECHNIQUES

6.1 GENERAL CONSIDERATIONS

The successful synthesis of transuranium nuclei in nuclear reactions involves a judicious balance of several factors. These factors can be grouped into two categories, "production" and "survival". The "production factors" determine the yield of the primary reaction products while the "survival factors" determine which primary product nuclei deexcite by particle emission which allows them to survive or which nuclei deexcite by fission which destroys them. Amongst the "production factors" are items such as the "starting material", the target nuclei, which must be available in sufficient quantity and suitable form. The transmuting projectile nuclei must be available in sufficient quantity also. The transmutation reaction must occur with adequate probability to insure a good yield of the product nucleus in a form suitable for further study. Equally important is that the product nuclei be produced with excitation energy and angular momentum distributions such that the product nuclei will deexcite by particle or photon emission rather than the disastrous fission process. The competition between particle emission and fission as deexcitation paths depends on excitation energy, angular momentum and the intrinsic stability of the product nucleus which is related to the atomic and mass numbers of the product.

Nuclear synthesis is analogous in some ways to inorganic or organic chemical syntheses with the synthetic chemist or physicist having to understand the reactions involved and the structure and stability of the intermediate species. While in principle, the outcome of any synthesis reaction is calculable, in practice such calculations are, for the most part, prohibitively difficult. Instead the cleverness of the scientists involved, their manipulative skills and the instrumentation available for their use determine the success of many synthetic efforts.

The nuclear reactions used to produce transuranium nuclei can be classified by the time scale and degree of interaction between the projectile and target nuclei. Following Weisskoff (Wei 57) (Figure 6.1), we can enumerate the reaction classes or mechanisms as follows. If the incident projectile nucleus interacts with the target nucleus, at its edge, there will be a partial reflection of the incident projectile corresponding to *shape elastic scattering*. The part of the projectile wave function that enters the nucleus undergoes absorption. The first step in the absorption process is a two body collision between a projectile and a target nucleon. If the struck target nucleon leaves the nucleus, a *direct reaction* occurs. If the

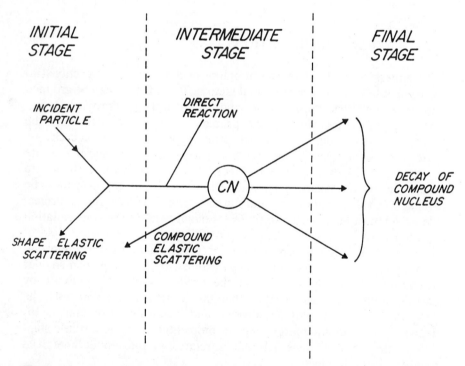

Figure 6.1 Representation, due to Weisskopf (Wei 57), of nuclear reactions in terms of successive collisions.

struck target nucleon does not leave the target nucleus, a more complicated series of events involving multiple collisions and collective effects can occur. (These can lead to *inelastic scattering* of the projectile and/or other direct reactions, such as *nucleon transfer*). Eventually the set of two body interactions becomes sufficiently complex that the *compound nucleus stage* of the reaction occurs. The compound nucleus does not remember how it was formed and its mode of decay is independent of its mode of formation. It may happen that the compound nucleus may reemit a projectile nucleus with the same center of mass energy as the incident projectile, in which case, *compound elastic scattering* occurs.

The nature of each of these reaction mechanisms is affected by the projectile size and charge. [Projectile nuclei can be single nucleons (n,p), simple light ions (d, t, α, etc.) or heavy ions ($Z > 2$, $A > 5$)]. Very heavy projectile nuclei (Kr, Xe, etc.) do not undergo compound nuclear reactions with actinide target nuclei while most lighter heavy ions (C,O) interact mostly via compound nuclear mechanisms for projectile energies below 10 MeV/A. In Sections 6.2 and 6.3, we discuss the details of these production (reaction) mechanisms with specific reference to the projectile size and charge.

6.2 COMPOUND NUCLEUS FORMATION

6.2.1 Neutrons (Cra 74)

Four of the transuranium elements (Np, Am, Es, Fm) were synthesized for the first time using compound nuclear reactions involving neutrons as projectile nuclei. Compound nuclear synthetic reactions involving neutrons account for all large-scale transuranium element production, with the largest terrestrial source of transuranium nuclei being the spent fuel from nuclear power reactors. There are two reasons for this: (a) the large numbers of neutrons available (in nuclear reactors) to induce reactions; and (b) the lack of a Coulomb barrier to inhibit neutron induced reactions.

6.2.1.1 *Qualitative Description.* The neutron capture paths used for synthesis of transuranium nuclei in nuclear reactors are shown in Figure 6.2. Note that the nuclides used as reactor fuel (^{235}U, ^{238}U, ^{239}Pu) lie on the same path, with the thorium cycle nuclides (^{233}Th and ^{233}U) joining this path also. Starting with ^{238}U, the main path goes to ^{239}Pu which undergoes multiple neutron capture to ^{243}Pu which decays to ^{243}Am which, in turn, undergoes capture and decay to ^{244}Cm. ^{244}Cm can undergo multiple

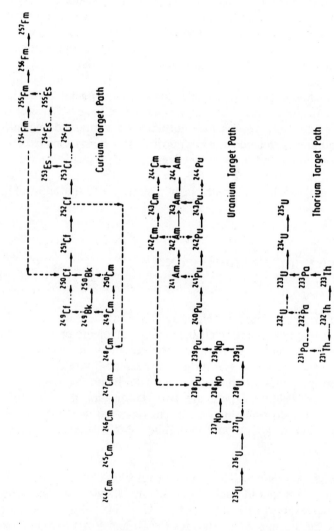

Figure 6.2 Neutron capture paths used to synthesize transuranium nuclei in nuclear reactors.

neutron captures to yield the Cm isotopes to ^{249}Cm while subsequent captures and decays lead to ^{249}Bk, ^{250}Bk, ^{250}Cf, ^{251}Cf, ^{252}Cf, and so on up to ^{257}Fm, where the chain ends due to the 380 μs spontaneous fission of ^{258}Fm. In power reactors, no appreciable concentrations of nuclides above ^{244}Cm are found.

To produce these heavier nuclides or "side chain" nuclides such as ^{238}Pu, one makes special reactor irradiations of separated nuclides. Typical facilities for such irradiations have included the High Flux Isotope Reactor (HFIR) at Oak Ridge accommodating 300 g of actinide target nuclides in a flux of 2–5 × 10^{15} n/cm^2 s or the plutonium production reactors at Savannah River.

Neptunium, the second most abundant transuranium element, exists largely as ^{237}Np, a by-product of the irradiation of natural uranium (70% of ^{237}Np results from the ^{238}U (n, 2n) reaction while 30% comes from ^{236}U (n, γ)). The principal use of ^{237}Np is as an irradiation target to produce ^{238}Pu.

Plutonium, the most abundant transuranium element, exists in ton quantities as ^{239}Pu due to production for use in nuclear weapons although its production in power reactors is growing. The plutonium isotopes, ^{240}Pu, ^{241}Pu, and ^{242}Pu are the results of reactor production as is ^{244}Pu.

The principal reactor-produced isotopes of americium are ^{241}Am and ^{243}Am. ^{241}Am results from the decay of the ^{241}Pu in reactor fuel while ^{243}Am is produced directly by the capture of two neutrons by ^{241}Am.

The nuclides ^{242}Cm and ^{244}Cm are produced by reactor irradiations of ^{241}Am and ^{243}Am as well as being reactor by-products. Other Cm nuclides result as intermediates in the production of ^{252}Cf from ^{244}Cm starting material. Isotopically pure ^{248}Cm can be obtained from the α-decay of ^{252}Cf.

The principal Bk nuclide made in reactor irradiations is ^{249}Bk, a by-product of ^{252}Cf production.

An important transuranium nuclide from a practical standpoint is ^{252}Cf, because of its convenient spontaneous fission half-life which also allows it to be used as a neutron source. The ^{252}Cf synthesis yields are shown in Figure 6.3. As one can see from examining Figure 6.3, neutron fluxes in excess of 10^{15} n/cm^2 s are needed to form significant amounts of ^{252}Cf. In addition, one should note that due to the low thermal neutron capture cross sections of ^{242}Pu and ^{244,246,248}Cm, one needs copious numbers of resonance neutrons for ^{252}Cf production.

Irradiation of ^{252}Cf targets will produce einsteinium and fermium with the principal product nuclides being ^{253}Es and ^{257}Fm.

The most efficient method of producing the heaviest transuranium nuclei is via a nuclear explosion such as that leading to the discovery of Es

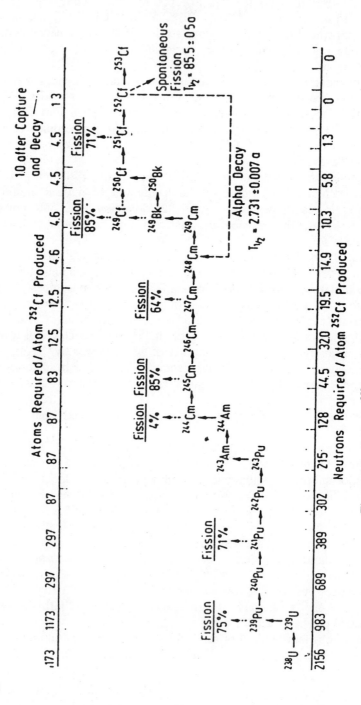

Figure 6.3 Path for ^{252}Cf production in a high flux reactor.

TABLE 6.1 Neutron Addition Paths to Transuranium Synthesis (Cra 74)

Neutron Addition Process	Neutron Flux (n/cm² s¹)	Reaction Time	Neutron Exposure (n/cm²)	Average Neutron Energy (keV)
High flux reactor	$\approx 5 \times 10^{15}$	0.5 years	$\approx 10^{23}$	2.5×10^{-5}
Stellar s process	$\approx 10^{16}$	$\approx 10^{3}$ years	$\approx 10^{26}$	≈ 10
Stellar r process	$\geq 10^{27}$	1–100 s	$>10^{27}$	≈ 100
Nuclear explosion	$>10^{31}$	$<10^{-6}$ s	$\approx 10^{25}$	≈ 20

and Fm. In Table 6.1, we compare various neutron capture paths for transuranium element synthesis.

The nuclear explosions have the highest neutron fluxes and shortest irradiation times. Some neutron moderation takes place in the explosions resulting in an average neutron energy of ~20 keV. Among the nuclides that are best produced by nuclear explosions is ^{250}Cm, a superheavy element synthesis starting material. Approximately 10^{20} atoms of ^{250}Cm were produced in a U.S. test such as Hutch (40 mol/cm^2 of neutrons). The principal problems of this method are product recovery, high cost, and the general undesirability of weapons testing.

6.2.1.2 Production Calculations.
The calculation of the production of transuranium nuclei in neutron induced reactions is usually approached from a semiempirical basis utilizing the large number of measured cross sections, et cetera. Following this approach, we can say that the rate of production of reaction products can be given as

$$R = N \int_0^\infty \Phi(E)\sigma(E)\, dE \tag{6.1}$$

where N is the number of target atoms, Φ the neutron flux in n/cm^2 s and σ the relevant cross section.

Microscopic considerations will give $\sigma(E)$ or one can simply refer to extensive tabulations of these quantities (Mug 81). A typical example of $\sigma(E)$ for a transuranium nuclide is shown in Figure 6.4. The neutron energy spectrum in a nuclear reactor $\Phi(E)$ is generally divided into three regions, a fission spectrum region ($E_n > 1$ MeV), a slowing-down or resonance spectrum (1 eV–1 MeV) and a thermal spectrum (with an average energy of ~0.025 eV) (Figure 6.5). The details of the neutron spectra and energy variation of the cross section are sufficiently complex within a given reactor to require the use of large computer codes to perform the calculations of expected transuranium nuclide activities. For thermal reactors, typical codes are CASMO or CPM while fast reactors are described by 3DB or MCMP.

Useful approximations to the computer calculations can be obtained by using fluxes integrated over the three energy regions discussed previously along with average cross sections for each group. For the thermal region, the neutron spectrum can be approximated as a Maxwell–Boltzmann distribution

$$\frac{n(E)}{n}\, dE = \frac{2\pi}{(\pi kT)^{3/2}} \exp\left(\frac{E}{kT}\right) dE \tag{6.2}$$

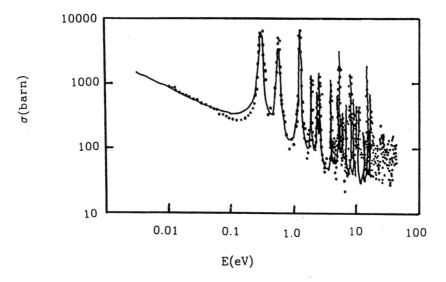

Figure 6.4 Total reaction cross section versus neutron energy for n + ^{241}Am.

where T is the temperature and E the neutron energy. In theory, the effective thermal cross section

$$\sigma_{\text{eff}} = \sigma_{2200}\left(\frac{293.2}{T}\right)^{1/2}\frac{\sqrt{\pi}}{2} \tag{6.3}$$

when multiplied by the total flux of thermal neutrons will give the reaction rate for thermal neutrons for a $1/v$ absorber (where σ_{2200} is the thermal cross section for monoenergetic neutrons of speed 2200 m/s). However, because many fissile nuclides are not $1/v$ absorbers, this reaction rate calculation should be done using a computer code, if possible.

For the resonance region, the neutron flux is given as

$$\Phi(E) = \beta/E \tag{6.4}$$

where β is a constant. This energy distribution can be combined with the Maxwell–Boltzmann energy distribution for thermalized neutrons, along with a correction for non-$1/v$ absorbers to give a simple "one-group" approximation for simplified thermal reactor calculations (Ben 81).

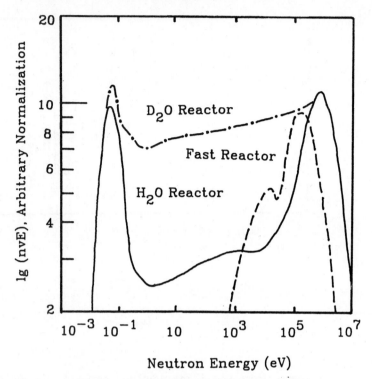

Figure 6.5 Neutron spectra in typical reactors.

In the fission energy region, the neutron spectra can be given as

$$n(E) = E^{1/2} \exp(-E/T) \qquad (6.5)$$

where $T \sim 1.2\text{--}1.3\,\text{MeV}$. Fast neutrons of this energy are only important in transuranium element synthesis because they can induce (n,2n) and (n,f) reactions.

6.2.2 Light Charged Particles (p,d,t,α,^{3}He)

6.2.2.1 *Energetics.* Compound nucleus reactions involving light charged particles were used for the first synthesis of the elements Pu, Cm, Bk, Cf, and Md. Such reactions are used routinely to synthesize small quantities of transuranium nuclei for studies of their nuclear properties, such as fission. Charged particle-induced reactions have the advantage of involving relatively well-understood interactions. As a consequence, the product nuclei are produced in well-defined states of energy and angular momentum. The particle fluxes of light charged particles from modern accelerators are usually copious, involving 10^{13}–10^{15} ions/s. The long ranges of light charged particles in matter allow the use of relatively thick targets in production reactions. It is possible to produce both n-rich and n-deficient product nuclei using these reactions.

Despite these advantages and the relative ease of conceiving a number of possible reactions that could lead to a desired transuranium product nucleus, there is one important consideration that limits the effectiveness of light charged particle-induced reactions in synthesizing transuranium nuclei. That consideration is the mutual Coulomb repulsion between the target nucleus and the charged particle projectile. That repulsion is characterized by a Coulomb barrier

$$B_C = \frac{Z_{\text{target}} Z_{\text{proj}} e^2}{R_{\text{proj}} + R_{\text{target}}} \tag{6.6}$$

where R_{proj}, R_{target} are the radii of the projectile and target nuclei with atomic numbers Z_{proj} and Z_{target}, respectively. (e^2 is the square of the electronic charge). A useful approximation is that

$$B_C \approx \frac{Z_{\text{proj}} Z_{\text{target}}}{A_{\text{target}}^{1/3}} \tag{6.7}$$

for light charged particle-induced reactions. While it is possible to induce reactions with projectiles whose center-of-mass (cm) kinetic energy is less than the Coulomb barrier height, B_C, such reactions have severely reduced probabilities. Therefore, it is common to assume the cm kinetic energy of the projectile, T_{cm}, must be at least equal to the Coulomb barrier height for compound nucleus reactions induced by light charged particles. The practical implication of this for the energetics of synthesis reactions is summarized in Table 6.2. In Table 6.2, we show typical values of the Coulomb barrier height, the binding energy of an added proton or alpha particle and the resulting excitation energy of the compound nucleus, E_{CN}, formed by the reaction of an incident projectile with $T_{\text{cm}} \approx B_C$ with a typical isotope of a given element. In contrast to neutron-induced reactions where $E_{\text{CN}} \sim 6\,\text{MeV}$, the product nuclei from these reactions are produced with much higher excitation energies. As a consequence, the yields of transuranium products of these reactions are reduced by several orders of magnitude relative to neutron-induced reactions due to fission/particle emission competition. The excitation energies, however,

TABLE 6.2 Simple Energetics of Charged Particle-Induced Reactions Involving Heavy Nuclei

Projectile	Average Value of Quantity	Z of Target Nuclei		
		90	95	100
Proton	Coulomb barrier (MeV)	12.6	13.2	13.6
	Binding energy (MeV)	5.2	5.4	4.0
	CN excitation energy (Mev)	17.8	18.6	17.6
Alpha	Coulomb barrier (MeV)	23.3	24.3	25.2
	Binding energy (MeV)	−4.6	−6.5	−6.2
	CN excitation energy (Mev)	18.7	17.8	19.0

are low enough to make the most likely particle to be emitted by the compound nucleus to be a neutron.

6.2.2.2 Cross Sections (Sat 80, Jac 70)

6.2.2.2.1 General Formalism. There is an extensive body of literature dealing with compound nucleus reactions induced by light charged particles. It is beyond the scope of this work even to summarize, in a cursory manner, this information. The reader is advised to consult the general references cited at the end of this chapter for appropriate guidelines for further study. It is most important that one understands that nuclear reactions leading to transuranium products can be understood using the same principles that one uses to treat reactions with other target nuclei and products. Other than the introduction of fission as a deexcitation mode for product nuclei, there is nothing unusual or special about synthesizing transuranium nuclei.

With that caveat in mind, one can use conventional nucleus reaction theory to treat compound nucleus reactions induced by light charged particles. We represent such compound nucleus reactions as two-step processes

$$a + A \rightarrow C \rightarrow b + B \tag{6.8}$$

where C is the compound nucleus. The decay of C should not depend upon the projectile nucleus a and the target nucleus A so that

$$\sigma_{\alpha\beta} = \sigma_{\alpha C} G_\beta^C = \sigma_{\alpha C} \Gamma_\beta^C / \Gamma^C \qquad (6.9)$$

where $\sigma_{\alpha\beta}$ denotes the cross section for the compound reaction with entrance channel $\alpha(a + A)$ and exit channel $\beta(b + B)$ and $\sigma_{\alpha C}$ denotes the cross section for compound nucleus formation. The relative probability G_β^C of decay into the β channel is taken as the ratio of the partial width Γ_β^C to the total width Γ^C. (Widths are used as convenient devices for expressing decay rates.) The reciprocity theorem states

$$\frac{\sigma(b \to a)}{p_a^2} = \frac{\sigma(a \to b)}{p_b^2} \qquad (6.10)$$

where $\sigma(a \to b)$ and $\sigma(b \to a)$ are the cross sections for the reactions shown in equation 6.8 and its inverse, respectively. (p_a is the momentum of a relative to A and p_b is the momentum of b relative to B.) Thus, ignoring spins,*

$$k_\alpha^2 \sigma_{\beta\alpha} = k_\beta^2 \sigma_{\alpha\beta} \qquad (6.11)$$

or

$$G_\beta^C / G_\alpha^C = (k_\beta^2 \sigma_{\beta C}) / (k_\alpha^2 \sigma_{\alpha C}) \qquad (6.12)$$

or since $\Sigma_\beta G_\beta^C = 1$

$$G_\beta^C = \frac{k_\beta^2 \sigma_{\beta C}}{\Sigma_\gamma k_\gamma^2 \sigma_{\gamma C}} \qquad (6.13)$$

Of course, the independence of formation and decay of the compound nucleus is subject to the constraints of conservation laws. Energy, total angular momentum, its projection on a space-fixed axis, and parity must

---

*For nuclei with spins, we have

$$\frac{k_\alpha^2 \sigma_{\alpha\beta}}{g_\alpha} = \frac{k_\beta^2 \sigma_{\beta\alpha}}{g_\beta}, \qquad G_\beta^C = \frac{k_\beta^2 \sigma_{\beta C}/g_\beta}{\Sigma_\gamma k_\gamma^2 \sigma_{\gamma C}/g_\gamma}$$

where the gs are statistical weight factors.

be conserved. Thus, in forming the implicit sums in equation 6.9, one must be careful to conserve E, J, π, and J_z.

Because the predicted excitation energies of the compound nuclei are relatively large, one is justified in using statistical models for calculating the cross sections for formation and decay of the compound nucleus. Thus, for the entrance channel, $a + A$

$$\sigma_{\alpha C} = \pi g_\alpha T_\alpha / k_\alpha^2 \tag{6.14}$$

where T_α is the transmission coefficient for a reacting with A. In general, we have the schematic expression

$$\sigma_{\alpha\beta} = \frac{\pi}{k_\alpha^2} \frac{g_\alpha T_\alpha T_\beta}{\Sigma_\gamma T_\gamma} \tag{6.15}$$

where proper accounting for conservation of J, π, and so on, is implied. Equation 6.15 is called the Hauser–Feshbach equation (Hau 52). For nuclei with spins, equation 6.15 can be written as

$$\sigma_{\alpha\beta} = \pi \lambda_\alpha^2 \Sigma_{J\pi} \frac{2J+1}{(2I+1)(2i+1)} \frac{\Sigma_{\ell s}(T_\ell^\alpha)_{J\pi} \Sigma_{\ell' s'}(T_{\ell'}^\beta)_{J\pi}}{[\Sigma_{\gamma \ell'' s''} T_{\ell''}^\gamma]_{J\pi}} \tag{6.16}$$

where I and i are the ground state spins of target and projectile nuclei, s is the channel spin, ℓ is the orbital angular momentum and $J\pi$ is the total angular momentum and parity. All summations must obey conservation laws (Tho 68). The sum in the denominator involves all channels open for compound nucleus decay. Since this summation could be very difficult if carried out explicitly, one usually evaluates the summation for discrete levels of known spin and parity up to some cutoff energy E_c and replaces the sum by an integral over a continuum of states for higher energies. Thus we have

$$[\Sigma_{\gamma \ell' s'} T_{\ell'}^\gamma] = [\Sigma_{\gamma \ell'' s''} T_{\ell''}^\gamma]_{E<E_C} + \int_{E_C}^{E_{max}} [\Sigma_{\gamma \ell' s'} T_{\ell'}^\gamma] \rho(U, J) \, dE \tag{6.17}$$

where $\rho(U, J)$ is the density of levels in the residual nucleus having spin J and excitation energy $E = U + \Delta$ where Δ is the pairing energy correction.

When equation 6.15 or 6.16 is used to predict the cross section for a particular reaction, one must substitute appropriate values of the transmission coefficients into the equation. For the capture or emission of neutrons or light charged particles, one is advised to use the best optical model values (Coo 81, Per 76). For the fission exit channel, one can use transmission coefficients for decay through specific states of the fission transition state nucleus (Beh 68) or a statistical model (Gro 76).

As discussed above, in transuranium element synthesis reactions with charged particles, one is generally involved with nuclei at sufficiently high excitation energies that the use of statistical models is justified. As indicated in equation 6.17, such models are used to describe the nuclear level densities. A formulation of the level density appropriate for use in equation 6.17 is

$$\rho(E, I) = \frac{2I + 1}{12} \sqrt{a} \left(\frac{\hbar^2}{2\mathcal{I}}\right)^{3/2} \frac{1}{(E - E_{\rm rot})^2} \exp 2(aE)^{1/2} \exp\left[\frac{-I(I + 1)}{2\sigma^2}\right] \quad (6.18)$$

where $\sigma = \mathcal{I}T/\hbar^2$. This formulation assumes a statistical distribution of equally spaced levels. Corrections can be made to equation 6.18 for the effect of nonuniform level spacings, pairing, deformation, et cetera (Sto 85). The parameters describing these various effects and the level spacing parameter a have been summarized by Gilbert and Cameron (Gil 65, Gil 65a) and Reffo (Ref 80).

For fission, the appropriate level densities are those of the fission transition state nucleus. For highly excited nuclei, the transmission coefficients in equation 6.17 are taken to be unity if $E > B_f$ and zero otherwise.

The implementation of the calculations implied by equations 6.16 and 6.17 is difficult and generally involves the use of a large computer code to perform the calculations. The most useful of these codes involves multistep calculations that trace the deexcitation of nuclei through multiple evaporation steps. These calculations work in two ways, using either a grid or a Monte Carlo approach. In the grid approach, a large grid in Z, A, J, and E^* is constructed and the population of each grid cell is traced. This approach has the advantage that the yields of improbable processes (such as the production of transuranium nuclei) can be calculated with precision. The Monte Carlo approach follows the decay of individual compound nuclei using Monte Carlo techniques. The accuracy of this approach depends on the number of "trials" in the calculation, that is, the amount of computer time used for a given calculation. Important examples of the grid approach are GROGI (Gro 67) and its "daughter" HIVAP (Rei 81) and MB-II (Bec 78). [The frequently used ALICE code (Bla 66, Bla72, Pla78) is of this gridded type but uses a Weisskopf–Ewing formulation rather than a Hauser–Feshbach analysis, so that angular momentum is not treated properly.] Examples of the Monte Carlo approach are DFF and its modifications (Dos 59, McG 85) and PACE (Gav 80). The use of any code is, of course, governed by the GIGO principle ("garbage in, garbage out") so that appropriate care must be taken in the choice of input parameters, et cetera.

For the reader who is more conservative or more comfortable with planning transuranium nuclide production using measured values of cross sections, there exist extensive tabulations of measured cross sections for transuranium nuclide production in light charged particle-induced reactions (Alo 74, Gor 80).

6.2.2.2.2 Survival of Transuranium Reaction Products. When considering the survival of transuranium nuclei produced in charged particle-induced compound nuclear reactions, often, *for simplicity*, one will emphasize the independence of formation and decay of the compound nucleus, as expressed in equation 6.9. When the compound nuclei are transuranium nuclei, the high Coulomb barrier against charged particle emission will generally cause the probability of charged particle emission (even in systems with high angular momentum J) to not be significant when compared to neutron emission, γ-ray emission, or fission (whose probability is enhanced by increasing angular momentum). Thus, it is usually sufficient when considering the deexcitation of moderately excited transuranium reaction products to only consider neutron emission and fission. Vandenbosch and Huizenga (Van 73) present a simple, frequently used form for the ratio of the fission width, Γ_f, to the neutron emission width, Γ_n, namely,

$$\frac{\Gamma_f}{\Gamma_n} = \frac{4A^{2/3}a_f(E - B_n)}{K_0 a_n[2a_f^{1/2}(E - E_f)^{1/2} - 1]} \exp[2a_n(E - B_n)^{1/2} - 2a_f(E - E_f)^{1/2}] \tag{6.19}$$

where A is the nuclear mass number; a_f, a_n are the level density parameters at the fission saddle point and ground state deformation, respectively; E, B_n, E_f are the nuclear excitation energy, neutron binding energy, and fission barrier height, respectively. The quantity K_0 is given by the equation

$$K_0^2 = \frac{T\Im_{\text{eff}}}{\hbar^2} \tag{6.20}$$

where T is the nuclear temperature and $\Im_{\text{eff}}$ the effective moment of inertia. Thus the ratio a_f/a_n along with the fission barrier heights (approximately known) and the neutron binding energy (well known) controls the competition between particle emission and fission (a_f/a_n is generally known to be 1.0–1.2). The effects of a finite nuclear angular momentum can be treated by replacing the fission barrier height, E_f, and the neutron binding energy, B_n, by

$$E_f^{\text{effective}} = E_f^r + E_H + R_{\text{sph}} \qquad (6.21)$$

$$B_n^{\text{effective}} = E_H + R_{\text{sph}} + B_n \qquad (6.22)$$

where

$$R_{\text{sph}} = \frac{\hbar^2}{2\mathfrak{I}_{\text{sph}}} J(J+1) \qquad (6.23)$$

and the quantities E_H and E_f^r are calculated from the rotating liquid drop model (Pla 73). Alternatively, straightforward rotating liquid drop model fission barriers can be used (Coh 74, Sie 86). Unfortunately, this simple calculational framework neglects a number of important quantities (discussed in Chapter 4) which are known to dramatically affect the fission probability. They include the double-humped fission barrier and its penetration, the role of nuclear symmetry in affecting the density of states at the fission barrier, the effect of angular momentum upon the shell effects which govern the ground state fission barrier heights, the "washing out" of the shell effects with increasing excitation energy, et cetera. A more sophisticated framework that considers some of these effects has been used by some (Bac 74, Gav 76, Gro 76). Moretto (Mor 72) has calculated fission barrier heights, level densities, as well as fission and neutron decay widths using microscopic models for a number of superheavy nuclei. Valuable as they are, such calculations are generally done neglecting angular momentum effects, collective enhancements of level densities, and so on. *In summary, there has been no completely correct treatment to date of the de-excitation of highly excited, high J species.* What is more serious perhaps, is that we have no universally accepted ideas about the importance of many of the aforementioned effects (collective enhancements, angular momentum dependence of shell effects, etc.) although Reisdorf (Rei 81) has made some interesting advances.

In view of this inability to approach the deexcitation of excited transuranium nuclei on sound theoretical grounds, many have reverted to the older semiempirical approach of Sikkeland and co-workers (Sik 68a, Sik 68b). Sikkeland et al. created a simple crude framework for treating compound nucleus reactions using the Jackson model (Jac 56) and an *energy-independent* Γ_n/Γ_f. The values were determined by fitting an extensive amount of data on survival probabilities in actinide nuclei. This leads to the empirical formula

$$\log[\Gamma_n/\Gamma_f] = 0.12\delta - 0.2767 + \begin{cases} 5.46 + 0.140N & \text{for } N < 153 \\ 19.23 + 0.050N & \text{for } N > 153 \end{cases} \quad (6.24)$$

where N is the neutron number of the excited nucleus and δ equals 1 for odd values of Z, otherwise it is zero. The approximate fraction of nuclei surviving fission, Γ_n/Γ_f, is computed at each evaporation step. Thus if $\Gamma_n/\Gamma_f = 0.1$, then after the evaporation of 5 neutrons, the approximate fraction surviving is $(0.1)^5 = 10^{-5}$. It is not clear, however, that the Sikkeland systematics are useful in treating deexcitation of nuclei with high angular momentum or nuclei with low excitation energies because of the neglect of the effect of E^* or J upon Γ_f/Γ_n. Alternative systematics of Γ_f/Γ_n that attempt to remedy these difficulties are available (Sim 83, Che 83, Kup 80, Kup 84, Plo 82).

6.2.3 Heavy Ions

Heavy ions ($Z > 2, A > 5$) are certainly charged particles and the general formalisms describing compound nucleus reactions induced by light charged particles apply to heavy ions. However certain aspects of heavy ion interactions are special. For example, the very short wavelength of an energetic heavy ion compared to its size means that heavy-ion interactions will be classical in character, a phenomenon not generally true of light charged particles. In addition, the very high angular momenta resulting from heavy ion interactions will decrease the fission barrier heights leading to decreased product survival. Finally, heavy ion reactions feature a larger number of reaction channels because of the high energies and angular momenta, thus emphasizing statistical phenomena.

6.2.3.1 *Energetics*. The minimum excitation energy of the compound system is achieved when $T_{cm} \sim B_C$ (or the fusion barrier B_{fu}). If one works through several examples, one finds, under these conditions, that for $A_{proj} < 80$ and $A_{proj} + A_{target} < 300$, the excitation energy of the completely fused species is $\sim 40\text{--}50$ MeV. For reactions like (Heavy Ion, xn), one finds $x = 3\text{--}5$ with a variance $(\langle x^2 \rangle - \langle x \rangle^2) \sim 0.3$. Thus the excitation functions $\sigma_{xn}(E)$ of (HI,xn) reactions are very narrow with FWHM $\sim 8\text{--}10$ MeV. Such reactions, with their high excitation energies, are called "hot fusion" reactions. While they can and were used to synthesize isotopes of elements 106 and lower Z elements, it is difficult to reach the heaviest elements when the losses due to fission of the compound nuclei are 10^5–10^{10}.

Oganessian and co-workers (Oga 74) were the first to point out that when projectiles with $A_{proj} > 40$ are used along with target nuclei with Z,A near the doubly magic ^{208}Pb, the excitation energy of the compound nucleus is at a minimum. (Figure 6.6) This process, known as "cold fusion", was used for the synthesis of elements 107-109. Unfortunately, the improved survival probabilities in these reactions are obtained at the expense of a decreased probability for compound nucleus formation due to the higher Coulomb forces between the interacting nuclei (Figure 6.7). In Figure 6.7, one also sees the general property of lower formation cross sections from (HI, xn) reactions compared to (α, xn) reactions. For example, ^{238}U(^{16}O,4n)^{250}Fm has $\sigma = 1.5$ μb while ^{250}Cf(α, 4n)^{250}Fm has $\sigma = 0.7$ mb.

6.2.3.2 Cross Sections. In the spirit of equation 6.9, we can write a separable expression for the probability of forming a compound nucleus with total angular momentum J as

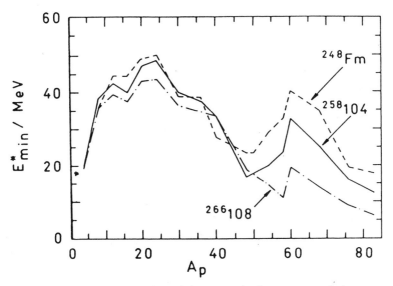

Figure 6.6 Dependence of the minimum excitation energy of the compound nucleus, E^*_{min}, on the projectile mass, for different projectile-target systems leading to the formation of the same compound nuclei ^{248}Fm, ^{258}104, and ^{266}108 (from Oga 85).

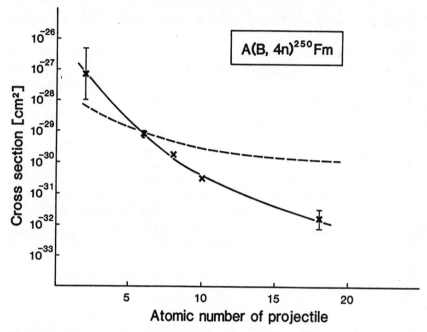

Figure 6.7 Cross sections for the fusion reaction $A + B \rightarrow {}^{250}\text{Fm} + 4n$. The dashed line is the expectation for unhindered fusion (see text) (Gäg 88).

$$\sigma_J = \pi \lambdabar^2 \frac{(2J + 1)}{(2I + 1)(2i + 1)} \Sigma_{\ell s} T_{\ell s J} \qquad (6.25)$$

where the symbols have the same meaning as in equation 6.16. If we neglect the spin of the target and projectile

$$\sigma_{fu} = \Sigma_J \sigma_J = \pi \lambdabar^2 \sum_{\ell=0}^{\infty} (2\ell + 1) T_\ell \qquad (6.26)$$

where σ_{fu} is the fusion cross section. (In the classical limit, where $T_\ell = 1$ for $\ell < L_{fu}$ and $T_\ell = 0$ for $\ell \geq L_{fu}$ where L_{fu} is the limiting angular momentum for fusion, we have

$$\sigma_{fu} = \pi \lambdabar^2 L_{fu}^2 \qquad (6.27)$$

and the average angular momentum of the compound nucleus is $2/3 L_{fu}$).

To calculate the complete fusion cross section, one must specify either T_ℓ (equation 6.26) or alternatively L_{fu} (equation 6.27). To do this, one has to pick a realistic potential for the projectile-target interaction. Modern calculations are usually based upon the use of the proximity potential (Blo 77), and the assumption of one-body friction (Bir 78, Bir 79). The procedures are described in the excellent monograph by Bass (Bas 80). Extensive tabulations of expected fusion cross sections for a large number of projectile-target combinations are available (Wil 80). Application of equation 6.26 along with appropriate estimates for Γ_n/Γ_f (equations 6.19 or 6.24) leads to an adequate parameterization of (HI,xn) data for production of the lighter transuranium nuclei with heavy ions (Figure 6.8).

To use heavy ions to synthesize the heaviest transuranium nuclei and any putative superheavy nuclei, one has to consider certain additional factors that both enhance and decrease the yield of transuranium products. These factors include fusion at low energies (subbarrier fusion) (Rei 86, Bas 80, Sto 80, Jah 82, Rei 82, Vaz 81, Sto 89) and a dynamical hindrance to fusion (for $Z_{\text{proj}} + Z_{\text{target}} > 80$) or a limit to fusion ($Z_{\text{proj}} + Z_{\text{target}} > 120$). While both effects are important in the synthesis of heavy and superheavy nuclei (and are thus discussed in Chapter 7), both effects are of general importance and interest in heavy ion-induced compound nucleus reactions and merit discussion here.

Up to now, we have asserted that the center of mass kinetic energy of the projectile T_{cm} should exceed the fusion barrier height.* However, quantum mechanics tells us that there is some finite probability for subbarrier fusion due to a penetration of the fusion barrier by tunneling. Conventionally, if we say

$$\sigma_{\text{fu}} = \pi \lambdabar^2 \sum_{\ell=0}^{\infty} (2\ell + 1) T_\ell \qquad (6.26)$$

---

*As a detail, we note the magnitude of the fusion barrier is usually taken (Bas 74) as

$$B_{\text{fu}} = \frac{Z_1 Z_2 e^2}{R_{12}} \left\{ \frac{R_{12}}{R_{12} + d_{\text{fu}}} - \frac{1}{x} \frac{d}{R_{12}} \exp\left(\frac{-d_{\text{fu}}}{d}\right) \right\}$$

where

$$R_{12} = r_0(A_1^{1/3} + A_2^{1/3})$$

and

$$x = \frac{e^2}{r_0 a_s} \frac{Z_1 Z_2}{A_1^{1/3} A_2^{1/3} (A_1^{1/3} + A_2^{1/3})}$$

The constants a_s, r_0, d, d_{int} are 17.0, 1.07, 1.35, and 2.70, respectively.

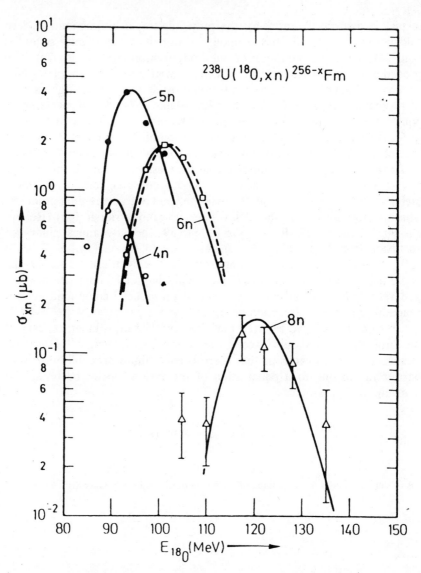

Figure 6.8 Excitation functions for the ^{238}U(^{18}O,xn)$^{256-x}$Fm reaction. The solid lines represent predictions based upon equations 6.26 and 6.24 (from Sik 67). The dashed line represents the prediction after a variation of the input data for the model.

we can calculate T_ℓ for the subbarrier case using the Hill–Wheeler penetrability

$$T_\ell = \{1 + \exp(2\pi(V_B(\ell) - E)/\hbar\omega\}^{-1} \qquad (6.28)$$

What one observes (Figure 7.5) is a substantial enhancement of the fusion probability over that expected from the penetration of two spherical nuclei (equations 6.26 and 6.28). The effects causing this enhanced penetrability are: (a) the static deformation of the target and/or projectile (Sto 80), (b) dynamic processes such as neck formation during fusion (Agu 89, Vaz 81); (c) vibrational excitations of the nuclear surface (Jah 82, Agu 89, Rei 82); or (d) nuclear structure effects (Rei 86). (The observation of a constant average angular momentum for fusion at subbarrier energies (Sto 89) may, however, decrease the survival probabilities of these species.) No simple calculational framework exists to treat such effects so that the simple one-dimensional barrier penetrabilities (equation 6.28) are altered by shifting V_B or $\hbar\omega$ to fit the data.

The dynamical hindrance to fusion in heavy nuclear systems is well established and represents a serious problem for heavy element synthesis efforts. This dynamical hindrance has its origin in the Coulomb repulsion between the fusing nuclei. When the values of $Z_{proj}Z_{target}$ are 1600–2000, a dramatic decrease in fusion probability occurs (Figure 6.9). Another example of this effect is the production of the evaporation residue ^{257}Rf by the reaction ^{249}Cf(^{12}C,4n) ($Z_pZ_t = 588$) and ^{208}Pb(^{50}Ti,1n) ($Z_pZ_t = 1804$). The cross section for the latter "cold fusion" reaction is a factor of ten less than the hot fusion cross section (Mün 88) due to dynamic limits on fusion probability.

Swiatecki has developed a schematic model that has been used widely to represent this dynamical limitation to fusion (Swi 82, Bjø 82). The model is illustrated in Figure 6.10. Three configurations in the dynamical evolution of a nucleus-nucleus collision are identified. They are: (a) the configuration where the colliding nuclei touch; (b) the "conditional" saddle configuration where the colliding nuclei have interpenetrated somewhat and are about to lose their identity; (c) the "unconditional" saddle configuration beyond which the product mononucleus has truly equilibrated and suffered "amnesia" about the collision partners from which it has formed (i.e., formed a compound nucleus). The energy required to make the nuclei touch is called the interaction barrier, V_B, while the extra radial energy above this barrier to make the two nuclei fuse to form a mononucleus is called the "extra push" energy. The extra energy above the barrier required to form the true compound nucleus is referred to as the "extra-extra push" energy. Within the framework of the schematic

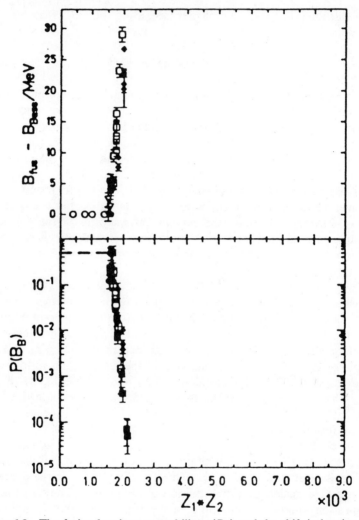

Figure 6.9 The fusion barrier penetrability $p(B_B)$ and the shift in barrier energy as a function of $Z_{proj}Z_{target}$ (from Arm 89).

model due to Swiatecki (or a later version (Blo 86)), the extra-extra push energy, E_{xx}, is given as

$$E_{xx} = 0 \quad \text{for } x_m \leq x_{th} \tag{6.29a}$$

$$E_{xx} = E_{ch}[a(x_m - x_{th}) + \text{higher powers of } (x_m - x_{th})]^2 \tag{6.29b}$$

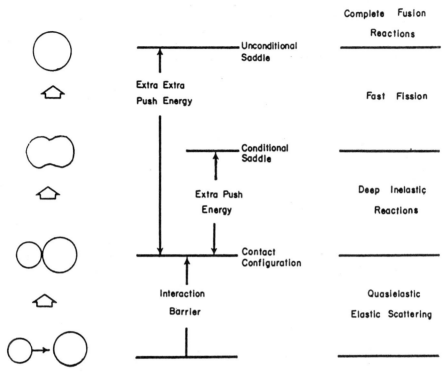

Figure 6.10 A schematic view of the stages of a heavy-ion reaction in the Swiatecki model. The nuclear shapes at various stages of the reaction are shown along with the projectile energies needed to get to the various stages and the reaction mechanisms associated with each stage.

where $x_m = 2/3x + 1/3x_e$. Here

$$x = (Z^2/A)/(Z^2/A)_{crit} \qquad (6.30)$$

$$x_e = (Z^2/A)_{eff}/(Z^2/A)_{crit} \qquad (6.31)$$

where Z, A refer to the fused system and

$$(Z^2/A)_{crit} = 50.883(1 - 1.7926I^2) \qquad (6.32)$$

$$(Z^2/A)_{eff} = 4Z_1Z_2/[A_1^{1/3}A_2^{1/3}(A_1^{1/3} + A_2^{1/3})] \qquad (6.33)$$

Here I is the relative neutron excess ($I = (A - 2Z)/A$) and the colliding nuclei have atomic and mass numbers Z_1, A_1, Z_2, A_2. E_{ch} is a constant.

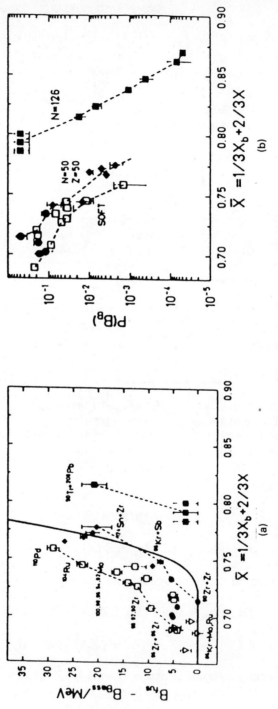

Figure 6.11 (a) The measured barrier shifts and (b) fusion probabilities as a function of fissility (from Arm 89).

Thus the behavior of the extra-extra push energy needed to make a compound nucleus (and thus a surviving transuranium product) is characterized by two dimensionless constants a and x_{th}, determined by fitting data. The functional form of equations 6.29–6.33 imply that when x exceeds x_{th} (~0.723), the value of E_{xx} eventually rises steeply with increasing x. This dependence implies a dynamic limit on fusion at $x \sim 0.8$. The large values of E_{xx} for $x \geq 0.8$ also imply vanishingly small yields of transuranium reaction products because not only is the fusion probability greatly reduced but the resulting high excitation energy lessens the product survival against fission. Swiatecki et al. (Swi 82, Bjø 82) also outlined a schematic model for calculating the reduction in fusion cross section due to dynamic hindrance. This schematic model was able to properly describe the shifts in the position of the fusion barrier leading to the production of evaporation residues and the magnitudes of the evaporation residue cross sections (Boc 82, Gäg 84, Lüt 84) for a large body of data.

Additional research (Gäg 88, Gäg 89, Arm 89) has pointed out that there are important nuclear structure effects on the fusion hindrance, or more correctly, the shift in the position of the observed fusion barrier height relative to the unhindered Bass barrier height (Bas 74). "Soft" collision partners show large ΔB ($= B_{fusion}^{expt} - B_{Bass}$) and small fusion probability, $p(B)$, for $x_m \sim 0.68$–0.76, while for closed shell collision partners, such effects are seen at higher x values (0.72–0.78). Lead-based systems (such as those used in the synthesis of elements 107–109) show the most resistance to barrier shifts and reduced fusion probability (Figure 6.11).

6.3 DIRECT REACTIONS

6.3.1 General Considerations, Light Charged Particles

In contrast to compound nucleus reactions are *direct reactions*, in which the incident projectile interacts primarily with surface nucleons. Direct reactions generally involve only a few nucleons in the target and projectile nuclei in the primary interaction. Direct processes occur more rapidly ($\sim 10^{-22}$ s) than compound nuclear processes ($\sim 10^{-16}$–10^{-18} s). Examples of direct reactions include stripping reactions such as (d,p) or (d,n), pickup reactions, such as (p,d), (d,^{6}Li) or with heavy ions, multinucleon transfer reactions. Both mechanisms can contribute to a given nuclear reaction such as inelastic scattering and, in fact, one can generally think of an entire range of reaction mechanisms (and time scales) between the extremes of "pure" direct and compound nucleus processes.

Direct reactions have many important applications in the study of heavy nuclei. Simple direct reactions such as (d,p) can be used to produce product nuclei with sharp, well-known values of spin, parity, and excitation energy. (These values can be measured by detecting the outgoing projectile remnants). Thus direct reactions are important nuclear spectroscopic tools. Direct inelastic scattering can be used to excite vibrational and rotational states while (d,p) reactions can be used to populate single particle states. Simple direct reactions can be used to study the states of the fission transition nucleus also (Van 73). From these studies, we have gained much of our current knowledge of fission barrier heights and the structure of the fission transition nucleus. Also, as discussed below, direct nucleon transfer reactions with heavy ions have become important in the synthesis of new transuranium nuclei.

The theory of these reactions can be approached from a simple semiclassical view given in elementary textbooks (Coh 71, Kra 88, Lef 68) or from a more complete, quantum mechanical viewpoint. The semiclassical approach gives one a feeling for the rudimentary physical phenomena involved, but is too crude to be of use. The complete theory of direct reactions (see, for example, Jac 70, Sat 80, Mar 70, Aus 70, Gle 75) is too detailed for this exposition, but we can sketch out the general outlines of the theory. The transition amplitude for the reacting system to go from an initial state $(a + A)$ to final state $(b + B)$ is given as a matrix element

$$M \int \Psi_B^* \Psi_b^* V \Psi_A \Psi_a \, dv \qquad (6.34)$$

The interaction V is a complicated function of many nuclear coordinates. A simple treatment of the problem assumes that Ψ_a and Ψ_b can be treated as plane waves (the plane wave Born approximation, PWBA). A more useful treatment uses the optical model to account for the distortion of the incoming and outgoing waves due to the nuclear interaction (the distorted wave Born approximation, DWBA). The use of the optical model allows for absorption as well as refraction of the waves. Nuclear shell model wave functions are used to describe the initial and final nuclear states. A volume interaction extending through the entire nuclear volume is employed as is a finite range force between nucleons. There are several, well-documented computer codes to do DWBA calculations (Mar 70, Aus 70, Sat 80, Gle 75). Using these codes can allow one to plan the synthesis of a given transuranium nuclide, or more likely, to understand the properties of such a synthesis by direct reactions involving light charged particles.

6.3.2 Heavy Ions

In principle, direct reactions involving heavy ions can be treated by the same general theoretical formalism used for light charged particle-induced reactions (Bas 80). However, the computational difficulties involved and the importance of multistep processes for many heavy ion-induced direct reactions have dictated special approaches to understanding these reactions. In utilizing these special approaches, we shall arbitrarily (but conventionally) divide the reactions into (a) single nucleon transfer and (b) multinucleon transfer. In considering (b), we shall restrict ourselves to considering reactions in which there is no strong perturbation of the entrance or exit channel trajectories because of the transfer process (so-called "quasielastic transfer"). The more complex deep inelastic transfer processes are considered in Section 6.4 below.

The transfer of a single nucleon from the projectile to the target nucleus (or vice versa) has been studied for many years. This transfer has been observed even at energies less than the Coulomb barrier height. The cross section for processes like (^{14}N, ^{13}N) are small, of the order of a few millibarns. The rudimentary theory of such processes is due to Breit (Bre 56). One envisions an overlap of the wave function of the projectile nucleon with that of the target nucleon, leading to a "tunneling" of the transferred nucleon. This overlap is thought to occur along the surfaces of both nuclei. More sophisticated theories (Hod 78, Ari 84, Asc 84) such as the DWBA with appropriate approximations (or the coupled channel Born approximation, CCBA) seem to account for the details of the experimental data adequately. Such single nucleon transfer reactions are used to study the structure of actinide nuclei.

It is also possible to observe direct multinucleon transfer processes. These processes are generally thought of as grazing collisions of projectile and target nuclei in which two or more nucleons are exchanged, producing an angular distribution of the projectile-like remnant involving a single bell-shaped peak at the grazing angle. The probability of such processes is substantial with cross sections being several hundred millibarns. Transfers of unusual clusters of nucleons, such as two protons, are possible. The localized spatial distribution of the reaction products is also reflected in a narrow window in ℓ space (Figure 6.12). The narrow distribution indicates there are severe "matching conditions" to conserve angular and linear momentum. The latter matching condition is known as the "optimum Q value" (Q_{opt}) (Bri 72, Bri 77). For reactions at the Coulomb barrier of the type $A(a,b)B$

$$Q_{\mathrm{opt}} = \left(\frac{Z_b Z_B}{Z_a Z_A} - 1\right) E_{\mathrm{cm}} \qquad (6.35)$$

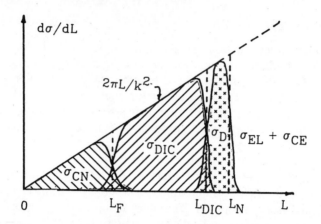

Figure 6.12 Partial reaction cross sections as a function of the scattering angular momentum L.

where E_{cm} is the cm energy of the projectile and Z_A, Z_B, Z_a, Z_b are the atomic numbers of the nuclei. Direct reactions among heavy ions will have large cross sections when ℓ and Q are in a narrow range close to l_{opt} and Q_{opt} (the L and Q windows). Invoking the concept of partial statistical equilibrium among the reacting nuclei, for which Bondorf et al. (Bon 71) proposed that the excitation energy E_{opt}^* (which mostly ends up in the heavy reaction partner) is given by

$$E^* = Q_{gg} - Q_{opt} \qquad (6.36)$$

(where Q_{gg} is the ground state reaction Q value), one can easily understand why proton transfer from projectile to target is favored over the reverse. (The case for target to projectile charge transfer leads to negative values of E^* or a mismatch indicating low cross sections.) Various models for Q_{opt} for transfer above the Coulomb barrier lead to different prescriptions than equation 6.35. These prescriptions are summarized and compared by Bass (Bas 80). It should be noted that these semiclassical considerations have been shown to be consistent with detailed DWBA calculations (Sie 71).

The relative yields of multinucleon transfer reactions can be correlated with Q_{opt} (see below). The prediction of the absolute values of the cross sections is a much more difficult matter especially for the transuranium nuclei where the fissionability of the product nuclei may impede their survival. Wilczynski et al. (Wil 80a, Wil 82) have proposed a simple semiclassical schematic sum-rule model to account for transfer reaction cross sections. In this model, in essence, one treats the transfer of a cluster from the projectile to the target in a manner similar to that employed for compound nucleus reactions (equation 6.9), that is,

$$\sigma(i) = \pi \lambdabar^2 \sum_{\ell=0}^{\ell_{max}} (2\ell + 1) \frac{T_\ell(i)p(i)}{\Sigma_j T_\ell(j)p(j)} \qquad (6.37)$$

where $\sigma(i)$ is the cross section for cluster i, $\lambdabar$ is the reduced wavelength ($=\hbar^2/2\mu E$) for the entrance channel, $T_\ell(i)$ and $p(i)$ are the transmission coefficients and probability of formation of cluster i. The cluster formation probability $p(i)$ is given as

$$p(i) = \exp(Q_{opt}/T) \qquad (6.38)$$

where T is an adjustable parameter. The transfer of each particular cluster is associated with a narrow range of ℓ values

$$T_\ell = \left[1 + \exp\frac{(\ell - \ell_{lim}(i))}{\Delta_\ell}\right]^{-1} \qquad (6.39)$$

where

$$\ell_{lim}(i) = \frac{(\text{mass of projectile})}{(\text{mass of cluster } i)} \ell_{cr}(\text{target} + \text{cluster}) \qquad (6.40)$$

The quantity ℓ_{cr} is the critical angular momentum (L_{fu} in equation 6.27) for fusion of the cluster and target nucleus.

One of the first and more careful studies of these heavy-ion "transfer" reactions involving production of transuranium nuclei was made by Hahn et al. (Hah 74). This remains one of the few studies in which kinematic measurements were attempted. Hahn et al. studied the excitation functions, recoil range distributions, and angular distributions of the heavy transuranium recoil products. In particular, they studied the characteristics of the production of ^{245}Cf and ^{244}Cf via the transfer reactions ^{239}Pu(^{12}C, α2n) and ^{239}Pu(^{12}C, α3n) and via the complete fusion reactions ^{238}U(^{12}C, 5n) and ^{238}U(^{12}C, 6n). As expected the complete fusion products were strongly forward focussed with their angular distributions peaked at 0°. They show Gaussian range distributions with mean ranges that increase with increasing projectile energy and whose values agree with the assumption of complete fusion. The same ^{244,245}Cf products when produced in the transfer reaction show angular distributions which peak near the grazing angle and show asymmetric range distributions whose mean value decreases with increasing projectile energy. The yields of ^{244,245}Cf are much larger in the transfer reactions compared with the complete fusion reactions. The yields of the transfer products are described by Hahn et al. with modest success using a modification of the semiempirical Sikkeland systematics (Sik 68a, Sik 68b). These calculations indicate the reason for the higher product yields in the transfer reaction is the relatively cold residual nucleus produced in this reaction compared to the complete fusion reaction.

Since this pioneering work by Hahn et al., there have been a number of experimental studies of multinucleon transfer reactions yielding transur-

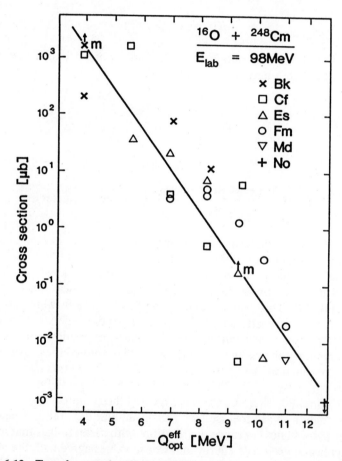

Figure 6.13 Transfer reaction cross sections versus Q_{opt} for the reaction of 98 MeV ^{16}O with ^{248}Cm (Gäg 88).

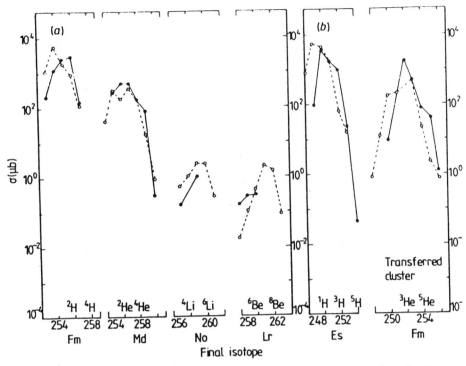

Figure 6.14 Measured (●) and calculated (○) production cross sections for the reactions (a) $^{16}O + ^{254}Es$ and (b) $^{18}O + ^{249}Cf$ at a bombarding energy of 97 MeV (Mag 87).

anium products (Lee 82, McF 82, Lee 83, Tan 84, Hof 85, Sch 86, Cha 87). Gäggeler has summarized and reviewed these and similar measurements (Gäg 88). He has shown that the results of these diverse studies show a simple and expected dependence of the relative cross sections for transuranium nuclide production on Q_{opt} (Figure 6.13). In Figure 6.13, one can also note the magnitude of the transfer reaction cross sections to produce nuclei that are 1–6 Z units above the target Z. The small cross sections preclude detailed multiparticle coincidence studies that could shed more light on the reaction mechanism(s) involved. Another relevant point concerning the data of Figure 6.13 is the values of the excitation energy E^* one would calculate for the transfer products using equation 6.36. They range from -17 MeV to $+18$ MeV with a mean value of ~ 4 MeV. Thus these transfer products are "cold", that is, they do not involve significant excitation energies and thus fission deexcitation of the primary reaction products is greatly reduced. Finally, one must note the remarkable agreement between the observed transfer reaction yields of the transuranium nuclei and the predictions of Magda et al. (Mag 87, Figure 6.14). The model used by Magda et al. was similar in principle to equation 6.37 but the probability $p(i)$ of forming a given cluster from the projectile were taken from a model due to Friedman (Fri 83) — which, in turn, is similar to equation 6.38.

6.4 DEEP INELASTIC TRANSFER

One of the benefits of attempts to synthesize new transuranium nuclei was the (re)discovery of a new nuclear reaction mechanism, that of deep inelastic scattering (Lef 73). In this reaction, projectile-like fragments were observed at angles other than the grazing angle and with small kinetic energies, indicative of a large kinetic energy loss during the collision. This nuclear reaction mechanism is unique to heavy ions and is intermediate between direct and compound nucleus mechanisms. We use the word rediscovery because this mechanism was first observed by Kaufman and Wolfgang several years prior (Kau 59, Kau 61) (Figure 6.15). What Kaufman and Wolfgang observed was a unique angular distribution of the projectile-like fragments. In addition to the expected direct reaction peak at the grazing angle, some products peaked at a forward angle. These products are the result of deep inelastic scattering. Unfortunately this work was largely neglected until the mid-1970s when attempts (Lef 73) were made to synthesize superheavy nuclei using the reaction of ^{84}Kr with ^{209}Bi. The results (Figure 6.16) showed, in addition to the expected peaks at the projectile and target mass numbers (at high energies), the presence of lower energy, highly inelastic events near the projectile and target mass numbers. This rediscovery was confirmed and extended by Wolf and coworkers (Wol 74). These works initiated an explosion of interest in this mechanism. (A comprehensive review article (Sch 84) is 611 pages in length, attesting to the prodigious amount of work that has occurred in this field.) The history of these developments is recounted in detail by Bromley (Bro 84a). Several comprehensive review articles have been written (Sch 77, Sch 84, Hui 85, Fre 84, Bas 80).

The rough characteristics of this mechanism are shown in Figure 6.17 (Wil 73). In this contour plot, the properties of the potassium ($Z = 19$) isotopes produced in the bombardment of thorium with 388 MeV argon ($Z = 18$) ions are shown. One observes a large peak in potassium yield at the grazing angle (34°) and nearly the beam energy, due to quasielastic transfer reactions. One also observes another peak at small angles (<15°) and low kinetic energies and a ridge of cross section extending from that

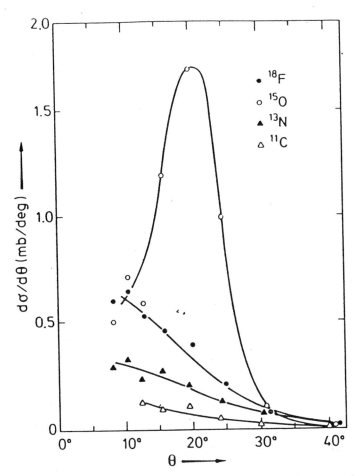

Figure 6.15 Differential cross sections for projectile-like fragments from the reaction of 101 MeV ^{16}O with Rh (from Kau 61).

peak to larger angles. These latter events are the deep inelastic events. A schematic explanation of these data is shown in Figure 6.17b. The deep inelastic events are attributed to "negative angle" scattering in which the nuclear forces deflect the incoming ions from the grazing trajectory. Plots such as Figure 6.17a are called "Wilczynski plots." Further investigations have shown the charge distribution of the fragments is essentially Gaussian, centered about the atomic numbers of the projectile and target nuclei. The variance of these charge distributions increase with increasing dissipation of kinetic energy in the interaction. The outcome of the reac-

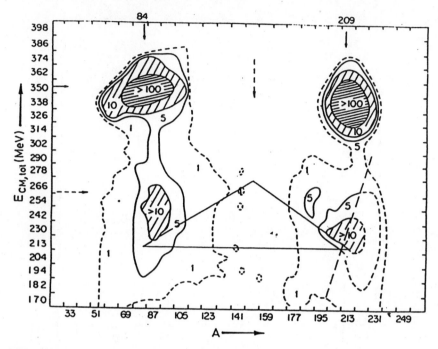

Figure 6.16 A contour plot of the relative fragment yield as a function of total kinetic energy and fragment mass number for the reaction of 5.95 MeV/nucleon ^{84}Kr with ^{209}Bi (Lef 73).

tions is sensitive to the $Z_{proj}Z_{target}$ product, that is, the Coulomb forces involved. At small values of $Z_{proj}Z_{target}$, the projectile-like fragments show forward-peaked angular distributions while larger values of $Z_{proj}Z_{target}$ lead to more sidewise-peaked distributions. The relative importance of this reaction mechanism increases with incresing values of $Z_{proj}Z_{target}$. It is the dominant reaction mechanism for the interaction of 10 MeV/nucleon Kr or heavier ions with heavy nuclei.

As remarked earlier, the deep inelastic reaction mechanism is thought to be intermediate between direct and compound nucleus reactions. That idea can be expressed in various ways. If τ represents the interaction time between projectile and target nuclei and τ_{rot} the rotational period of any dinuclear complex formed in the interaction, then direct quasielastic transfer processes (QET) are characterized by $\tau \ll \tau_{rot}$. If $\tau \gg \tau_{rot}$, compound nucleus reactions occur. If $\tau \sim \tau_{rot}$, then deep inelastic scattering occurs. (This view of these reactions is in the spirit of Figure 6.1). Another way to express the same general ideas is in terms of the trajectories of

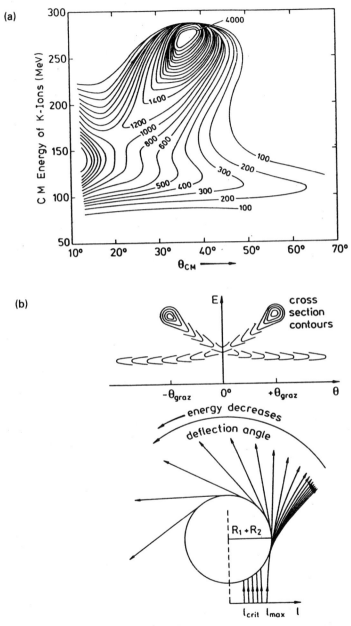

Figure 6.17 (a) Contour plot of the differential cross section for the ^{232}Th (^{40}Ar,K) reaction at 9.48 MeV/nucleon. (b) Schematic representation of the reaction mechanism responsible for (a) (from Wil 73).

the ions (Figure 6.18). Here the sequence (direct reactions, deep inelastic reactions, compound nuclear reactions) is marked by decreasing impact parameters, "harder" collisions, and longer reaction times. Classical trajectory calculations then allow one to further parameterize these gradations in terms of the angular momentum involved (Figure 6.12). Here the deep inelastic collisions (DIC) are associated with a range of ℓ values intermediate between the low ℓ values associated with complete fusion and the high ℓ values associated with QET reactions.

A detailed treatment of the models for deep inelastic collisions is beyond the scope of this work (see Sch 84, Hui 85, for further information).

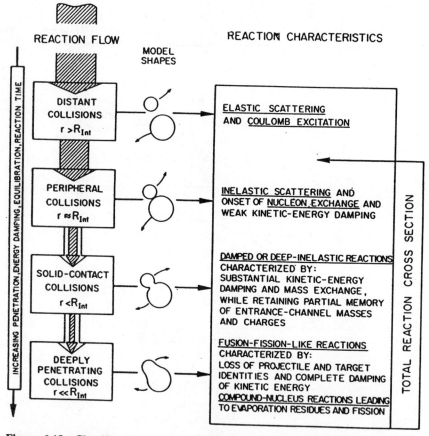

Figure 6.18 Classification of heavy ion reaction mechanisms based upon impact parameter (Sch 84). The open and hatched arrows depict the reaction flow for light and heavy ion reactions, respectively.

In general, one abandons the quantum mechanical description used for the simpler direct reactions in favor of using statistical concepts that utilize classical treatments of certain microscopic degrees of freedom. The most successful of these models have been based upon transport theories. The dissipation of kinetic energy in the interaction (i.e., the loss of kinetic energy through transfer from relative motion to intrinsic excitations) has been successfully depicted in terms of a nucleon in the projectile (target) interacting with the mean field of the other nucleus. This is called "one-body dissipation" to distinguish it from the nucleon-nucleon collisions of high energy reactions, which are termed "two-body dissipation". These one-body transport models (Ran 78, Ran 79) do an excellent job of describing the variances of the charge and mass distributions of the (projectile-like) fragments.

The width of these (Z, A) distributions is the feature that motivates the use of these reactions to synthesize new (transuranium) nuclei, especially on the n-rich side of stability. The widths of these distributions increase with increasing total kinetic energy dissipation in the reaction (Figure 6.19) and are appreciable. Since there is partial statistical equilibrium in the dinuclear complex during the reaction, the widths of these distributions should be determined by the product of the level densities of the final fragments, that is,

$$\text{prob} = \rho_3 \rho_4 \sim \exp(\Delta E^*/T) \tag{6.41}$$

where

$$\Delta E^* = Q_{gg} + \Delta E_C + \Delta E_{rot} - \delta(p) - \delta(n) \tag{6.42}$$

The temperature T is taken as the same for both fragments, ΔE^* is the gain in fragment excitation energy during the rotation due to the changes in Coulomb and rotational energy, ΔE_C and ΔE_{rot}, respectively. The last two terms in equation 6.42 denote the pairing energies of the transferred protons and neutrons, respectively. Thus the potential energy surfaces describing the reaction will be expected to play an important role in describing the relative fragment yields (Kra 79, Fre 84, Moo 86, Wel 87).

The first use of deep inelastic reactions to synthesize new transuranium nuclei was the work of Wolf et al. (Wol 77). Wolf et al. (Wol 77) measured the yields of Am, Cm, Cf, Es, and Fm nuclei produced in the interaction of 7.2 MeV/nucleon ^{40}Ar and ^{84}Kr, and 8.3 MeV/nucleon ^{136}Xe with thick ^{238}U targets. Using a semiempirical diffusion model to calculate the primary fragment distributions (from the initial deep inelastic transfer reaction) and semiempirical values of Γ_f/Γ_n, these workers were able to fit the

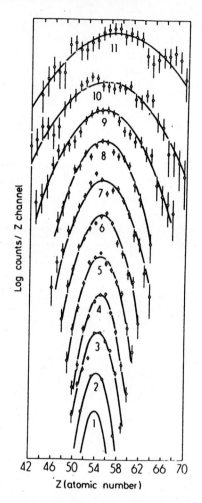

Figure 6.19 Charge distributions of the Xe-like fragment for the reaction of 1130 MeV ^{136}Xe with ^{209}Bi shown for various total kinetic energy loss values. The total kinetic energy loss increases from near zero for curve 1 to ~300 MeV for curve 11 (from Sch 76).

Cf and Es isotopic distributions from the ^{84}Kr + ^{238}U and ^{136}Xe + ^{238}U reactions. Interestingly enough, they also predicted a peak cross section for producing element 106 from the ^{253}Es + ^{136}Xe reaction to be 10^{-32} cm^2 and the cross section for producing element 110 in the same reaction to be ~10^{-36} cm^2.

The most significant use and understanding of deep inelastic transfer

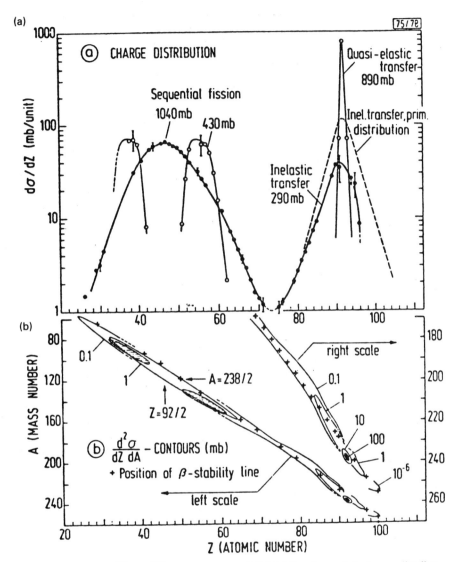

Figure 6.20 The ^{238}U + ^{238}U reaction at ≤7.5 MeV/nucleon: (a) charge distribution for quasielastic transfer and sequential fission at low excitation energies (○), and for damped collisions with the associated sequential fission process (●) and the reconstructed primary fragment distribution (----); (b) independent cross section isopleths in a Z-A plane (from Sch 78).

reactions to produce transuranium nuclei has been in the studies of the ^{238}U + ^{238}U reaction at the Unilac at GSI. The first realization of the possible potential of this reaction for transuranium nuclide synthesis was in the work of Hildenbrand, Freiesleben, and co-workers (Hil 77, Fre 79). Radiochemical studies by Schädel et al. (Sch 78) and Gäggeler et al. (Gäg 81) confirmed these results. In Figure 6.20, we show the product yields as a function of Z and A for the reaction of 7.5 MeV/nucleon ^{238}U ions with thick ^{238}U targets. The distribution of target-like fragments from the deep inelastic reaction can be seen to peak at $Z = 91$ rather than $Z = 85$ (as found in the Xe + U reaction (Ott 76) or $Z = 79$ (as found in the Kr + U reaction (Kra 74)). Thus the "goldfinger" (as this feature was dubbed in Kra 74) had become the "protactinium finger." Reconstruction of the primary target-like fragment distribution led to an estimation of the production cross section of $Z = 70$ fragments in this reaction of 10^{-28} cm^2 which, by symmetry, must also be the estimate of the primary fragment yield of the superheavy element 114 in this reaction.

The yields of the heavy actinides were measured (see Figure 6.21) and were found to show similar variations with Z and A as observed in the yields of the same species formed in the Xe + U reaction. The yields of the transuranium nuclei in the U + U reaction are much greater than those observed in the Xe + U reaction.

Herrmann (Her 78) pointed out that if one compares the Po and Fm product distributions (see Figure 6.22) from this reaction (both have $|Z - 92| = 8$), one observes that the center of the Fm distribution is $\sim 3.7\, A$ units from the most probable primary fragment mass number A_p but the center of the Po distribution is $\sim 9.4\, A$ units from A_p. Thus, one concludes that the Fm yields result from the low excitation energy tails of the primary fragment distributions. The fact that the transuranium element distributions have the same general shape in the U + U and Xe + U reactions can be understood in terms of the fact that despite changes in the primary distributions with projectile Z, A, and E, only those few nuclei in the low E^*, low J tails of the primary distributions will survive fission. The principal advantage of the U + U reaction was thought to be that because of the generally broader primary product distributions, the number of nuclei in the tails of the distributions would increase enormously.

Freiesleben et al. (Fre 79) and Schädel et al. (Sch 78) attempted, using very simple phenomenological models and semiempirical estimates of Γ_f/Γ_n, to predict the yields of the transuranium nuclei in these reactions. The general shapes and centroids of the isotopic distributions are well reproduced in the calculations, but the calculated yields are 1–2 orders of magnitude higher than the observed yields. Gäggeler (Gäg 88) has

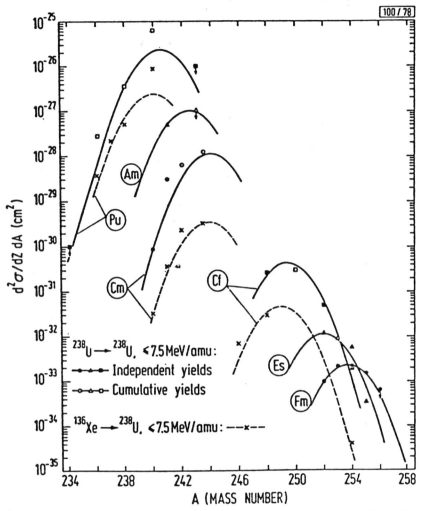

Figure 6.21 Production cross sections of transuranium nuclei in the ^{238}U + ^{238}U and ^{136}Xe + ^{238}U reactions (from Sch 78).

reexamined this question using the semiempirical model of Wollersheim et al. (Wol 82), which accurately represents the U + U reaction along with the Sikkeland systematics for Γ_f/Γ_n. He comes to the same conclusion as (Fre 79, Sch 78) but is able to attribute the failure of the calculation to the deficiencies of the Sikkeland systematics for use in this context.

A more fundamental approach to treating the initial deep inelastic

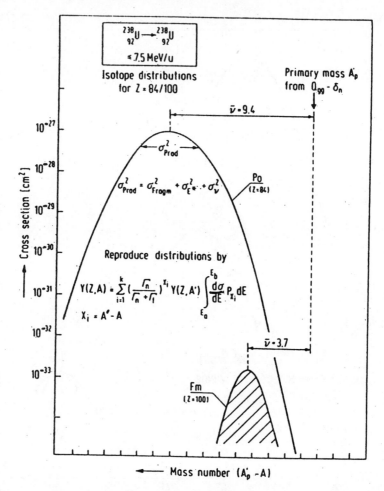

Figure 6.22 Cross section distribution of polonium and fermium isotopes representing the lightest and heaviest complementary reaction products observed in the damped collision of ^{238}U with ^{238}U (from Her 78). Reprinted, with permission, from M.A.K. Lodhi, *Superheavy Elements*, Copyright, 1978, Pergamon Press.

transfer reaction has been taken by Nörenberg and collaborators (Ayi 76a, Ayi 76b, Wol 78, Wol 78a, Ayi 78, Rie 79a, Wol 77b, Rie 79b). Riedel and Nörenberg (Rie 79b) use a semiphenomenological model to predict the primary fragment distributions in the reaction of ^{238}U with ^{238}U. Comparison of the calculated primary product distributions with the measured distributions indicates the necessity of assuming the transur-

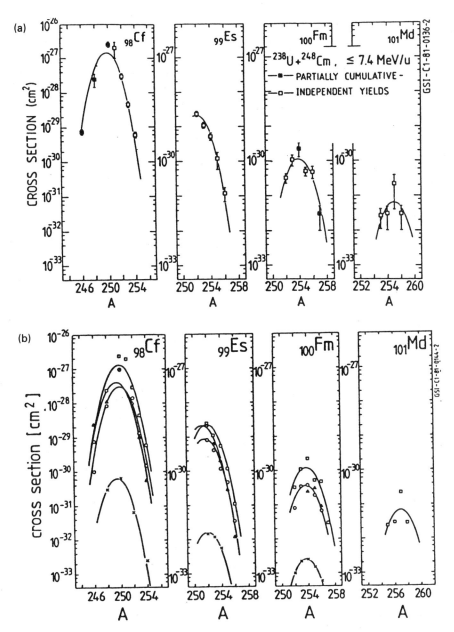

Figure 6.23 (a) Cross sections for the formation of heavy actinides in the reaction of 7.4 MeV/nucleon ^{238}U projectiles with thick ^{248}Cm targets. (b) Intercomparison of formation cross sections for heavy actinides from the reaction of ≤ 7.4 MeV/n ^{238}U + ^{248}Cm (□); ≤ 5.6 MeV/n ^{48}Ca + ^{248}Cm (△); ≤ 6.4 MeV/n ^{136}Xe + ^{248}Cm (○) and ≤ 7.5 MeV/n ^{238}U + ^{238}U (×).

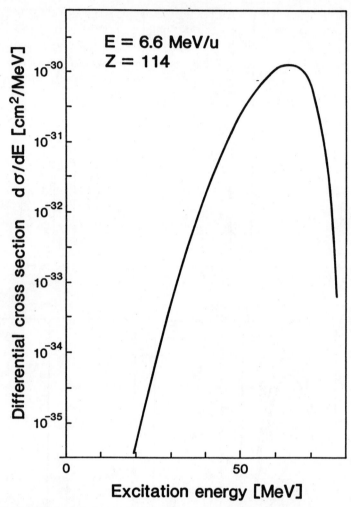

Figure 6.24 Calculated primary differential cross sections $d\sigma/dE^*$ for $Z = 114$ in the reaction of 6.9 MeV/nucleon ^{238}U with ^{238}U (from Gäg 88).

anium products only arise from the low energy tails of the initial distributions. This work also indicated that large gains in the transuranium product yields would be obtained with heavier projectiles and targets.

Gäggeler et al. (Gäg 81) and Schädel et al. (Sch 82) followed up on this suggestion. They measured the yields of transuranium nuclei in the reaction of 7.4 MeV/nucleon ^{238}U with ^{248}Cm. Their results are shown in

Figure 6.23 along with the results of other investigations of deep inelastic transfer reactions with ^{248}Cm. The shapes and centroids of the isotopic distributions are similar in all cases, but the magnitudes of the yields are greatest in the U + Cm reaction. For example, the U + Cm reaction gives $\sim 10^4$ times more Cf and 10^3 times more Fm than the U + U reaction.

Attempts to synthesize superheavy nuclei using these reactions (U = U, U + Cm) have failed (see Chapter 7). The fundamental problem seems to be the relationship implied by Figure 6.19. The nuclei far from the target mass are associated with high total kinetic energy loss events, that is, where large amounts of total kinetic energy have been dissipated into internal excitation energy. The resulting primary product excitation energy spectrum expected for $Z = 114$ (Gäg 88, Figure 6.24) is peaked at too high an energy and is too narrow to allow a significant yield, after deexcitation, of heavy transuranium nuclei.

GENERAL REFERENCES

The literature concerning nuclear reactions is extensive. At the risk of neglecting some important works, the authors cite the following general references as being especially valuable in designing and understanding nuclear synthetic reactions.

Elementary Level

Lef 68   M. Lefort, *Nuclear Chemistry* (Van Nostrand, London, 1968).

Intermediate Level

Mar 68   J.B. Marion and F.C. Young, *Nuclear Reaction Analysis* (North-Holland, Amsterdam, 1968)

Mar 70   P. Marmier and E. Sheldon, *Physics of Nuclei and Particles*, Vols. I and II, (Academic Press, New York, 1970)

Muk 87   K.N. Mukhin, *Experimental Nuclear Physics*, Vol. I (Mir, Moscow, 1987)

Sat 80   G.R. Satchler, *Introduction to Nuclear Reactions* (Wiley, New York, 1980)

Advanced Level

Alo 74   J.R. Alonso in *Gmelin Handbuch der Anorganischen Chemie*, Vol. 7B, Part A1, II, (Springer, Berlin, 1974) p. 28.

Bas 80   R. Bass, *Nuclear Reactions with Heavy Ions* (Springer-Verlag, Berlin, 1980)

Bro 84   *Treatise on Heavy-Ion Science*, Vol. I–IX, D.A. Bromley, Ed. (Plenum Press, New York, 1984).

Van 73   R. Vandenbosch and J.R. Huizenga, *Nuclear Fission*, (Academic Press, New York, 1973).

Tho 68   T.D. Thomas, Annu. Rev. Nucl. Sci. **18**, 343 (1968).

REFERENCES

Agu 89   C.E. Aguiar, A.N. Aleixo, V.C. Barbosa, L.F. Canto, and R. Donangelo, Nucl. Phys. **A500**, 195 (1989).

Ari 84   A. Arima and S. Kubono in *Treatise on Heavy-Ion Science*, Vol. I, D.A. Bromley, Ed. (Plenum Press, New York, 1984).

Arm 89   P. Armbruster, GSI Report 89-65, August, 1989.

Asc 84   R.J. Asciutto and E.A. Seglie, in *Treatise on Heavy-Ion Science*, Vol. I, D.A. Bromley, Ed. (Plenum Press, New York, 1984).

Aus 70   N. Austern, *Direct Nuclear Reaction Theories* (Wiley, New York, 1970).

Ayi 76a   S. Ayik, B. Schürmann, and W. Nörenberg, Z. Phys. **A277**, 299 (1976).

Ayi 76b   S. Ayik, B. Schürmann, and W. Nörenberg, Z. Phys. **A279**, 145 (1976).

Ayi 78   S. Ayik, G. Wolschin, and W. Nörenberg, Z. Phys. **A286**, 271 (1978).

Bac 74   B.B. Back, H.C. Britt, O. Hansen, B. Ledoux, and J.D. Garrett, Phys. Rev. **C10**, 1948 (1974).

Bas 74   R. Bass, Nucl. Phys. **A231**, 45 (1974).

Bas 80   R. Bass, *Nuclear Reactions with Heavy Ions* (Springer, Berlin, 1980).

Bec 78   M. Beckerman, Phys. Lett. **78B**, 17 (1978).

Bec 78a   M. Beckerman and M. Blann, Phys. Rev. **C17**, 1615 (1978).

Beh 68   A.N. Behkami, J.H. Roberts, W.D. Loveland, and J.R. Huizenga, Phys. Rev. **171** (1968).

Ben 81   M. Benedict, T.H. Pigford, and H. Levi, *Nuclear Chemical Engineering*, 2nd edn. (McGraw-Hill, New York, 1981).

Bir 78   J.R. Birkelund, J.R. Huizenga, J.N. De, and D. Sperber, Phys. Rev. Lett. **40**, 1123 (1978).

Bir 79   J.R. Birkelund, L.E. Tubbs, J.R. Huizenga, J.N. De, and D. Sperber, Phys. Rep. **56**, 107 (1979).

Bjø 82   S. Bjønholm and W.J. Swiatecki, Nucl. Phys. **A391**, 471 (1982).

Bla 66   M. Blann, Nucl. Phys. **80**, 223 (1966).

Bla 72   M. Blann and F. Plasil, Phys. Rev. Lett. **29**, 303 (1972).

Blo 77   J. Blocki, J. Randrup, W.J. Swiatecki, and C.F. Tsang, Ann. Phys. **105**, 427 (1977).

Blo 86	J.P. Blocki, H. Feldmeier and W.J. Swiatecki, Nucl. Phys. **A459**, 145 (1986).
Boc 82	R. Bock et al., Nucl. Phys. **A388**, 334 (1982).
Bon 71	J.P. Bondorf, F. Dickmann, D.H.E. Gross, and P.J. Siemens, J. Phys. Colloq. **C6**, 145 (1971).
Bre 56	G. Breit and M.E. Ebel, Phys. Rev. **103**, 679 (1956).
Bri 72	D.M. Brink, Phys. Lett. **40B**, 37 (1972).
Bri 77	D.M. Brink, *Physics with Heavy Ions and Mesons*, Vol. 1 (North-Holland, Amsterdam, 1978) p. 1.
Bro 84a	D.A. Bromley in Bro 84.
Cha 87	R.M. Chastelier, R.A. Henderson, D. Lee, K.E. Gregorich, M.J. Nurmia, R.B. Welch, and D.C. Hoffman, Phys. Rev. **C36**, 1820 (1987).
Che 83	E.A. Cherepanov, A.S. Iljinov, and M.V. Mebel, J. Phys. G. **9**, 931 (1983).
Coh 71	B.L. Cohen, *Concepts of Nuclear Physics* (McGraw-Hill, New York, 1971).
Coh 74	S. Cohen, F. Plasil and W.J. Swiatecki, Ann. Phys. (N.Y.) **82**, 557 (1974).
Coo 81	J. Cook, Atom. Data Nucl. Data Tables **26**, 19 (1981).
Cra 74	J.L. Crandall in *Gmelin Handbuch der Anorganischen Chemie*, Vol. 7B, Part A1, II, (Springer, Berlin, 1974) p. 1.
Dos 59	I. Dostrovsky, Z. Fraenkel, and G. Friedlander, Phys. Rev. **116**, 683 (1959).
Fre 79	H. Freiesleben et al., Z. Phys. **A292**, 171 (1979).
Fre 84	H. Freiesleben and J.V. Kratz, Phys. Rep. **106**, 1 (1984).
Fri 83	W.A. Friedman, Phys. Rev. **C27**, 569 (1983).
Gäg 81	H. Gäggeler et al., Proc. Fourth Intl. Conf. on Nuclei Far From Stability, Helsingør, Denmark, June 1981, CERN 81-09, (1981) p. 763.
Gäg 84	H. Gäggeler et al., Z Phys. **A316**, 291 (1984).
Gäg 86	H. Gäggeler et al., Phys. Rev. **C33**, 1983 (1986).
Gäg 88	H.W. Gäggeler, Paul Scherrer Institut Report PSI-Berecht 14, October 1988.
Gäg 89	H.W. Gäggeler, Nucl. Phys. **A502**, 561c (1989).
Gav 76	A. Gavron, H.C. Britt, E. Konecny, J. Weber and J.B. Wilhelmy, Phys. Rev. **C13**, 2374 (1976).
Gav 80	A. Gavron, Phys. Rev. **C21**, 230 (1980).
Gil 65	A. Gilbert and A.G.W. Cameron, Can. J. Phys. **43**, 1446 (1965).
Gil 65a	A. Gilbert, F.S. Chan, and A.G.W. Cameron, Can. J. Phys. **43**, 1248 (1965).

Gle 75   N.K. Glendenning in *Nuclear Spectroscopy and Reactions*, Part D, J. Cerny, Ed. (Academic Press, New York, 1975).

Gor 80   V.M. Gorbachev, Y.S. Zamyatnin, and A.A. Lbov, *Nuclear Reactions in Heavy Elements* (Pergamon, Oxford, 1980).

Gro 67   J.R. Grover and J. Gilat, Phys. Rev. **157**, 802 (1967).

Gro 76   H.R. Groening and W.D. Loveland, Phys. Rev. **C10**, 697 (1976).

Hah 74   R.L. Hahn, P.F. Dittner, K.S. Toth, and O.L. Keller, Phys. Rev. **C10**, 1889 (1974).

Hau 52   W. Hauser and H. Feshbach, Phys. Rev. **87**, 366 (1952).

Her 78   G. Herrmann in *Superheavy Elements*, M.A.K. Lodhi, Ed. (Pergamon, New York, 1978) p. 24.

Hil 77   K. D. Hildenbrand et al., Phys. Rev. Lett. **39**, 1065 (1977).

Hod 78   P.E. Hodgson, *Nuclear Heavy-Ion Reactions*, (Clarendon Press, Oxford, 1978).

Hof 85   D.C. Hoffman et al., Phys. Rev. **C31**, 1763 (1985).

Hui 85   J.R. Huizenga and W. U. Schröder, in *Semiclassical Descriptions of Atomic and Nuclear Collisions*, J. Bang and J. de Boer, Eds. (North-Holland, Amsterdam, 1985) p. 255.

Jac 56   J.D. Jackson, Can. J. Phys. **34**, 767 (1956).

Jac 70   D.F. Jackson, *Nuclear Reactions* (Methuen, London, 1970).

Jah 82   U. Jahnke, H.H. Rossner, D. Hilscher, and E. Holub, Phys. Rev. Lett. **48**, 17 (1982).

Kau 59   R. Kaufman and R. Wolfgang, Phys. Rev. Lett. **3**, 232 (1959).

Kau 61   R. Kaufman and R. Wolfgang, Phys. Rev. **121**, 206 (1961).

Kra 74   J.V. Kratz, J.O. Liljenzin, A.E. Norris, and G.T. Seaborg, Phys. Rev. Lett. **33**, 502 (1974).

Kra 79   J.V. Kratz, W.Brüchle, G. Franz, M. Schädel, I. Warnecke, G. Wirth, and M. Weis, Nucl. Phys. **A332**, 477 (1979).

Kra 88   K.S. Krane, *Introductory Nuclear Physics* (Wiley, New York, 1988).

Kup 80   V.M. Kupriyanov, K.K. Istekov, B.I. Fursov, and G.N. Smirenkin, Yad. Fiz. **32**, 355 (1980); trans. Sov. J. Nucl. Phys. **32**, 184 (1980).

Kup 84   V.M. Kupriyanov, G.N. Smirenkin, and B.I Fursov, Yad. Fiz. **39**, 556 (1984); trans. Sov. J. Nucl. Phys. **39**, 352 (1984).

Lee 82   D. Lee, H.R. von Gunten, B. Jacak, M. Nurmia, Y. Liu, C. Luo, G.T. Seaborg, and D.C. Hoffman, Phys. Rev. **C25**, 286 (1982).

Lee 83   D. Lee, K.J. Moody, M.J. Nurmia, G.T. Seaborg, H.R. von Gunten, and D.C. Hoffman, Phys. Rev. **C27**, 2656 (1983).

Lef 73   M. LeFort, C. Ngô, J. Peter, and B. Tamain, Nucl. Phys. **A216**, 166 (1973).

Lüt 84   K. Lützenkirchen, J.V. Kratz, W. Brüchle, H. Gäggeler, K. Sümmerer, and G. Wirth, Z. Phys. **A317**, 55 (1984).

Mag 87	M.T. Magda, A. Pop, and A. Sandulescu, J. Phys. G **13**, L127 (1987).
McF 82	R.M. McFarland, Lawrence Berkeley Laboratory Report LBL-15027, September 1982.
McG 85	P.L. McGaughey, W. Loveland, D.J. Morrisssey, K. Aleklett, and G.T. Seaborg, Phys. Rev. **C31**, 896 (1985).
Moo 86	K.J. Moody, D. Lee, R.B. Welch, K.E. Gregorich, G.T. Seaborg, R.W. Lougheed, and E.K. Hulet, Phys. Rev. **C33**, 1315 (1986).
Mor 72	L.G. Moretto, Nucl. Phys. **A180**, 337 (1972).
Mug 81	S.F. Mughabghab et al., *Neutron Cross Sections* (Academic, New York, 1981).
Mün 88	G. Münzenberg, Rep. Prog. Phys. **51**, 57 (1988).
Oga 74	Y.T. Oganessian et al., Soviet Phys. JETP Letters, **20**, 265 (1974).
Oga 81	Y.T. Oganessian, Actinides 81, Asilomar, CA, September 1981.
Oga 85	Y.T. Oganessian and Y.A. Lazarev, *Treatise on Heavy Ion Science*, Vol. 4, D.A. Bromley, Ed. (Plenum, New York, 1985).
Ott 76	R.J. Otto, M.M. Fowler, D. Lee and G.T. Seaborg, Phys. Rev. Lett. **36**, 135 (1976).
Per 76	C.M. Perey and F.G. Perey, Atom. Data Nucl. Data Tables **17**, 1 (1976).
Pla 73	F. Plasil and W.J. Swiatecki, as quoted in Van 73, p. 244.
Pla 75	F. Plasil and M. Blann, Phys. Rev. **C11**, 508 (1975).
Pla 78	F. Plasil, Phys. Rev. **C17**, 823 (1978).
Plo 82	M. Ploszajczak and M.E. Faber, Phys. Rev. **C25**, 1538 (1982).
Ran 78	J. Randrup, Nucl. Phys. **A307**, 319 (1978).
Ran 79	J. Randrup, Nucl. Phys. **A327**, 490 (1979).
Ref 80	G. Reffo in *Theory and Applications of Moment Methods in Many Fermion Systems* (Plenum Press, New York, 1980).
Rei 81	W. Reisdorf, Z. Phys. **A300**, 227 (1981).
Rei 82	W. Reisdorf et al., Phys. Rev. Lett. **49**, 1811 (1982).
Rei 86	W. Reisdorf, Proc. Int. Nucl. Phys. Conf. Harrogate (AIP, UK, 1986) p. 205.
Rie 79a	C. Riedel, G. Wolschin, and W. Nörenberg, Z. Phys. **A290**, 47 (1979).
Rie 79b	C. Riedel and W. Nörenberg, Z. Phys. **A290**, 385 (1979).
Sch 76	W.U. Schröder et al., Phys. Rev. Lett. **36**, 514 (1976).
Sch 77	W.U. Schröder and J.R. Huizenga, Annu. Rev. Nucl. Sci. **27**, 465 (1977).
Sch 78	M. Schädel et al., Phys. Rev. Lett. **41**, 469 (1978).
Sch 82	M. Schädel et al., Phys. Rev. Lett. **48**, 852 (1982).
Sch 84	W.U. Schröder and J.R. Huizenga in *Treatise on Heavy-Ion Science*, Vol. 2, D.A. Bromley, Ed., (Plenum Press, New York, 1984).

Sch 86	M. Schädel et al., Phys. Rev. **C33**, 1547 (1986).
Sie 71	P.J. Siemens, J.P. Bondorf, D.H.E. Gross, and F. Dickmann, Phys. Lett. **36b**, 24 (1971).
Sie 86	A.J. Sierk, Phys. Rev. **C33**, 2039 (1986).
Sik 67	T. Sikkeland, Ark. Fys. **36**, 539 (1967).
Sik 68a	T. Sikkeland, A. Ghiorso, and M. Nurmia, Phys. Rev. **172**, 1232 (1968).
Sik 68b	T. Sikkeland, J. Maly, and D.F. LeBeck, Phys. Rev. **169**, 1000 (1968).
Sim 83	M.H. Simbel, Z. Phys. **A313**, 311 (1983).
Sto 80	R.G. Stokstad, W. Reisdorf, K.D. Hildenbrand, J.V. Kratz, G. Wirth, R. Lucas, and J. Poiton, Z. Phys. **A295**, 269 (1980).
Sto 85	R.G. Stokstad in *Treatise on Heavy-Ion Science*, Vol. 3, D.A. Bromley, Ed. (Plenum Press, New York, 1985) p. 83.
Sto 89	R.G. Stokstad, D.E. Digregorio, K.T. Lesko, B.A. Harmon, E.B. Norman, J. Pouliot, and Y.D. Chan, Phys. Rev. Lett. **68**, 399 (1989).
Swi 82	W.J. Swiatecki, Nucl. Phys. **A376**, 275 (1982).
Tan 84	S. Tanaka, K.J. Moody, and G.T. Seaborg, Phys. Rev. **C30**, 911 (1984).
Vaz 81	L.C. Vaz, J.M. Alexander, and G.R. Satchler, Phys. Rep. **69**, 373 (1981).
Wei 57	V.F. Weisskopf, Rev. Mod. Phys. **29**, 174 (1957).
Wel 87	R.B. Welch, K.J. Moody, K.E. Gregorich, D. Lee, and G.T. Seaborg, Phys. Rev. **C35**, 204 (1987).
Wil 73	J. Wilczynski, Phys. Lett. **47B**, 484 (1973).
Wil 80	W.W. Wilcke, J.R. Birkelund, H.J. Wollersheim, A.D. Hoover, J.R. Huizenga, W.U. Schröder, and L.E. Tubbs, Atom. Data Nucl. Data Tables **25**, 391 (1980).
Wil 80a	J. Wilczynski, K. Siwek-Wilczynski, J. van Driel, S. Gonggrijp, D.C.J.M. Hageman, R.V.F. Janssens, J. Lukasiak, and R.H. Siemssen, Phys. Rev. Lett. **45**, 606 (1980).
Wil 82	J. Wilczynski et al., Nucl. Phys. **A373**, 109 (1982).
Wol 74	K.L. Wolf, J.P. Unik, J.R. Huizenga, J. Birkelund, H. Freiesleben, and V.E. Viola, Phys. Rev. Lett. **33**, 1105 (1974).
Wol 77	K.L. Wolf, J.P. Unik, E.P. Horwitz, C.A.A. Bloomquist, and W. Delphin, Bull. Am. Phys. Soc. **22**, 67 (1977).
Wol 77b	G. Wolschin, Nukleonika **22**, 1165 (1977).
Wol 78	G. Wolschin and W. Nörenberg, Z. Phys. **A284**, 209 (1978).
Wol 78a	G. Wolschin and W. Nörenberg, Phys. Rev. Lett. **41**, 691 (1978).
Wol 82	H. Wollersheim, W.W. Wilcke, J.R. Birkelund, and J.R. Huizenga, Phys. Rev. **C25**, 338 (1982).

7

SUPERHEAVY ELEMENTS

7.1 INTRODUCTION

In the 50 years following the discovery of the first transuranium elements, neptunium and plutonium, scientists synthesized and identified (or "discovered") 15 others ranging in atomic number from 95 through 109. These 17 elements represent an addition of nearly 20% to the elemental building blocks of nature. In this chapter, we explore the possibility of extending this still further.

Up until about 1970, the experimental data on the stability of the transuranium elements seemed to indicate that a practical limit to the periodic table would be reached at about element 108 (Figure 7.1). By simple extrapolation of the experimental data known at that time, one would have concluded that at about element 108, the half-lives of the longest lived isotopes of the elements would become so short ($< 10^{-6}$ s), owing to decay by spontaneous fission, as to preclude their production and study. However, between 1966 and 1972, a number of theoretical calculations based upon newly developed theories of nuclear structure showed that in the region of proton number $Z = 114$ and neutron number $N = 184$, the nuclei have particularly stable spherical ground states.* This

---

*In analogy to the filling of neutron shells in nuclei, one might have expected special stability associated with $Z = 126$ rather than $Z = 114$. However, the influence of the Columb force causes a shift in the proton single particle states relative to the neutron states, causing a modest gap in the proton level scheme at $Z = 114$, in addition to the larger gap near $Z = 126$.

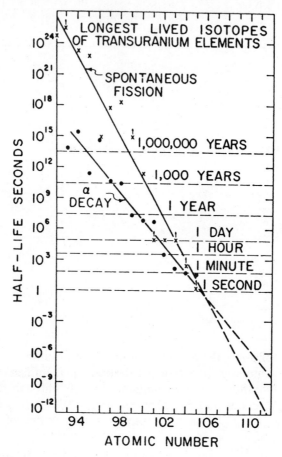

Figure 7.1 The half-life of the longest-lived isotope of a given element plotted against atomic number, as known in 1970.

stabilization is due to the complete filling of proton and neutron shells and is analogous to the stabilization of the chemical elements, such as the noble gases, which is due to the filling of electronic shells in these atoms. Of even more interest was the prediction that some of these "superheavy" nuclei have halflives of the order of the age of the universe, thus stimulating efforts to find these "missing" elements in nature. The superheavy elements (SHE) were predicted to form an "island" of relative stability around $Z = 114$ and $N = 184$, separated from the "peninsula" of known nuclei by a sea of instability (Figure 7.2).

Following this initial period of optimism about the existence and ac-

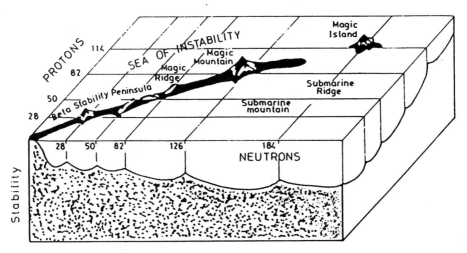

Figure 7.2 An allegorical representation (circa 1970) of the stability of nuclei, showing a peninsula of known nuclei and an island of superheavy nuclei (predicted to be relatively stable) in a sea of instability.

cessibility of the superheavy nuclei, a worldwide effort was launched to "jump the gap" between the peninsula of known nuclei and the predicted island of stability by combining two nuclei in a nuclear reaction, thus synthesizing superheavy nuclei in the laboratory. In the two decades that have passed since the original optimistic predictions were made, many attempts to synthesize superheavy nuclei have failed, the theoretical predictions about the nature and shape of the superheavy "island" have changed, and some striking successes in the synthesis of isotopes of elements 107, 108, and 109 have occurred.

In this chapter we attempt to review and clarify the situation and to indicate why this search for new elements continues. One fact should be emphasized from the outset: while the various theoretical predictions about the superheavy nuclei differ as to the expected half-lives and regions of stability, all theoretical predictions are in agreement: superheavy nuclei can exist. Thus the search for superheavy nuclei remains as a unique, rigorous test of the predictive power of modern theories of the structure of nuclei.

7.2 PROPERTIES OF THE SUPERHEAVY ELEMENTS

The general nuclear properties of unknown nuclei, such as their masses, decay modes, et cetera, are calculated using the macroscopic-microscopic approach (see Chapter 4) in which the potential energy of the nucleus as a function of shape is calculated as the sum of a macroscopic term and microscopic term. The macroscopic term, which is usually calculated using a refined version of the liquid drop model, gives the average smooth variation of nuclear properties with particle number and shape. The microscopic term, which (for the heaviest elements) is a 0.5% correction to the macroscopic term, accounts for the nonuniform distribution of single particle levels in the nucleus. The microscopic term will lower the ground state mass of closed shell nuclei owing to the increased stability of these "magic" nuclear configurations. There are two major parts to the microscopic term: a shell correction and a pairing correction. They are both determined from a set of single particle levels that is used as an input to the calculation. The shell correction is calculated using the Strutinsky method (Str 67) while the pairing correction is made using the BCS approximation (see, for example, Nix 72).

The crucial factor in these calculations is the nature of the set of assumed single particle levels. Some years ago, Chasman used a set of realistic single particle levels that had been fitted to the known levels of the actinide nuclei to calculate the microscopic correction term for the superheavy elements (Cha 78). He found special stability associated with $Z = 114$ and $N = 164$, 178, and 184. But he also found that because of the uncertainties in our knowledge of the single particle level schemes and the methods of calculation used in estimating the microscopic term, the spontaneous fission half-lives of the SHE were uncertain by a factor of $10^{\pm 7}$–$10^{\pm 10}$, which is in agreement with a previous estimate made by Bemis and Nix (Bem 77). With this information as a background, it is not surprising that as our best theoretical prescriptions for the single particle levels have changed, so too have our estimates of the nature of the superheavy island.

In Figure 7.3, we show the results of three calculations of the properties of the superheavy nuclei, as performed in 1972, 1976, and 1989. While differences between these predictions (Fis 72, Ran 76, Pat 89, Sob 89) are understandable, they are frustrating to the experimentalists who must try to "hit" a moving target whose characteristics are subject to change. The early calculations emphasized that the superheavy island was centered on $N = 184$, with either a gradual slope to lesser values of N (1972 calculation) or a steep drop towards lower N, crashing into the sea of instability (1976 calculations), with the 1972 calculations predicting half-lives of the order of the age of the universe, which the later calculations failed to confirm. The significance of the 1976 calculations which included the effects of axially asymmetric nuclear deformations was that they forced experimentalists to attempt to assemble composite nuclei with $N \sim 184$ — a very difficult task — in their attempts at synthesizing SHE in the laboratory. The calculations in 1989 showed the most stable nuclei to have $N = 182$–184

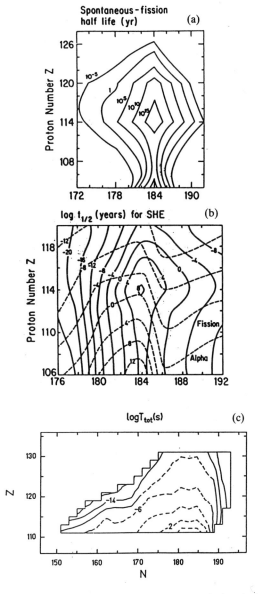

Figure 7.3 Contour plots of predicted spontaneous fission half-lives for SHE as calculated (a) in 1972 (Fis 72), (b) in 1976 (Ran 76) and (c) contour plot of predicted total half-lives for SHE as calculated in 1989 (Pat 89). ———, spontaneous fission; − − − −, alpha decay.

($t_{1/2} \sim 12$ days)* with a larger region of nuclei with measurable half-lives ($t_{1/2} > 10^{-6}$ s) extending to $N \sim 160$. The nuclei near ^{294,296}112 are predicted to be spherical while those with $N < 166$ (e.g., ^{274,276}112) are predicted to be deformed (Sob 89). Unlike the allegorical representation of Figure 7.2, these latest calculations (Figures 1.2 and 7.3c) predict that there is no vast sea of instability separating the known nuclei and the nuclei with $Z \geq 110$. These recent calculations along with others (Loj 88) have also confirmed the earlier work on the special stability associated with $N \sim 162$–164 (which is due to the gap between the $j_{15/2}$ and $d_{3/2}$ neutron levels) for $Z = 108$–111.

Since most of the data used to deduce the single particle levels that are employed in the calculation come from fits to data from the rare earth and actinide regions, it would be reassuring to find some experimental evidence that the recent calculations have some measure of accuracy in the transactinide region. In Figure 7.4, we show some such evidence in the form of a comparison between current theoretical calculations and experimental data for the alpha decay energies of the $N = 157$ isotopes (nuclei with the same neutron number but different atomic numbers). The agreement between the calculations and data is generally excellent.

Once a superheavy nucleus has been formed, one would like to find a unique signature for its existence amongst the many other products of possible synthesis reactions. The higher Z of the superheavy nuclei should lead to increased total kinetic energy of the fission fragments during spontaneous fission (~ 50 MeV higher than for $Z = 92$) (Sch 72) and a very large number of neutrons ($\bar{\nu} \sim 10$–14 for $Z = 114$ compared to 2.4 for ^{235}U(n, f)). [This latter conclusion about $\bar{\nu}$ is sensitive to the fragment shapes and could be as low as $\bar{\nu} = 5$–6 (Hof 78).] In order to identify the atomic number of a superheavy nucleus, one would try to detect any emitted alpha particles (with $8 < Q_\alpha < 10$ MeV), especially those in time correlation with particles emitted in the decay of known daughter, granddaughter, and so on, nuclei, or use some other signature such as the energy of characteristic x-rays.

One further consequence of our current thinking about the properties of the SHE is that, contrary to the belief of some concerning the initial optimistic picture of the late 1960s and early 1970s, we do not now believe that it would be fruitful to search for the existence of the superheavy elements in nature: we think that the half-lives of these elements are short compared to the age of the universe. Also, it seems unlikely that the r-

---

*Figure 7.3c shows the contours of total half-life and emphasizes the point that α-decay is expected to be the stability limiting decay mode for these nuclei, thereby shifting the region of maximum stability to $Z = 110$–112.

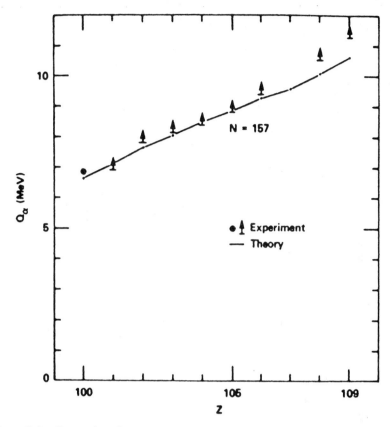

Figure 7.4 Comparison between experimental (Arm 85) and calculated (Möl 86) values of Q_α, the alpha decay energy, for $N = 157$ isotones. The experimental Q_α values are lower limits for most of the nuclides because the transitions are believed to populate excited states of the daughter nuclei rather than the ground state. The length of the arrows shows the expected uncertainty in the Q_α values.

process (rapid neutron capture in supernovae) of heavy element nucleosynthesis will lead to the production of superheavy nuclei (Mat 76).

Among the most interesting aspects of superheavy elements are their predicted chemical properties (Kel 77). (This discussion presumes the lifetime of some superheavy species is long enough to make the discussion of chemical behavior meaningful.) Predictions of the ground state electron configurations of these elements have been made using Dirac–Fock and Dirac–Fock–Slater calculations (Table 3.3). These calculations indicate that the 6d shell will be complete at element 112. The relativistic 7p orbital is predicted to be split into $7p_{1/2}$ and $7p_{3/2}$ orbitals with the energy splitting between these orbitals being such that the filled $7p_{1/2}^2$ orbital will act as a closed shell and additional $7p_{3/2}$ electrons will act as electrons outside of a closed shell. As an example of this effect, element 115 (eka-bismuth) is predicted to have its valence electrons in the configuration $7p_{1/2}^2\ 6p_{3/2}$ with a stable +1 oxidation state in contrast to the stable +3 oxidation state of its homolog bismuth. Similarly, Pitzer (Pit 75) has pointed out that because of relativistic effects elements 112 (eka-mercury) and 114 (eka-lead) may, in fact, be very noble, that is, volatile gases or liquids.

Other chemical properties of the superheavy elements can be predicted on the basis of these electron configurations along with a judicious use (Kel 77) of Mendeleev-like extrapolations of the smooth trends of these properties among the members of the particular group in the periodic table (see earlier comments).

Chemical separation methods for identifying the atomic numbers of any superheavy nuclei produced in laboratory syntheses have been divised on the basis of these predicted chemical properties. Separations based upon the ion-exchange behavior of the bromide complexes, the predicted strong affinity to sulfur or sulfur compounds, and the predicted high volatility and ease of reduction have been used.

7.3 LABORATORY SYNTHESIS OF SUPERHEAVY ELEMENTS

The laboratory synthesis of the heaviest elements is a formidable challenge to experimentalists. The predicted cross sections for heaviest element formation by heavy ion bombardment are less than 10^{-8} of the total reaction cross section, corresponding to the production of less than one atom per day of irradiation. The synthesis process, as discussed in Chapter 6, can be broken down into two related steps: (a) the initial formation of a composite nucleus with the appropriate values of Z and A; and (b) the deexcitation of that composite species predominantly by the emission of neutrons (in competition with the much more probable fission process which will destroy the nucleus). Understanding how this overall process might occur is difficult because the surviving products in a reaction frequently arise from the poorly characterized low excitation energy (E^*), low angular momentum (J) tails of the primary product distributions.

The nuclear reactions used in superheavy element synthesis are heavy ion reactions. A smaller projectile nucleus is made to collide with a larger target nucleus to form a composite species. Because all successful first elemental syntheses have involved complete fusion reactions, we shall generally restrict our attention to this type of reaction in which the projectile nucleus completely amalgamates with the target nucleus forming a composite system. The probability of fusion of the projectile and target nuclei depends on their specific properties, their' interaction potential and any dissipative processes that occur during fusion (see Chapter 6). For the low energies involved in SHE synthesis where the fusion and reaction cross sections are the same, a simple classical model of fusion gives for the fusion cross section

$$\sigma_{fus} = \pi R_B^2 (1 - (V(R_B)/E_{cm})) \tag{7.1}$$

where E_{cm} is the center of mass energy of the projectile, and the nuclear radius R_B and interaction potential $V(R_B)$ are given by

$$R_B = 1.4(A_1^{1/3} + A_2^{1/3}) \text{ (fm)}$$

$$V(R_B) = Z_1 Z_2 e^2 / R_B$$

Here Z_1 and Z_2 are the atomic numbers, and A_1 and A_2 the mass numbers of the projectile and target nuclei, respectively. These equations can be translated into angular momentum space giving

$$\sigma_{fus} = \pi \lambdabar^2 \Sigma (2\ell + 1) = \pi \lambdabar^2 (\ell_{max} + 1)^2 \tag{7.2}$$

where $\lambdabar$ is the reduced wavelength of the projectile and ℓ_{max}, the highest partial wave to fuse, is the highest partial wave at energy E_{cm} that just overcomes the reaction barrier, $V(R_B)$. This simple classical model can be modified to account for barrier penetration and the effects of nuclear dissipation during the collision (Bir 79). The results of calculations based upon these modifications give a good general description of fusion cross sections for nuclei of light and intermediate masses.

For the heavy target nuclei involved in attempts to synthesize SHE, additional phenomena significantly affect the ability of nuclei to fuse. The first of these effects involves the enhancement of the fusion probability at low projectile energies, that is, subbarrier fusion (see Chapter 6). Two processes appear to lead to this enhancement: (a) the static deformation of one (or both) of the reacting nuclei which creates a greatly preferred (and highly favorable) orientation of the colliding nuclei (Sto 80), and (b) dynamic deformations of the nuclei due to coupling of the vibration modes of the reacting nuclei which also lead to enhanced fusion cross sections (Rei 82). In Figure 7.5, we show a case of subbarrier fusion compared to "normal" fusion. The significance of subbarrier fusion in the synthesis of heavy nuclei is that it allows experimenters to combine two nuclei with reasonable probability at low projectile energies, thus leading to composite

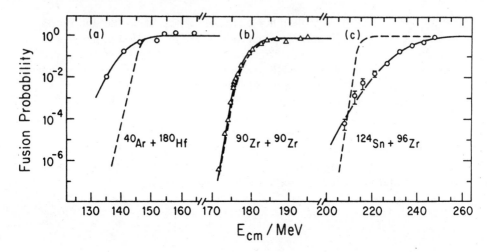

Figure 7.5 Three examples of excitation functions for complete fusion reactions. In each case, the dashed line represents the behavior for "normal", that is, unhindered or unenhanced fusion: (a) subbarrier enhancement of fusion probability; (b) "normal" fusion; (c) dynamic limitation of fusion due to "extra push" effects (from Arm 85). Reproduced, by permission, from the Annual Review of Nuclear and Particle Science, Vol. 35, Copyright, 1985 by Annual Reviews, Inc.

systems with low excitation energies and thus relatively higher survival probabilities.

Unfortunately, there is a competing dynamical process in the reaction of heavy nuclei that lowers the fusion probability: Coulomb repulsion between the reacting heavy nuclei hinders their fusion. If the values of the atomic numbers of the projectile and target nuclei Z_p and Z_t, respectively, become large enough ($Z_p + Z_t > 120$), then fusion is virtually impossible. For lesser values ($120 \geq (Z_p + Z_t) \geq 80$), fusion is significantly hindered. Asymmetric projectile-target combinations do lead to higher fusion probabilities than symmetric systems (see Chapter 6). A typical example of an "extra push effect" is shown in Figure 7.5c. In the synthesis of transactinide nuclei, these "extra push" effects are predicted to occur for $Z_{proj} \geq 18$ (Ar).

Because of the intrinsically low probability of fusion reactions leading to new elements and the complex interplay of subbarrier enhancement and dynamical limitations of fusion probability, it is difficult to make theoretical estimates of fusion probability that are reliable enough to serve as a guide for experimental efforts. With this in mind, Armbruster has developed a semiempirical representation of fusion probability which uses the Swiatecki model as a basis. This representation takes the form of the equation

$$p(V_B) = 0.5 \exp[-a(x_{mean} - b)] \qquad (7.3)$$

where $p(V_B)$ is the probability of fusion at the s-wave ($\ell = 0$) fusion barrier $V_B (=\sigma_{fus}(V_B)/\pi \lambdabar^2)$, x_{mean} is the mean fissility of the composite system (Arm 85) and the coefficients a and b are determined to be 71 and 0.72, respectively, from fitting experimental data. Qualitatively, the more fissile the fusing system is, the harder it will be to form a stable mononucleus. The fissility of a nucleus is proportional to Z^2/A (the ratio of the repulsive Coulomb energy to the attractive surface energy.) The mean fissility, x_{mean}, is taken to be the geometric mean of the fissility of the colliding ions, x_{eff}, and the composite system, x_{cn}. Formally,

$$x_{mean} = (x_{eff} x_{cn})^{1/2}$$

$$x_{eff} = 4A_1A_2(Z_1 + Z_2)^2/(A_1A_2)^{1/3}(A_1^{1/3} + A_2^{1/3})(A_1 + A_2)^2(Z^2/A)_{crit}$$

$$x_{cn} = (Z_1 + Z_2)^2/(A_1 + A_2)(Z^2/A)_{crit}$$

$$(Z^2/A)_{crit} = 50.883(1 - 1.7826\{(N_1 + N_2 - Z_1 - Z_2)/(A_1 + A_2)\}^2). \qquad (7.4)$$

This representation of the fusion probability which combines the effect of both enhancements and limitations to σ_{fus} is shown in Figure 7.6. The maximum fusion probabilities are observed in light projectile-heavy target reactions with the decrease in fusion probability with increasing target and projectile atomic numbers becoming more severe as Z_{proj} increases.

Thus our best current estimates (Figure 7.6) lead us to believe that superheavy composite systems having appropriate values of Z and A can be formed in nuclear reactions with adequate probability. The next task is to determine whether or not such composite systems will survive. As noted earlier, the composite systems are excited and the question is whether they can get rid of the excitation energy by the relatively benign process of particle evaporation, as opposed to the more probable and destructive fission process. The lower the excitation energy of the composite species (that is the "cooler" it is), the more likely it is to survive. The minimum excitation energy of various heavy composite systems formed in nuclear reactions is shown in Figure 7.7. One can see that for $Z_{target} \sim 82$ (Pb), the special stability of the target nuclei leads to low values of Q and E^*. This realization, the "cold fusion" mechanism (Oga 75), has been used in the synthesis of elements 107, 108 and 109. Also, the use of a

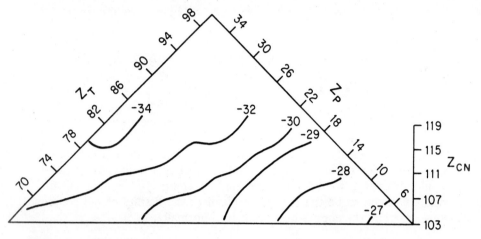

Figure 7.6 Plot of the contours of $\log_{10}\sigma_{\text{fus}}$ ($\ell = 0$, $E = V_B$) in cm^2 for various values of Z_p and Z_t according to the semiempirical formula of Armbruster (Arm 85). The actual fusion cross sections (neglecting survival probabilities) may include other partial waves besides $l = 0$. For a given maximum P value contributing to fusion, $P_{\max}$, these $l = 0$ cross section values should be multiplied by $(l_{\max} + 1)^2$.

doubly magic projectile, such as ^{48}Ca, can lower E^* significantly (Figure 7.7).

Successful synthesis of a new heavy or superheavy element will require selection of a product of maximum stability (Figure 7.3) and then juggling of the competing considerations of minimum excitation energy (Figure 7.7) and reasonable probability of composite nucleus formation (Figure 7.6) to pick the projectile-target combination that appears most promising. In this treatment, it is necessary to assess the probable outcome of the competition between fission and neutron emission in the deexcitation of the composite species. Given the excitation energies, neutron binding energies, and fission barriers for the nuclei involved, detailed procedures that are accurate to an order of magnitude exist for use in making such estimations (Oga 85, Chapter 6). A very crude approximation to these calculations for the heaviest nuclei assumes that for each 8–10 MeV of excitation energy (that is, the average excitation energy removed by an evaporated neutron), the survival probability drops by a factor of about 100. Thus a species excited to 15–20 MeV will have survival probability of 10^{-4}. This arithmetic reflects the fact that at each stage of the deexcitation of an excited nucleus, 100 nuclei "die" due to fission for every nucleus that emits a neutron.

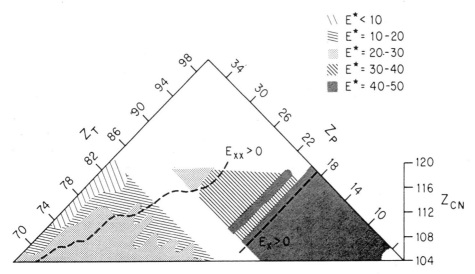

Figure 7.7 Plot of the minimum excitation energy E^* (MeV) of the composite species formed from a given target-projectile combination. "Extra push" effects are not considered in the computation of E^* although contours of E_x (extra push energy) > 0 and E_{xx} (extra extra push energy) > 0 are shown. The nuclear masses used are those of Möller and Nix (Möl 81) and Möller et al. (Möl 86) with droplet masses where not otherwise available taken from Myers and Swiatecki (Mye 66). The reactions are assumed to occur at a projectile energy (cm system) of 0.96 V_B (Swi 82).

The excitation energy of the composite species must be kept as low as possible, to reduce the number of times that the species will undergo the rigours of struggling for survival against fission. But there are even more compelling reasons for requiring a minimum excitation energy since as the excitation energy rises, the chances of survival at each step in the deexcitation process decrease. This is due to the fact that if the excitation energy is high enough, the nuclear "shell effects" that stabilize any potential superheavy nucleus against fission "washout", that is, the fission barriers vanish. There is some disagreement as to when this occurs; while some estimates are as low as $E^* = 15$ MeV for spherical superheavy nuclei (Arm 85), most estimates of this energy are in the range 30–50 MeV (Mus 78, Oga 85). Nonetheless, it is clearly important to achieve the minimum excitation energy of the composite species. For similar reasons, it is important to minimize the angular momentum of any intermediate nuclei and thus, in this chapter, we have restricted our attention to the production of $\ell \sim 0$ product nuclei.

To see how all those factors act in element synthesis, we survey the successful efforts to synthesize elements 106 and 108. Ghiorso et al. (Ghi 74) at Berkeley used the nuclear reaction

$$95 \text{ MeV } {}^{18}\text{O} + {}^{249}\text{Cf} \rightarrow {}^{263}106 + 4n$$

to synthesize the 0.9 ± 0.2 s $^{263}106$. This reaction produced a composite species of $^{267}106$ at an excitation energy of 40 MeV. For this system, one calculates x_{mean} to be about 0.605 with a corresponding $\ell = 0$ fusion cross section of about 5×10^{-28} cm². The survival probability of the composite species should be $\sim 10^{-8}$ leading to an estimated production cross section of $\sim 5 \times 10^{-36}$ cm² (the measured cross section is 3×10^{-34} cm²). The 0.9 s $^{263}106$ was detected by following its α-decay to the known $^{259}104$, which in turn decays to the known ^{255}No. Oganessian et al. (Oga 74) at Dubna used the "cold fusion" approach to attempt the synthesis of this element with the reactions

$$280 \text{ Mev } {}^{54}\text{Cr} + {}^{207,208}\text{Pb} \rightarrow 106 + xn$$

The formed composite species $^{261,262}106$ would be "colder" than those in the Ghiorso et al. work, with an excitation energy of ~ 22–23 MeV. The predicted $\ell = 0$ fusion cross section would be about 3×10^{-30} cm² with a survival probability of 10^{-4} for the emission of two neutrons, giving a predicted formation cross section of about 3×10^{-34} cm² (measured cross section 10^{-33} cm²). A spontaneous fission activity with a half-life of 4–10 ms was detected and, based upon knowledge of known nuclear reactions, this activity was assigned to $^{259}106$. We now know that this assignment was incorrect in that the observed spontaneous fission activities were mainly due to the daughters of element 106, $^{256,255}104$, and not element 106 (Dem 84). The isotope $^{260}106$ (which may also have been produced in the work of Oganessian et al.) is now known to have a half-life of ~ 4 ms with a partial half-life for decay by spontaneous fission of ~ 7 ms (Mün 85).

Following the pioneering work of Oganessian, which showed that a minimum excitation energy of the composite species occurs when one of the two reacting nuclei has $Z \sim 82$, and the use of these ideas by the Darmstadt and Dubna groups to synthesize (or attempt to synthesize) element 107 (Mün 81, Oga 75), Münzenberg et al. (Mün 84) produced 1.8 ms $^{265}108$ through the reaction

$$292 \text{ MeV } {}^{58}\text{Fe} + {}^{208}\text{Pb} \rightarrow {}^{265}108 + n$$

The excitation energy of the $^{266}108$ composite was about 20–23 MeV. One estimates that the $\ell = 0$ fusion cross section was 5×10^{-31} cm², with a survival probability of 10^{-4}, giving rise to a predicted production cross section of 5×10^{-35} cm^2 (measured cross section is 2×10^{-35} cm^2). Clearly if it were not for the enhanced survival probability associated with the low excitation energy, this elemental synthesis would have been impossible. (The survival probability can be estimated to be $\sim 10^{-4}$ greater than in the synthesis of element 106 using the actinide target.) Typical reactions with actinide targets that could be used to synthesize element 108 have predicted production cross sections that are at least an order of magnitude lower.

One important result from this recent heavy element research is the finding that, contrary to the trends shown in Figure 7.1, the half-lives of the known isotopes of elements 107, 108, and 109 are roughly similar ($t_{1/2} \sim 1$ ms) and furthermore, these nuclei have α-decay half-lives that are shorter than their spontaneous fission halflives. This result suggests the fission barrier heights of the most stable isotopes of elements 106–110 are approximately constant ($B_f \sim 5$–6 MeV). This constancy of the barrier height is thought to arise from microscopic shell corrections, since the macroscopic liquid-drop barriers are predicted to decrease by a factor of four in this Z range.

Since the first predictions of the stability of the SHE, there have been more than 25 reported attempts to synthesize SHE in the laboratory. All of these attempts have ended in failure for a variety of reasons which are discussed elsewhere (Sea 87, Kra 83). The most intensively studied synthesis reaction is the ^{48}Ca + ^{248}Cm reaction. The composite nucleus ^{296}116 was expected to have an excitation energy of ~ 30 MeV, giving rise to a survival probability of 10^{-6}. If the $\ell = 0$ fusion cross section was 3×10^{-30} cm^2, then the predicted formation cross section should have been $\sim 3 \times 10^{-36}$ cm^2. The experimental upper limits for superheavy element formation for this reaction are 10^{-35}–10^{-34} cm^2, at least an order of magnitude greater than the predicted cross section (Arm 85a).

But the situation may even be worse. In nuclear reactions that produce the "shell-stabilized" light actinides such as ^{216}Th (where stabilization is due to the $N = 126$ shell), the survival probabilities of the spherical thorium products were smaller than predicted by two orders of magnitude while the deformed products showed nearly "normal" survival probabilities. This observation, which has been called "a dark cloud on the superheavy element horizon" (Bjø 82) has been interpreted as meaning either (a) that there is some problem in calculating the survival probabilities for spherical nuclei, (b) that the nuclear shells that stabilize spherical nuclei, like the superheavy nuclei, are ineffective at excitation energies greater than 15 MeV (Arm 85), or (c) that there is—contrary to the numerical estimates of the Swiatecki model—a significant "extra-extra push" energy needed in these cases to go beyond the deformed mononucleus to the spherical compound nucleus (Sie 86). This extra-extra push energy has been estimated to be of such a magnitude as to cause the compound nucleus excitation energy to be 50 MeV, which would result in a negligibly small survival probability. Thus, whatever the cause, we suspect one must pay an additional penalty beyond that considered in our normal estimates of survival probability when one forms a spherical shell-stabilized heavy nucleus.

Alternative superheavy synthesis reactions, such as ^{48}Ca + ^{244}Pu or ^{48}Ca + ^{243}Am offer some improvement in the nominal fusion probability over the ^{48}Ca ^{248}Cm reaction and lead to products of similar properties (Figure 7.3c) but should suffer from the same catastrophes during the deexcitation of any composite species formed. The superheavy element synthesis reaction, ^{48}Ca + ^{254}Es, which forms a more neutron-rich composite species, would have a decreased fusion probability relative to the ^{48}Ca + ^{248}Cm reaction (approximately four times less), but might afford increased product survival owing to the lower excitation energy ($E^* = 25$ MeV). (Upper limits of values of the production cross sections for spontaneously fissioning superheavy elements of 3×10^{-31} cm^2 have been measured for this reaction (Lou 85).)

Up to now, we have restricted our attention to the production of the SHE in complete fusion reactions. Another nuclear reaction mechanism has also been used (unsuccessfully) in attempts to synthesize superheavy nuclei. This mechanism — deep inelastic transfer — involves inelastic scattering of a large projectile nucleus by a target nucleus whereby there is a large transfer of nucleons from the projectile to the target nucleus. Spurred on by studies of the ^{238}U + ^{238}U reaction in which 20 or more protons appeared to be transferred from the projectile to the target nucleus with moderate excitation energies, hopes were raised that from the tails of the E^* and J distributions of the heavy nuclei produced in these reactions, a sufficient number of surviving SHE could be made (Rei 82). To date, the upper limits on superheavy element formation in the ^{238}U + ^{238}U and the more favourable ^{238}U + ^{248}Cm reactions are 10^{-35}–10^{-34} cm^2 (Gäg 80, Kra 86). Further attempts at superheavy element synthesis using this approach appear unlikely to be successful owing to the relatively high excitation energies and deformed character of the intermediate species.

7.4 WHAT'S NEXT?

By now, the reader should have an understanding of why those who would synthesize SHE are frustrated. The best efforts to synthesize these elements appear to have fallen short by one or more orders of magnitude.

We believe that the superheavy elements exist, but the method of making them in sufficient quantities as to be observable has eluded us. There are several natural roadblocks in our journey to the superheavy island; we seem to have exhausted the obvious or easy routes. If we are to continue using complete fusion reactions to synthesize superheavy nuclei, our element production rates must improve by one or more orders of magnitude.

Faced with this situation, most workers in this field have lowered their sights to the more modest (yet still difficult) goal of synthesizing the elements immediately beyond element 109. These efforts have focused first on the production of $Z = 110$, $N \sim 162$–164, a species calculated to have special stability due to nuclear shell effects (Figure 7.3c). The overall half-life of this species is calculated to be approximately 10–100 ms (Möl 87), a lifetime which is very compatible with existing equipment for studying and identifying shortlived nuclei. Furthermore, this species is calculated to be deformed in its ground state (with a large β_4 deformation as well as a prolate deformation, giving rise to a sausage-shaped nucleus), thus possibly avoiding some of the difficulties encountered in forming a spherical superheavy nucleus in the ^{48}Ca ^{248}Cm reaction. Armbruster (Arm 85) has suggested the best cold fusion reaction to synthesize this element is

$$^{64}\text{Ni} + {}^{208}\text{Pb} \rightarrow {}^{271}110 + n$$

If the projectile energy is chosen to be near the interaction barrier, the excitation energy of the $^{272}110$ composite is 10–15 MeV, allowing the 1n reaction. But the dynamical hindrance to fusion is large and the overall formation cross section is predicted to be about 10^{-36} cm^2 (Figure 7.8). This corresponds to a yield, with current experimental techniques, of one atom per three weeks of irradiation. A further difficulty is the possibility that the product may decay by electron capture (Moo 89). An upper limit for the production of element 110 in this reaction was found to be 5×10^{-36} cm^2 (Mün 86). Another cold fusion reaction,

$$^{59}\text{Co} + {}^{209}\text{Bi} \rightarrow {}^{266}110 + 2n$$

(suggested by Ghiorso) has a slightly higher fusion probability but produces a more excited product with a lower survival probability and, because of the lower value of N, the product is predicted (Loj 88) to have a considerably shorter $t_{1/2}(\sim 10^{-6}$ s).

It is also possible to use a reaction with an actinide target and a lighter projectile to synbthesize element 110. For example, the reactions

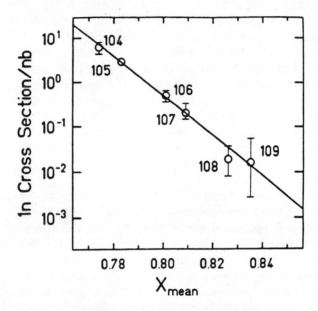

Figure 7.8 Variation of element formation cross sections for reactions involving the emission of one neutron as a function of the parameter, x_{mean} (from Arm 85). Reproduced with permission, from the Annual Review of Nuclear and Particle Science, Volume 35, Copyright 1985 by Annual Reviews, Inc.

$$^{23}\text{Ni} + {}^{254}\text{Es} \rightarrow {}^{273}110 + 4n$$

$$^{26}\text{Mg} + {}^{249}\text{Cf} \rightarrow {}^{271}110 + 4n$$

should not suffer any significant dynamical hindrance to fusion. The survival probabilities of the intermediate species are low (10^{-8}) due to high excitation energies and the overall formation cross sections again are of the order of 10^{-36} cm^2. Another set of possible reactions employing heavier projectiles and lighter actinide targets, such as

$$^{40}\text{Ar} + {}^{236}\text{U} \rightarrow {}^{272}110 + 4n$$

$$^{44}\text{Ca} + {}^{232}\text{Th} \rightarrow {}^{271}110 + 5n$$

offer alternative paths to element 110. These reactions have the advantage of employing readily available projectile and target nuclei and have the same or somewhat improved survival probabilities as the previously discussed reactions. Possible problems that one might encounter in using these reactions involve the possibility of getting full momentum transfer events (complete fusion) without forming a compound nucleus (Les 83, Sch 89). As described in Section 2.15, Oganessian and co-workers (Oga 88) claim to have discovered an isotope of element 110 ($t_{1/2} \sim 10$ ms; decay by SF using such reactions while Ter-Akop'yan et al. (Ter 79) have set an upper limit of 5×10^{-35} cm^2 for SHE production in the related ^{48}Ca + ^{232}Th reaction.

One interesting question that has been discussed recently is whether we have already discovered SHE. Certainly, nuclear shell effects are already playing a crucial and important role in the stability of elements 102–108 (namely, for this range of Z, the fission barrier, (B_f), is approximately constant (~ 5–6 MeV) while the macroscopic liquid-drop model fission barrier heights decrease by a factor of five). If "superheavy element" is synonymous with "element whose stability is determined primarily by a nuclear shell effect" (Boh 74), then the answer is "yes—SHE have been discovered."

However, we believe that the term "superheavy element" connotes an element whose lifetime is strikingly longer than its neighbors in the chart of the nuclides. Thus we conclude that SHE cannot be said to have been discovered. If the measured half-life of element 110 is as long as predicted, this criterion would begin to be satisfied and we might be justified in referring to this as part of an "islet of the superheavy island."

In summary, it appears that the superheavy island has so far successfully resisted all of our attempts to land on it. Because the superheavy elements represent a unique and important test of our knowledge of nuclear structure and the dynamics of the reactions used to synthesize them, we do not want to give up. The problem of synthesizing them is more difficult than we imagined in the 1960s when this research began. But it has been possible to overcome extraordinary difficulties in the synthesis and study of elements 107, 108, and 109. We understand many of the reasons for our past failures. If improvements of orders of magnitude in the production and detection of superheavy reaction products are achieved, our current theoretical understanding of the superheavy elements and their properties provides a realistic basis for optimism. We therefore think it will be possible, eventually, to synthesize and identify the superheavy elements.

REFERENCES

Reviews

Arm 85 P. Armbruster, Annu. Rev. Nucl. Part. Sci. **35**, 135 (1985).
Arm 88 P. Armbruster and G. Münzenburg, Spektrum der Wissenschaft, September 1988.
Arm 89 P. Armbruster and G. Münzenburg, Sci. Am. *May*, 66, 1989.
Kra 83 J.V. Kratz, Radiochim. Acta **32**, 35 (1983).
Kum 89 K. Kumar, *Superheavy Elements* (Hilger, Bristol, 1989).
Mün 88 G. Münzenberg, Rep. Prog. Phys. **51**, 57 (1988).
Oga 85 Y.T., Oganessian and Y.A. Lazarev, *Treatise on Heavy Ion Science*, Vol. 4, D.A. Bromley, Ed. (Plenum Press, New York, 1985).
Sea 87 G.T. Seaborg and W. Loveland, Contemp. Phys. **28**, 33 (1987).

Research Papers

Arm 85a P. Armbruster, et al., Phys. Rev. Lett. **54**, 406 (1985).
Bem 77 C.E. Bemis and J.R. Nix, Comments Nucl. Part. Phys., **7**, 65 (1977).
Bir 79 J.R. Birkelund, L.E. Tubbs, J.R. Huizenga, J.N. De, and D. Sperber, Phys. Rep., **56**, 107 (1979).
Bjø 82 S. Bjørnholm and W.J. Swiatecki, Nucl. Phys., **A391**, 471 (1982).
Boh 74 A. Bohr, Physica Scripta **10A**, 52 (1974).
Cha 78 R.R. Chasman, in *Superheavy Elements*, M.A.K. Lodhi, Ed. (Oxford, Pergamon, 1978), p.387.
Dem 84 A.G. Demin, S.P. Tretyakov, V.K. Utyonkov, and I.V. Shirokovsky, Z.Phys., **A315**, 197 (1984).
Fis 72 E.O. Fiset and J.R. Nix, Nucl. Phys., **A193**, 647 (1972).
Gäg 80 H.W. Gäggeler, et al., Phys. Rev. Lett., **45**, 1824 (1980).
Gäg 84 H.W. Gäggeler, et al., Z. Phys., **A316**, 291 (1984).
Ghi 74 A. Ghiorso, J.M. Nitschke, J.R. Alonso, C.T. Alonso, M. Nurmia, G.T. Seaborg, E.K. Hulet, and R.W. Lougheed, Phys. Rev. Lett. **33**, 1490 (1974).
Hof 78 D.C. Hoffman, in *Superheavy Elements*, M.A.K. Lodhi, Ed. (Oxford, Pergamon, 1978), p. 89.
Hui 78 J.R. Huizenga, Lawrence Berkeley Laboratory Report LBL-7701 (1978).
Kel 70 O.L. Keller, J.L. Burnett, T.A. Carlson and C.W. Nestor, J. Phys. Chem. **74**, 1127 (1970).
Kel 77 O.L. Keller and G.T. Seaborg, Ann. Rev. Nucl. Sci. **27**, 139 (1977).
Kra 86 J.V. Kratz, et al., Phys. Rev., **C33**, 504 (1986).
Les 83 K.T. Lesko, S. Gil, A. Lazzarini, V. Metag, A. Scamster and R. Vandenbosch, Phys. Rev. **C27**, 2999 (1983).

Loj 88	Z. Lojewski and A. Baran, Z. Phys. **A329**, 161 (1988).
Lou 85	R.W. Lougheed, et al., Phys. Rev. **C32**, 1760 (1985).
Mat 76	G.J. Mathews and V.E. Viola, Jr., Nature, **261**, 382 (1976).
Möl 81	P. Möller and J.R. Nix, J.R., At. Data Nucl. Data Tables, **26**, 165 (1981).
Möl 86	P. Möller, G.A. Leander, and J.R. Nix, Z. Phys. **A323**, 41 (1986).
Möl 87	P. Möller, J.R. Nix, and W.J. Swiatecki, Nucl. Phys. **A469**, 1 (1987).
Moo 89	K.J. Moody, private communication (1989).
Mün 81	G. Münzenberg, S. Hofmann, F.P. Hessberger, W. Reisforf, K.H. Schmidt, J.H.R. Schneider, P. Armbruster, C.C. Sahm, and B. Thuma, Z. Phys. **A300**, 107 (1981).
Mün 84	G. Münzenberg et al., Z. Phys. **A317**, 235 (1984).
Mün 85	G. Münzenberg et al., Z. Phys. **A322**, 227 (1985).
Mün 86	G. Münzenberg et al., GSI Report 86-1 (1986) p. 29.
Mus 78	M.G. Mustafa in *Superheavy Elements*, M.A.K. Lodhi, Ed. (Pergamon, Oxford, 1978) p. 284.
Mye 66	W.D. Myers and W.J. Swiatecki, Nucl. Phys. **81**, 1 (1966).
Nix 72	J.R. Nix, Annu. Rev. Nucl. Sci., **22**, 65 (1972).
Oga 74	Y.T. Oganessian et al., Soviet. Phys., JETP Lett. **20**, 265 (1974).
Oga 75	Y.T. Oganessian, A.S. Iljinov, A.G. Demin, and S.P. Tretyakov, Nucl. Phys. **A239**, 353 (1975).
Pat 89	Z. Patyk, J. Skalski, A. Sobiczewski, and S. Cwiok, Nucl. Phys. **A502**, 591c (1989).
Pit 75	K.S. Pitzer, J. Chem. Phys. **63**, 1032 (1975).
Ran 76	J. Randrup, S.E. Larsson, P. Möller, A. Sobiczewski, and A. Lukasiak, Phys. Scr. **10A**, 60 (1976).
Rei 82	W. Reisdorf et al., Phys. Rev. Lett. **49**, 1811 (1982).
Rie 79	C. Riedel and W. Norenberg, Z. Phys. **A290**, 385 (1979).
Sch 82	M. Schädel et al., Phys. Rev. Lett. **48**, 852 (1982).
Sch 72	H.W. Schmitt and U. Mosel, Nucl. Phys. **A186**, 1 (1972).
Sch 89	E. Schwinn, B. Cramer, G. Ingold, V. Jahnke, D. Hilscher, M. Lehmann, and H. Rossner, Nucl. Phys. **A502**, 551c (1989).
Sie 86	A.J. Sierk, Los Alamos National Laboratory Report LA-UR-86-1175, March 1986.
Sob 89	A. Sobiczewski, Z. Patyk, and S. Cwiok, Phys. Lett. **B224**, 1 (1989).
Sto 80	R.G. Stokstad, W. Reisdorf, K.D. Hildenbrand, J.V. Kratz, G. Wirth, R. Lucas, and J. Poitou, Z. Phys. **A295**, 269 (1980).
Str 67	V.M. Strutinsky, Nucl. Phys. **A95**, 420 (1967).
Swi 82	W.J. Swiatecki, Nucl. Phys. **A376**, 275 (1982).
Ter 79	G.M. Ter-Akop'yan, H. Bruchertseifer, G.V. Buklanov, O.A. Orlova, A.A. Pleve, V.I. Cherpigin, and Choy Val Sek, Yad. Fiz **29**, 608 (1979); trans. Sov. J. Nucl. Phys. **29**, 312 (1979).

8

PRESENCE IN NATURE

8.1 NATURAL ABUNDANCES

The transuranium elements are called "the man-made elements" and their half-lives are short compared to the age of the universe. Therefore it may seem paradoxical to discuss their "natural abundance." This apparent paradox may be resolved if we consider three ideas: (a) the interaction of neutrons with existing deposits of uranium can lead to the formation of small quantities of such nuclides as ^{239}Pu and ^{237}Np; (b) the radionuclide ^{244}Pu cannot be extinct because of its long half-life, 80×10^6 years; and (c) the activities of man, such as nuclear weapons testing, the nuclear power industry, and the accidental destruction of transuranium radionuclide power sources have led to the input of substantial quantities of transuranium nuclides to the environment.

The solar system abundances for the actinide elements are taken from analyses of the primordial class of meteorites, the carbonaceous chondrites. Based upon isotope dilution data for C1 and C2 chondrites, a value of the thorium/uranium ratio of 3.64 is generally adopted (Bur 82). As discussed above, because of its long half-life, ^{244}Pu cannot be extinct, although the total amount of primordial ^{244}Pu on earth is less than 10 g. Direct measurement of this primordial ^{244}Pu in terrestrial materials is very difficult although an interesting result (of ^{244}Pu concentration in a rare earth mineral of 1 part in 10^{18}) has been obtained (Hof 71). In general, however, a compelling case for the existence of ^{244}Pu (an alpha and spontaneous fission emitter) in the early solar system is made by analysis of the products of its spontaneous fission, in particular, the xenon isotopes. One observes an overabundance of fission tracks relative to ^{238}U spontaneous fission in meteorite mineral grains rich in thorium, uranium, and the rare earth elements. Correlated with these tracks is a large excess of unshielded xenon isotopes not due to uranium fission. The relative yields of these unshielded xenon isotopes agrees with the measured xenon yields from the spontaneous fission of ^{244}Pu. The best estimate of the ^{244}Pu/^{238}U ratio in the early (4.6 aeons ago) solar system is 0.005–0.007 (Jon 87). Evidence for ^{244}Pu on the moon has been established by analysis of samples returned from the Apollo missions.

As early as 1942, only a short time after the first synthesis of ^{238}Pu and ^{239}Pu, ^{239}Pu was found in extremely small concentrations in the uranium mineral Canadian pitchblende (Sea 48). The half-life of ^{239}Pu of less than 25,000 years precludes a primordial origin for this nuclide. It must have been formed by the capture of neutrons by ^{238}U, resulting in ^{239}U which in turn decays by β^- emission through ^{239}Np to ^{239}Pu. The neutrons originate from either the spontaneous fission of ^{238}U, the slow neutron induced fission of ^{235}U, (α, n) reactions on elements of low atomic number or fission or spallation reactions induced by cosmic rays. The best estimate of the "natural" ^{239}Pu/^{238}U ratio is $\sim 3 \times 10^{-12}$ (KSM 86).

The amounts of the naturally occurring transuranium nuclides are so small that for all practical purposes any significant amounts found in nature are due to the activities of man.

8.2 NUCLEOSYNTHESIS OF THE TRANSURANIUM ELEMENTS

Despite the general absence of primordial transuranium elements in nature, it is interesting to consider what the primordial concentrations of these elements were and how they were formed. It is thought that the heaviest elements are produced during the final explosive stages of stellar evolution via rapid multiple neutron-capture reactions (*the r-process*) (B^{2}FH 57). "In this process, a nucleus captures neutrons via (n, γ) reactions whose rates are fast as compared with β-decay rates. Neutron captures continue until the rate for the (γ, n) reaction balances that for the (n, γ) reaction. At that point, the nucleus waits until β-decay allows it to again capture neutrons. The resulting path that the r-process follows is typically ten neutrons richer than the line of β-stability. The process terminates either when the neutron density and temperature fall low enough that the (n, γ) and (γ, n) reactions cease, or when fission occurs and prevents buildup to heavier nuclei." (Sch 82).

The site for the occurrence of the r-process is not clear. Among the

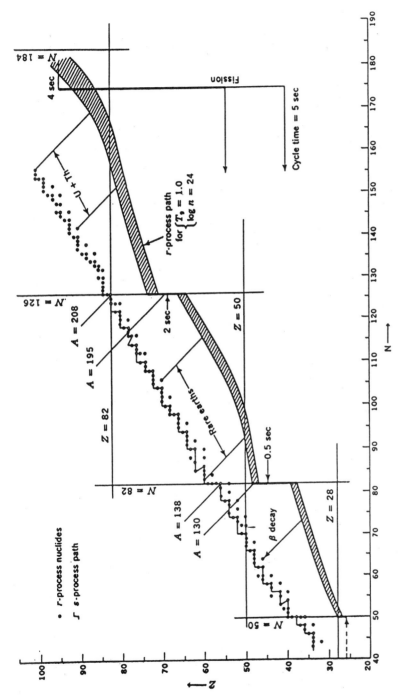

Figure 8.1 Neutron capture paths for the s-process and the r-process. The s-process follows a path in the NZ plane along the line of beta stability and is equivalent to the neutron capture process occurring in nuclear reactors. The neutron-rich progenitors to the stable r-process nuclei, which are here shown as small circles, are formed in a band in the neutron-rich area of the NZ plane, such as the shaded area shown here. After the synthesizing event the nuclei in this band beta-decay to the stable r-process nuclei (See 65). Reprinted courtesy of P.A. Seeger and *The Astrophysical Journal*, published by the University of Chicago Press, Copyright, 1965, The American Astronomical Society.

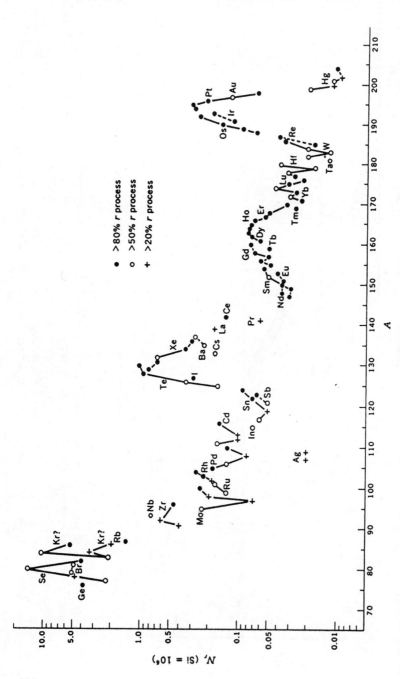

Figure 8.2 The solar system r-process abundances. Isotopes of a given element are joined by lines, solid lines for even Z and dashed lines for odd Z. The most important characteristics are the three main abundance peaks and the broad hump in the region of the rare earths (See 65). Reprinted courtesy of P.A. Seeger and *The Astrophysical Journal*, published by the University of Chicago Press. Copyright, 1965, The American Astronomical Society.

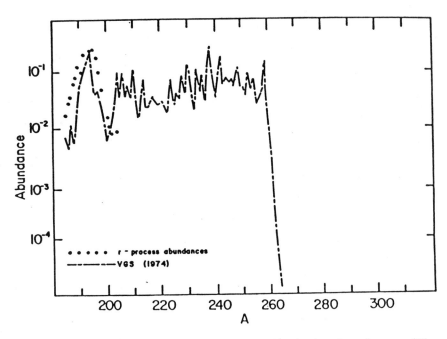

Figure 8.3 r-Process abundances before decay for the heaviest elements (Vio 78). (···, measured, ---, calculated using masses of Viola, Swant, and Graber (Vio 74).) Reprinted, with permission, from M..K. Lodhi, *Superheavy Elements*, Copyright, 1978, Pergamon Press.

possibilities are various regions of supernovae, neutron star-black hole collisions, and so on.

The neutron capture paths for the s- (slow) and r-process are shown in Figure 8.1 (See 65). The r-process nucleosynthesis path crosses the magic numbers $N = 50$, 82, and 126 at values of $A \sim 80$, 130, 195 leading to peaks in the abundance distributions at these mass numbers (Figure 8.2). The r-process path terminates eventually due to fission, which becomes faster than β-decay. The exact point at which this termination occurs is a sensitive function of the masses and fission barriers of the nuclei involved. Most calculations indicate that the termination occurs at $Z \sim 94$, $N \sim 170$–175 with the abundance pattern for the heaviest elements being similar to that shown in Figure 8.3. The last r-process nucleosynthetic event occurred $\sim 10^8$ years before the solar system was formed (Was 82).

8.3 SOURCES AND DISTRIBUTION OF MAN-MADE TRANSURANIUM ELEMENTS IN THE ENVIRONMENT

The transuranium nuclides which may be of interest as anthropogenic environmental contaminants are shown in Table 8.1. Of these, plutonium is the most prevalent, while ^{237}Np is the second most prevalent transuranium element. ^{241}Am results primarily from nuclear weapons testing; it is produced from the decay of ^{241}Pu. ^{244}Cm is least important because very little is present due to nuclear weapons testing fallout, although it is significant in nuclear power and plutonium production reactor wastes.

Plutonium is clearly the most significant transuranium element in the environment. The plutonium in the environment is due primarily to atmospheric testing of nuclear weapons, secondarily to the disintegration upon reentry of satellites equipped with ^{238}Pu power sources and lastly, to the processing of irradiated fuel and fuel fabrication in the nuclear power industry and the plutonium production program. During the period from 1950-1963, about 4.2 tons of plutonium (mostly a mixture of ^{239}Pu and ^{240}Pu) was injected into the atmosphere as a result of nuclear weapons testing. Because of the high temperatures involved, most of this plutonium was thought to be in the form of a refractory oxide. Most of this plutonium has been redeposited on the earth with concentrations being highest at the midlatitudes. Of the 350,000 Ci of ^{238}Pu and ^{239}Pu originally present in the atmosphere, about 1000 Ci remained in 1989. Approximately 9.7×10^6 Ci of ^{241}Pu were also injected into the atmosphere during weapons testing. When this completely decays, a total of $\sim 3.4 \times 10^5$ Ci of ^{241}Am will be formed. There is an additional ~ 1.4 tons of plutonium deposited in the ground (1989) due to surface and sub-surface nuclear weapons testing. Approximately 16,000 Ci of ^{238}Pu were injected into the

TABLE 8.1 Nuclear Properties of Long-lived Transuranium Elements which Serve as Environmental Contaminants

Eement	Isotope	Emission	Half-life
Neptunium	^{237}Np	α	2.1×10^6 years
Plutonium	^{238}Pu	α	87.7 years
	^{239}Pu	α	24,110 years
	^{240}Pu	α	6563 years
	^{241}Pu	β	14.4 years
Americium	^{241}Am	α	432.7 years
Curium	^{244}Cm	α	18.1 years

atmosphere when a satellite containing an isotopic power source disintegrated over the Indian Ocean in 1964. The amount of plutonium in the environment due to fuel reprocessing is small.

Over 99% of the plutonium released to the environment ends up in the soil, and in sediments. The global average concentration of plutonium in soils is 5×10^{-4}–2×10^{-2} pCi/g dry weight with most of the plutonium being near the soil surface. The concentrations of plutonium in water are quite low with an average concentration being $\sim 10^{-4}$ pCi/l. (Greater than 96% of any plutonium released to an aquatic ecosystem ends up in the sediments. In these sediments, there is some translocation of the plutonium to the sediment surface due to the activities of benthic biota). Less than 1% (and perhaps closer to 0.1%) of all the plutonium in the environment ends up in the biota. The concentrations of plutonium in vegetation range from 10^{-5} to 2%, with concentrations in litter and animals ranging from 10^{-4} to 3% and 10^{-8} to 1%, respectively. None of these concentrations has been observed to cause any discernible effect.

8.4 TRANSPORT AND FATE IN THE ENVIRONMENT
(Wat 83, Dah 76, KSM 86, All 82, Cle 83, Tra 76)

Most transuranium elements in the environment exist in a strongly adsorbed state on surface soils. The physical transport of these elements in the environment is mainly due to wind and water. Focussing on processes involving water, we find that the solubility (mobility) of these elements is highly dependent on the chemistry of the element and the chemical environment to which it is exposed.

Americium and curium remain in the +3 oxidation state over the normal range of environmental conditions. The chemistry of neptunium and plutonium is more complex because they display multiple oxidation states in aqueous solutions within the range of natural concentrations. Neptunium may exist as either Np (IV) or Np (V) and plutonium may be Pu (III), Pu (IV), Pu (V), or Pu (VI). For plutonium, Pu (III) is unstable to oxidation at environmental acidities and so the other three states are observed with Pu (IV) being the most stable in waters that contain significant amounts of organic material. In the absence of organic reductants, Pu (V) is the stable state. (Humic materials cause a slow reduction of Pu (V) to Pu (IV)). Under reducing conditions, neptunium should be present as Np (IV) and behave like Pu (IV); under oxidizing conditions, NpO_2^+ will be the stable species. In marine waters, Pu (IV) and the transplutonium elements will tend to undergo hydrolysis to form insoluble hydroxides and oxides. However, these elements can also form strong complexes with inorganic anions (OH^-, CO_3^{2-}, HPO_4^{2-}, F^- and SO_4^{2-}) and organic complexing agents that may be present. The speciation and solubility of these elements are largely determined by hydrolysis and formation of carbonate, fluoride, and phosphate complexes. Stable soluble species involve Pu (V, VI) and Np (V), although under most conditions the actinides will form insoluble species that concentrate in sediments.

Pu (IV), which forms highly charged polymers, strongly sorbs to soils and sediments. Other actinide III and IV oxidation states also bind by ion exchange to clays. The uptake of these species by solids is in the same sequence as the order of hydrolysis: Pu > Am (III) > U (VI) > Np (V). The uptake of these actinides by plants appears to be in reverse order of hydrolysis Np (V) > U (VI) > Am (III) > Pu (IV), with plants showing little ability to assimilate the immobile hydrolyzed species. The further concentration of these species in the food chain with subsequent deposit in man appears to be minor. Of the ~4 tons of plutonium released to the environment in atmospheric testing of nuclear weapons, the total amount fixed in the world population is less than 1 g [of this amount, most (99.9%) was inhaled rather than ingested].

REFERENCES

All 82 B. Allard, in *Actinides in Perspective*, N.M. Edelstein, Ed. (Pergamon, Oxford, 1982) p. 553.

B^2FH 57 E.M. Burbidge, G.R. Burbidge, W.A. Fowler, and F. Hoyle, Rev. Mod. Phys. **29**, 547 (1957).

Bar 82 C. A. Barnes, D.D. Clayton and D.N. Schramm, Eds. *Essays in Nuclear Astrophysics* (Cambridge, 1982).

Bur 82 D.S. Burnett, M.I. Stapanian, and J.H. Jones in Bar 82, pp 141-158.

Cle 83 J.M. Cleveland, T.F. Rees, and K.L. Nash, Science **222**, 1323 (1983).

Dah 76 R.C. Dahlman, E.A. Bondietti, and L.D. Eyman, in Fri 76.

Fri 76 A.M. Friedman, Ed. *Actinides in the Environment* (ACS, Washington, 1976).

Han 80 W.C. Hanson, Ed. *Transuranic Elements in the Environment* (USDOE, Washington, 1980).

Hof 71 D.C. Hoffman, F.O. Lawrence, J.L. Mcwherter, and F.M. Rourke, Nature **234**, 132 (1971).

Jon 87 J.H. Jones and D.S. Burnett, Geochim. Cosmochim. Acta **51**, 769 (1987).

KSM 86 J.J. Katz, G.T. Seaborg, and L. Morss, Eds., *The Chemistry of the Actinide Elements* (Chapman, London, 1986).

Sch 82	D.N. Schramm in Bar 82, pp 325–354.
Sea 48	G.T. Seaborg and M.L. Perlman, J. Am. Chem. Soc. **70**, 1571 (1948).
See 65	P.A. Seeger, W.A. Fowler, and D.D. Clayton, Astrophys. J. Suppl. **11**, 121 (1965).
Tra 76	*Transuranium Nuclides in the Environment* (IAEA, Vienna, 1976).
Vio 74	V.E. Viola, Jr., J.A. Swant, and J. Graber, Atom. Data Nucl. Data Tables **13**, 35 (1974).
Vio 78	V.E. Viola, Jr. and G.J. Mathews in *Superheavy Elements*, M.A.K. Lodhi, Ed. (Pergamon, New York, 1978) 499.
Was 82	G.J. Wasserburg and D.A. Papanastassiou, in *Essays in Nuclear Astrophysics*, C.A. Barnes, D.D. Clayton and D.N. Schramm, Eds. (Cambridge, 1982) pp 77–140.
Wat 80	R.L. Watters, D.N. Edgington, T.E. Hakonson, W.C. Hanson, M.H. Smith, F.W. Whicker, and R.E. Wilding in Han 80.
Wat 83	R.L. Watters, T.E. Hakonson, and L.J. Lane, Radiochim. Acta **32**, 89 (1983).

9

PRACTICAL APPLICATIONS

9.1 NUCLEAR POWER

An important practical application of uranium and the transuranium elements is the production of electric power from nuclear fission. While a detailed discussion of the subjects of nuclear reactors, the nuclear fuel cycle, and the disposal of radioactive waste is generally beyond the scope of this book, a few comments seem in order.

A typical nuclear reactor system is shown in Figure 9.1. The reactor consists of a core of fissionable material (containing UO_2, enriched to 3.3% ^{235}U) in which the chain reaction takes place. The energy released in the fission process, which is primarily in the form of the kinetic energy of the fission fragments, is absorbed as heat in the core. Also present in the core is the water moderator which slows down the fission neutrons to thermal energies where their probability of inducing another fission is greatest. A reflector helps to prevent neutrons from escaping from the core. The heat is removed from the core by a coolant and the chain reaction is controlled by rods of neutron absorbing material inserted into the core. In the pressurized water reactor (PWR), the heat energy produces steam for the turbine through the use of a heat exchanger, whereas in a boiling water reactor (BWR), the steam is produced for direct use in the turbine.

Reactors are classified as thermal or fast or intermediate according to the energy of the neutrons inducing nuclear fission in the core. A thermal

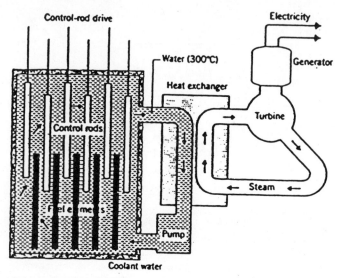

Figure 9.1 Schematic diagram of a pressurized-water nuclear power plant.

reactor is one in which the neutrons are in thermal equilibrium with the reactor materials. In a typical thermal power reactor, the neutron speeds are about 3000 m/s. At these speeds, the neutrons are readily absorbed by many materials. The reactor fuel in thermal reactors is primarily ^{235}U although ^{239}Pu is also used following its production by neutron absorption by ^{238}U. Ordinary water can act as both a moderator and coolant in thermal reactors. It is also possible to moderate thermal reactors using graphite or heavy water and cool them by the flow of a gas such as carbon dioxide or helium. In a fast reactor the average speed of the neutrons is about 15 million m/s. At these high speeds the probability of a neutron being absorbed or causing fission is lower. One fission event results in the release of about 200 MeV of energy, or about 3.20×10^{-4} erg which corresponds to 3.20×10^{-11} W-s. Thus 3.1×10^{10} fissions/s produces 1 W of power as heat. The fission of 1 g of uranium or plutonium per day liberates 0.96×10^3 kW, or about 1 MW. This is the energy equivalent of 3 tons of coal or about 600 gallons of fuel oil per day.

One remarkable fact about nuclear reactors is that they can produce their own fuel. If the total worldwide installed nuclear capacity is 4.4×10^5 MWe, one can estimate that about 140 tons of ^{239}Pu are produced each year. In such converter reactors the primary power source is the fission of ^{235}U. The more abundant uranium isotope, ^{238}U, which is always present, even in enriched uranium fuel, is converted to ^{239}Pu which could, in principle, be used to fuel another reactor. In the plutonium production reactors, such as those that were operated at Hanford, Washington, the ratio of ^{239}Pu produced to ^{235}U burned was about 0.8.

In "breeder reactors", which operate with a uranium blanket, it is possible to generate more plutonium than the plutonium that is used up as fuel in the core. The best candidate for a breeder reactor is one fueled with ^{239}Pu and operating with fast neutrons, a so-called "fast breeder" reactor. The smaller fission cross sections associated with the fast neutrons (as compared with thermal neutrons) lead to higher fuel concentrations in the core and higher power densities which in turn create significant heat transfer problems. These are compensated to some degree by a reduction of fission product poisons during reactor operation. Breeder reactors operating currently use 20% PuO_2–80% UO_2 fuel with liquid sodium as a coolant. (In a fast-breeder reactor, water cannot be used as a coolant because it will moderate the neutrons. Thus liquid sodium metal is a commonly chosen coolant.)

At the time of this writing (1989) recent developments in reactor technology have been directed towards producing new types of nuclear reactors—two types of advanced passive light water reactors (APLWR), the liquid metal reactor (LMR), and modular high temperature gas-cooled reactor (MHTGR). Passive safety features can be thought of as characteristics of a reactor which, without intervention of the human operator, will tend to shut a reactor down, keep it in a safe configuration, and prevent release of radiation to the public. These features fall into two broad categories—features which are designed to prevent accidents from taking place and those which mitigate the effects of potential accidents if they do happen.

The most developed is the Advanced Passive Light Water Reactor (APLWR), of which two designs are being developed. These two reactors are similar in that they are both smaller (600 MWe) in size, run at lower temperatures and with larger water inventories than current (1989) light water reactors, and have passive emergency core-cooling systems which utilize gravity, thus, eliminating the necessity for the large electrically driven emergency pumps common to the current generation light water reactors. They both use passive natural circulation for the removal of decay heat from the core. The two current (1989) designs are the AP-600 (Advanced Passive 600 MWe), a pressurized water reactor, and the SBWR (Simplified Boiling Water Reactor).

As of 1988, more than 430 nuclear power plants were operating around the world, generating more than 310,000 MW of electricity in 26 countries. Some countries depend vitally on the electricity generated by nuclear power. In 1988, France generated 70% of its electricity from nuclear power plants, Belgium 66%, South Korea 47%, Taiwan 41%, Sweden 47%, Finland 36%, and Japan 23%. Bulgaria generated 36% of its electricity from nuclear power, Hungary 49%, and Czechoslovakia 27%. Furthermore, although the United States was not a leader in percentage, it had the largest total electric output from nuclear power: 95,000 MWe from 108 plants, generating 20% of the U.S. electric power.

The chemical steps often used to recover the transuranium elements from irradiated fuel have been described in Chapter 3 as the PUREX process (Plutonium Uranium Recovery by EXtraction).

The important actinides in irradiated uranium fuel are uranium, neptunium, plutonium, americium and curium. The paths for the production of these radionuclides in a nuclear reactor are shown in Figure 9.2. Neutron capture by ^{235}U leads to the formation of ^{237}Np which in turn can, by (n, γ) and $(n, 2n)$ reactions, form ^{238}Pu and ^{236}Pu. ^{237}Np is an important, long-lived component of radioactive waste. ^{238}Pu is the largest contributor to the total alpha activity of plutonium in irradiated fuel. The daughters of ^{232}U produced by the decay of ^{236}Pu emit high energy gamma rays and are of concern in shielding during recycling plutonium. The largest amount of the plutonium present in irradiated fuel is as ^{239}Pu, formed by neutron capture by ^{238}U followed by beta decay of ^{239}U and ^{239}Np. Neutron capture by ^{239}Pu leads to ^{240}Pu, ^{241}Pu, ^{242}Pu, and ^{243}Pu. Decay of ^{241}Pu and ^{243}Pu lead to the formation of ^{241}Am and ^{243}Am. Neutron capture by ^{241}Am ultimately leads to the production of ^{242}Cm, which is the most intense alpha activity in spent fuel. Some heavier curium isotopes are also produced, as shown in Figure 9.2. The curium isotopes present in spent fuel decay back to plutonium, thus acting as the greatest source of the α-active ^{238}Pu in the high level waste from fuel reprocessing.

In a yearly operating cycle of a typical (1000 MWe) pressurized water reactor, the spent fuel at discharge contains about 25 MT of uranium and about 250 kg of plutonium (1 MT (one metric ton) = 1000 kg). Some 40% of the energy produced in the course of a nuclear fuel cycle comes from ^{239}Pu. (Since about 20% of the electricity generated in the United States in 1987 came from nuclear power plants, about twice as much electricity was generated from the synthetic element plutonium (200×10^9 kWh in 1987) as was generated from oil-fired electrical generating plants (110×10^9 kWh in 1987). Table 9.1 shows the typical activity levels of various actinides in such spent fuel.

The ingestion toxicities (expressed as water dilution volumes) of the actinides in spent fuel are shown in Figure 9.3. During the first 10 years, the toxicities are controlled by the fission products. After that the controlling activities are those of ^{241}Am, ^{240,239}Pu, and then ^{237}Np. The

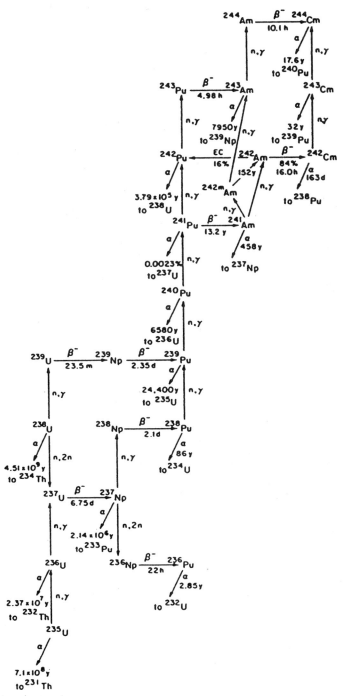

Figure 9.2 Nuclide chains producing plutonium, neptunium, americium, and curium in nuclear power reactors (from Ben 81).

TABLE 9.1 Actinides in Discharge Uranium Fuel[a]

Radionuclide	Half-life	kg/year	Ci/year
^{234}U	2.47×10^5 years	3.14	1.94×10^1
^{235}U	7.1×10^8 years	2.15×10^2	4.61×10^{-1}
^{236}U	2.39×10^7 years	1.14×10^2	7.22
^{237}U	6.75 days	9.15×10^{-7}	7.47×10^1
^{238}U	4.51×10^9 years	2.57×10^4	8.56
Total		2.60×10^4	$\alpha 3.56 \times 10^1$
			$\beta 7.47 \times 10^1$
^{237}Np	2.14×10^6 years	2.04×10^1	1.44×10^1
^{239}Np	2.36 days	2.05×10^{-6}	4.78×10^2
Total		2.04×10^1	$\alpha 1.44 \times 10^1$
			$\beta 4.78 \times 10^2$
^{236}Pu	2.85 years	2.51×10^{-4}	1.34×10^2
^{238}Pu	87.7 years	5.99	1.01×10^5
^{239}Pu	24,110 years	1.44×10^2	8.82×10^3
^{240}Pu	6563 years	5.91×10^1	1.30×10^4
^{241}Pu	14.4 years	2.77×10^1	2.81×10^6
^{242}Pu	3.73×10^5 years	9.65	3.76×10^1
Total		2.46×10^2	$\alpha 1.23 \times 10^5$
			$\beta 2.81 \times 10^6$
^{241}Am	432.7 years	1.32	4.53×10^3
^{242m}Am	141 years	1.19×10^{-2}	1.16×10^2
^{243}Am	7370 years	2.48	4.77×10^2
Total		3.81	$\alpha 5.01 \times 10^3$
			$\beta 1.16 \times 10^2$
^{242}Cm	163 days	1.33×10^{-1}	4.40×10^5
^{243}Cm	28.5 years	1.96×10^{-3}	9.03×10^1
^{244}Cm	18.1 years	9.11×10^{-1}	7.38×10^4
^{245}Cm	8500 years	5.54×10^{-2}	9.79
^{246}Cm	4700 years	6.23×10^{-3}	1.92
Total		1.11	$\alpha 5.14 \times 10^5$
Total		2.63×10^4	$\alpha 6.42 \times 10^5$
			$\beta 2.81 \times 10^6$

[a]Uranium-fueled 1000-MWe pressurized water reactor, 150 days after discharge (Ben 81).

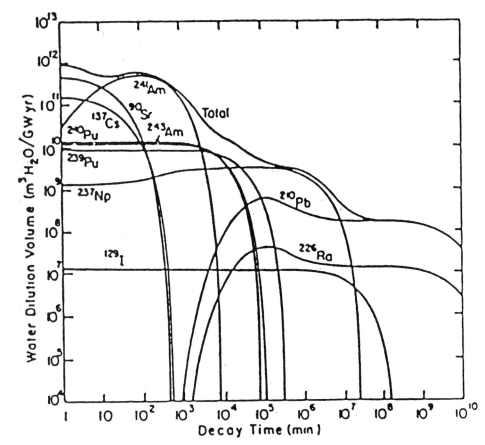

Figure 9.3 Principal contributions to the long term ingestion toxicity of spent fuel (fuel from uranium-fueled pressurized water reactor). The water dilution volume is the volume of water with which the radionuclides must be diluted so that drinking the water will result in an accumulated dose rate of 0.5 rem/year (from Cho 81). Note the abscissa in this figure is incorrectly labeled. The units should be years and not minutes.

ultimate, very long term toxicity is governed by the ^{238}U decay products, ^{226}Ra and ^{210}Pb. These estimates of the toxicity of spent fuel can be put in perspective relative to the hazards associated with other components of the nuclear fuel cycle (Figure 9.4). Without reprocessing, the toxicity of the spent fuel is the most important contribution to the toxicity associated with the fuel cycle for millions of years. Reprocessing only changes

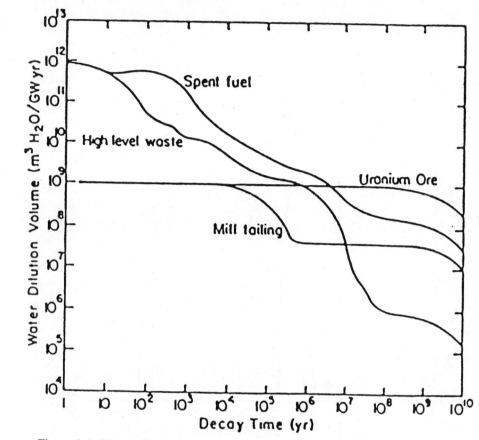

Figure 9.4 Water dilution volume for high level reprocessing waste spent fuel, uranium ore and mill tailings, all on the basis of 1 GW(e)year of electricity produced (Cho 81).

this situation modestly. The long-term toxicity of either the spent fuel or the high level waste from reprocessing is due to the transuranium nuclei.

There has been much attention given to the idea of chemically isolating the actinides from radioactive waste and burning them in a nuclear reactor. This effort will be enormous, requiring removal of more than 99.5% of the actinides for even the most modest reduction in toxicity. Transmuting the separated actinides will ultimately result in an increase in the residual transplutonium content of the reactor with its attendant problems. For example, it has been estimated that ^{252}Cf concentrations in irradiated fuel under such a program will increase until the neutron flux from the ^{252}Cf will be 10^{12} n/s per 1000 kg of heavy metal in the fuel (Ben 81).

Another important use of nuclear reactors is to supply power for motion for ships. Such shipboard propulsion systems are usually pressurized water reactors with energy conversion based on the steam-turbine cycle. Shipboard reactors are generally more compact than similar installations ashore. Special attention has to be given to maintenance, protection from collision, leakage, et cetera. The primary use of nuclear power has been in submarines and other naval vessels although a few nonnaval marine installations such as the Soviet icebreaker *Lenin* and the United States demonstration merchant ship NS *Savannah* have occurred.

The United States has built and launched (as of 1988) an impressive number of nuclear powered ships—a total of 155 attack and missile submarines, 5 aircraft carriers (with a total of 24 reactors) and 9 guided missile cruisers (18 reactors)—and has built and operated 9 land prototype reactors. Of this total (as of 1988), 22 submarines have been decommissioned (or are nonoperational) and 1 land prototype reactor is no longer operational.

9.2 NUCLEAR WEAPONS

The first nuclear weapon test at Alamogordo, New Mexico on 16 July, 1945 involved the use of a transuranium element, plutonium. As remarked earlier, the first public announcement of the existence of plutonium was the use of a nuclear weapon in World War II. Thus nuclear weapons and the transuranium elements have been inexorably linked since then. While a full discussion of nuclear weapons is beyond the scope of this book, some comments about the operating principles of such devices and their connection to the transuranium elements are desirable.

The techniques used to produce a "nuclear explosion" (i.e., an essentially instantaneous, self-perpetuating nuclear chain reaction) are very complex. A nuclear explosion must utilize a high energy neutron spectrum (fast neutrons). This results basically from the fact that, for an explosion to take place, the nuclear chain reaction must be very rapid—of the order of microseconds. Each generation in the chain reaction must occur within about 0.01 μs (a "shake") or less. The energy release takes place over many generations although 99.9% of the energy release occurs within the last 7 generations, that is, in a time of the order of 0.1 μs. The rapid time scale of this reaction requires the use of fast neutrons. The process by which a neutron is degraded in energy is time consuming and largely eliminates the possibility of an explosion. This also explains why power reactors that operate with a slow or thermal neutron spectrum cannot undergo a nuclear explosion, even if the worst accident is imagined. In the case of reactors that operate with higher energy neutrons, a nuclear explosion is also precluded because of the geometrical arrangement of the fissionable material and the rearrangement of this material if an accident occurs.

The explosive ingredients of fission weapons are limited, in practice, to ^{239}Pu and ^{235}U, because these are the only nuclides that are reasonably long-lived, capable of being produced in significant quantities, and also capable of undergoing fission with neutrons of all energies, from essentially zero or thermal to the higher energies of the secondary neutrons emitted in fission. Other nuclides—as, for example, ^{238}U or ^{232}Th—can undergo fission with some of these higher energy neutrons, but not with those of lower energy. It is not possible to produce a self-sustaining chain reaction with these nuclides, since an insufficient fraction of the neutrons produced in the fission reaction has an appropriate energy to induce, and hence perpetuate, the fission reaction.

Fission weapons currently use ^{239}Pu or highly enriched ^{235}U (usually greater than 90%) although, in principle, enrichments as low as 10% are usable. Fission weapons utilizing ^{239}Pu have higher yield to weight ratios and can be made with smaller sizes and weights. One problem in plutonium-based weapons is the presence of ^{240}Pu, whose high spontaneous fission rate can present problems with the preinitiation of the weapon. Preinitiation of the weapon is defined as the initiation of the nuclear chain reaction before the desired degree of supercriticality (see below) is achieved. The neutrons emitted during the spontaneous fission of ^{240}Pu can cause such a preinitiation which will decrease the yield of the weapon and increase the uncertainty in that yield. To prevent this preinitiation, weapons grade plutonium contains less than 7% ^{240}Pu while ordinary reactor grade plutonium may contain more than 19% ^{240}Pu. The ^{240}Pu content of plutonium can be regulated by controlling the time ^{238}U is left in the reactor for generating ^{239}Pu. Many United States fission weapons contain both ^{239}Pu and ^{235}U as a tradeoff between the higher efficiency of using ^{239}Pu and the greater availability of ^{235}U. If the conditions are such that the neutrons are lost at a faster rate than they are formed by fission, the chain reaction is not self-sustaining. The escape of neutrons occurs at the exterior of the ^{239}Pu (or ^{235}U) mass undergoing fission, and thus the rate of loss by escape will be determined by the surface area. On the other hand, the fission process, which results in the formation of more neutrons, takes place throughout the whole of the material; the rate of growth of neutron population is therefore dependent upon the mass. If the quantity of ^{239}Pu (or ^{235}U) is small, that is, if the ratio of the surface area to the volume is large, the proportion of neutrons lost by escape to those producing fissions will be so great that the propagation of a nuclear fission chain, and hence the production of an explosion, will not be possible. But as the size of the piece of ^{239}Pu (or ^{235}U) is increased and the relative loss of neutrons is thereby decreased, a point is reached at which the chain reaction can become self-sustaining. This is referred to as the "critical mass" of the fissionable material.

The critical mass of a bare sphere of normal density ^{235}U metal has been reported to be 52 kg while the same number reported for certain phases of plutonium metal is about 10 kg (Coc 84). However, the critical mass may be lowered in a number of ways. Use of a reflector can lower the critical mass by a factor of 2–3. Compression of the material to increase its density will also lower the value of the critical mass, with the critical mass being approximately proportional to the inverse square of the density. Most nuclear weapons employ only a fraction of the critical mass (at normal density). Because of the presence of stray neutrons in the atmosphere or the possibility of their being generated in various ways, a quantity of ^{239}Pu (or ^{235}U) exceeding the critical mass would be likely to melt or possibly explode. It is necessary, therefore, that before detonation a nuclear weapon should contain no single piece of fissionable material that is as large as the critical mass for the given condition. In order to produce an explosion, the material must then be made supercritical, that is, made to exceed the critical mass, in a time so short as to completely preclude a subexplosive change in the configuration, such as by melting.

Two general methods have been described for bringing about a nuclear explosion, that is to say, for quickly converting a subcritical system into a supercritical one.

In the first procedure, two or more pieces of fissionable material, each less than a critical mass, are brought together very rapidly in the presence of neutrons to form one piece that exceeds the critical mass. This may be achieved in some kind of gun-barrel device, in which a high explosive is used to blow one subcritical piece of fissionable material from the breech end of the gun into another subcritical piece firmly held in the muzzle end. The first nuclear weapons had a mass of ^{235}U in the form of a sphere with a plug removed from its center. The plug was then fired into the center of the sphere creating a supercritical assembly. This technique is largely of historical interest.

The second method makes use of the fact that when a subcritical quantity of an appropriate isotope, that is, ^{239}Pu (or ^{235}U), is strongly compressed, it can become critical or supercritical. The reason for this is that compressing the fissionable material, that is, increasing its density increases the rate of production of neutrons by fission relative to the rate of loss by escape. The surface area (or neutron escape area) is decreased, while the mass (upon which the rate of propagation of fission depends) remains constant. A self-sustaining chain reaction may then become possible with the same mass that was subcritical in the uncompressed state.

In a fission weapon, the compression may be achieved by encompassing the subcritical material with a shell of chemical high explosives, which is imploded by means of a number of external detonators, so that a uniform inwardly directed "implosion" wave is produced. The implosion wave creates overpressures of millions of pounds per square inch in the core of the weapon, increasing the density by a factor of two. A simple estimate may be made (Kra 88) to show that the resulting assembly should have a size of 10 cm, the mean free path of a fast neutron in ^{235}U or ^{239}Pu. The implosion technique is used in modern nuclear weapons.

In both methods, high density, heavy metals are used to surround the fissionable material, thereby reducing or preventing the escape of neutrons from the reacting assembly. To contain the fissionable material and insure that a large enough fraction of the nuclei undergo fission before the expansion of the exploding material causes subcriticality, the fissile material is surrounded by a heavy metal case which acts as a tamper (and a neutron reflector.)

In a thermonuclear or hydrogen bomb, a significant fraction of the energy release occurs by nuclear fusion rather than nuclear fission. The hydrogen isotopes, ^{2}H (deuterium, D) and ^{3}H (tritium, T), can be made to fuse, as

$$^2H + {}^3H \rightarrow {}^4He + n + 17\,\text{MeV}$$

To initiate such a D-T fusion reaction requires temperatures of 10–100 million degrees. Relatively large amounts of deuterium/tritium and/or lithium deuteride can be heated to such temperatures by a fission explosion. (Tritium is generated in situ by the neutron bombardment of ^{6}Li during the fusion reaction by the reaction $^6\text{Li} + n \rightarrow {}^3H + {}^4He + n + 17\,\text{MeV}$ thus making the overall fusion reaction $^6\text{Li} + {}^2H \rightarrow 2\,{}^4He + 21.78\,\text{MeV}$.)

The energy release can be enhanced further by using the high energy neutrons released in the fusion reactions to induce fission in the abundant isotope, ^{238}U. Thus, we have fission-fusion and fission-fusion-fission weapons, which can give rise to explosions of much greater energy than those from simple fission weapons.

In a typical modern multistage thermonuclear weapon, the radiation from a fission explosion is used to transfer energy and compress a physically separate component containing the fusion material. The fissile material is referred to as the primary stage while the fusion material is called the secondary stage. A third stage can be added in which the fast neutrons from the fusion reaction are used to initiate the fission of ^{238}U. In modern multistage thermonuclear weapons, comparable energy release is said to come from fission and fusion reactions (Coc 84).

A published schematic diagram of the operation of a modern multistage thermonuclear weapon is shown in Figure 9.5 (Mor 81). The fission stage is similar to the implosion weapon used over Nagasaki but is only 12 inches in diameter. The chemical explosives are arranged in a soccer ball configuration with 20 hexagons and 12 pentagons forming a sphere. Detonator wires are attached to each face.

In this example (Mor 81), the fusion reaction must take place before the expanding fireball of the exploding fission trigger destroys the fusion materials (that is, in a time scale of less than 100 shakes). This is accomplished through the use of x- and γ-radiation to transmit the energy of the fission reaction. The x- and γ-radiation travels about a hundred times faster than the exploding debris from the fission reaction to the fusion assembly. As shown in Figure 9.5, the thermonuclear weapon in this example is a 3–4 ft long cylinder with an 18-in diameter with the fission stage located near one end and the fusion stage near the other. The x- or γ-radiation is directed to a tamper of polystyrene foam which surrounds the fusion assembly. The radiation energy is absorbed by the polystyrene foam which is transformed into a highly energized plasma which compresses the fusion fuel assembly.

The "neutron bomb" or "enhanced radiation" weapon is a thermonuclear weapon in which the energy release in the form of heat and blast is minimized and the lethal effects of the high energy neutrons generated in fusion are maximized. This is reported to be done by the elimination of the ^{238}U components of the weapon (Coc 84). The suggested net effect of this is that the instantaneously incapacitating radius (dose of 8000 rad) of a neutron bomb is about the same as a fission weapon with ten times the yield. The instantaneously incapacitating radius for a one kiloton neutron bomb is thus about 690 m.

Nuclear weapon yields are measured in units of kilotons of TNT (1 kiloton of TNT = 10^{12} calories = the explosive energy release from 60 g of fissile material (Gla 67)). The first nuclear explosive device which was detonated at Alamogordo, New Mexico had a yield of about 20 kT as did the Fat Man bomb dropped over Nagasaki, Japan (both fueled by ^{239}Pu). The Little Boy bomb dropped over Hiroshima, Japan had a yield of 12–15 kT (fueled by ^{235}U). The efficiency of the plutonium based devices

314 PRACTICAL APPLICATIONS

Monsanto makes the electrically fired detonators surrounding the primary . . .

which set off the chemical high-explosive charges, made by Mason and Hanger–Silas Mason, that surround a hollow spherical tamper made of beryllium and uranium-238. The tamper, manufactured by Union Carbide, is liquefied by the implosive shock wave and driven inward toward the softball-sized fissionable core of the primary.

The core is compressed to supercriticality by the tamper, and a beam of high energy neutrons is fired from outside the casing by a high-voltage vacuum tube made by General Electric. The neutrons start a fission chain reaction in the plutonium-239 "pit" made by Rockwell.

The chain reaction spreads to a layer of uranium-235 surrounding the pit, and the heat and pressure of fission ignite a hydrogen fusion reaction in the "booster" charge of tritium and deuterium gas, supplied by Du Pont. Fusion adds neutrons to the fission reaction, speeding it up and raising its temperature.

The energy of the fission reaction races away from the primary in the form of x-rays which are momentarily trapped by the bomb's metal casing . . .

focused through a paper honeycomb shield, and absorbed by a special polystyrene foam "channel filler" made by Bendix, which serves as a thermal explosive encasing the secondary.

The exploding styrofoam compresses the secondary, which is filled with lithium-6 deuteride. A "spark plug" of uranium-235 or plutonium-239, running down its center, is compressed to supercriticality, and a second fission chain reaction thus begins to supply neutrons which convert lithium-6 into tritium.

The nuclear explosion of the spark plug generates the temperatures and pressures needed to fuse the newly created tritium with deuterium, showering the casing of the secondary with high-energy neutrons created by fusion. The neutrons cause uranium-238 in the casing of the secondary (called the "pusher") to undergo fission. The lithium and uranium parts are made by Union Carbide.

Figure 9.5 Schematic view of the operation of a hydrogen bomb according to Mor 73. From THE SECRET THAT EXPLODED by Howard Morland and Peter Garrison, Copyright 1981 by Howard Morland and Peter Garrison. Reprinted by permission of Random House, Inc.

was about 17% while the uranium-based device had an efficiency of about 1.3%. The smallest nuclear weapons have been reported to have weight that is about 0.5% of the Fat Man bomb (10,800 lb) and a total size of 25–30 in in length and 10–12 in in diameter, with explosive yields about 0.25 kT (Coc 84). Modern thermonuclear weapons with yields above 100 kT have yield/weight ratios of 1–3 kT/kg which is far from the theoretical maximum of 80 kT/kg (Coc 84).

9.3 RADIONUCLIDE POWER SOURCES

The fuels for radionuclide power sources should involve easily shielded, weakly penetrating radiation, reasonably long half-life, a specific power of 0.2 W/g or more, good corrosion resistance, insolubility in water, low cost and reasonably available material (Ma 83). Among the transuranium elements, oxides of the α-emitting nuclides ^{238}Pu ($t_{1/2}$ = 88 years), ^{242}Cm ($t_{1/2}$ = 163 days) and ^{244}Cm ($t_{1/2}$ = 18.1 years) are useful fuels. A few grams to kilograms of such nuclides, in appropriately shielded containers, provide intense sources of heat with power levels up to hundreds of watts, since the alpha particles are stopped very easily and their energy converted into heat. Using thermoelectric devices without moving parts, it is possible to convert this heat into usable electricity.

Such power sources are small, light weight, and rugged. One of their uses is in the SNAP (Space Nuclear Auxiliary Power) units which have been used to power satellites or more importantly, to power remote sensing instrument packages (Ato 87). SNAP sources (fueled by ^{238}Pu) served as the power sources for instrument packages on the five Apollo missions, the Viking unmanned Mars lander, and the Pioneer and Voyager probes to Jupiter, Saturn, Uranus, Neptune, and beyond.

Early versions of the SNAP power sources were designed to burn up into submicron-sized particles upon reentry into the earth's atmosphere. Following the controversy surrounding the burnup of a SNAP 9A (^{238}Pu) power source in 1963, the design concept was changed to require the power source to remain intact during reentry and after impact. The success of this design change was shown in the intact recovery of the SNAP-19 (^{238}Pu) heat sources from the ocean floor following an abortive launch of a weather satellite in 1968. In addition to space applications, radionuclide power sources have been used as terrestrial sources of energy wherever compact and long-lived sources, not requiring much maintenance, are needed such as in cardiac pacemakers. In the early 1970s ^{238}Pu batteries were used as the power sources for cardiac pacemakers. Over 3500 units were implanted and most remain functioning. Due to a lower cost and easier construction, the plutonium power sources have largely been supplanted by lithium batteries despite the shorter lifetime (~10 years) of the lithium battery compared to the plutonium battery (~30 years).

9.4 INDUSTRIAL APPLICATIONS OF THE TRANSURANIUM NUCLIDES (Fol 86)

Because of their short ranges in matter, α-emitting radionuclides can be used to measure and control the thickness of very thin samples (i.e., approximately 1 mg/cm^2 or less). Thus a number of α-emitting transuranium elements can be used in such devices. In addition, ^{241}Am, because of its soft 60 keV γ-ray and its relatively long half-life, can be used in thickness gauges designed to measure and control larger thicknesses. A common application of this type is the use of ^{241}Am to measure and control metal sheet and foil thicknesses, up to thicknesses of 5000 g/m^2, in cold rolling mills. Source strengths in such applications are of the order of 10–40 GBq. ^{241}Am sources are also used to measure and control the thickness of plate glass during manufacture, steel plate during hot rolling, and the wall thickness of steel pressure bottles.

The high specific ionization of α-emitting radionuclides such as the transuranium elements makes them useful in devices designed to analyze gases by ionization. A common example of this is the smoke detector in which ion currents generated by passing alpha particles (typically from ^{241}Am sources) through a reference cell of clean air and another cell containing the air to be tested are compared. When combustion occurs, heavily ionized atoms are produced. When these atoms enter the detector, they collide with the ions responsible for the ambient current and actually decrease the current. Unlike photoelectric detectors, these detectors do not detect the visible aerosol particles, but rather the invisible ions from combustion.

An unusual use of the transuranium elements involves their use in quantitative analysis of surfaces by Rutherford backscattering. When alpha particles from a transuranium nucleus strike a surface, they can be backscattered (through a large angle). In doing so, they lose energy with that energy loss being given as

$$\Delta E = E_\alpha \frac{4m_\alpha/M}{[1+(m_\alpha/M)]^2}$$

where E_α is the incident alpha energy, m_α is the α-particle mass and M

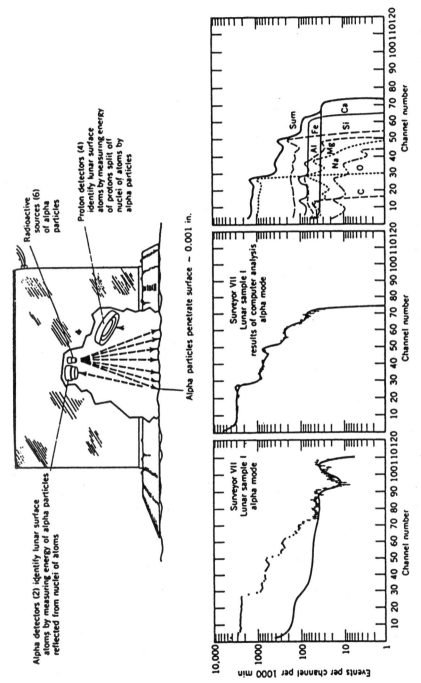

Figure 9.6 (Top) The α-particle scattering experiment from the Surveyor Moon Lander (from Tur 73). Bottom: Data from the α-scattering experiment on the Moon. The figure on the left shows the raw data; the solid line is background from a naturally occurring alpha source. The data with background subtracted is shown in the center, and the analysis showing the constituents is on the right. The composition is dominated by oxygen and silicon, as are earth rocks (from Pat 69).

is the mass of the scattering atom. By measuring the loss in α-particle energy, ΔE, one can determine the mass (and identity of the scattering atom). While the technique is most sensitive for light atoms, measurable values of ΔE (~ 0.5 MeV) can result from scattering from heavier nuclei ($m_\alpha/M \sim 0.02$). In Figure 9.6 we show the use of this technique to study the composition of the lunar surface (Tur 73).

Among the transuranium elements which have important practical uses, ^{252}Cf is one of the most important (Jan 83). ^{252}Cf has found widespread application because of its unique properties as a neutron source. A small fraction of all ^{252}Cf decays by spontaneous fission with the emission of neutrons. As a consequence, a source of ^{252}Cf emits 2.311×10^{12} n/s per g. Because of this high neutron specific activity, ^{252}Cf neutron sources are quite compact. (A 10^9 n/s source has a small heat output, of the order of 1.6 W.)

The most important industrial use of ^{252}Cf is as a startup source for nuclear reactors. To calibrate the instruments in a nuclear reactor core and to aid in the approach to criticality, a small radioactive neutron source is brought near the core. Such sources are called startup sources. ^{252}Cf neutron sources have replaced the use of Pu-Be neutron sources because of their small physical size, low heat generation, and lower gas generation.

The second most important use of ^{252}Cf is in nuclear reactor fuel rod scanners. The scanners use activation by the neutrons emitted by ^{252}Cf to measure the uniformity of the fissile material in the fuel rods and the total content of fissile material. Another important use of ^{252}Cf is as a neutron source for neutron activation analysis. ^{252}Cf sources used in this application have typical source strengths of 10^9–10^{10} n/s. The emitted neutrons are fast neutrons and usually are thermalized for most applications by the use of a water moderator. Typical thermal neutron fluxes are two orders of magnitude less than the primary, fast flux. As such, the thermal neutron fluxes obtainable from ^{252}Cf are small by comparison to those available from nuclear reactors but the sources offer some advantages for certain applications of neutron activation analysis. For example, the portability of the ^{252}Cf source allows it to be used in field studies for in situ activation analysis without disturbing the material being analyzed. An example of this application is borehole logging in geology. The source can be lowered into an existing borehole and the formation outside the borehole can be analyzed by radiation detectors mounted near the source. Another common example of the use of ^{252}Cf sources in industry is to do continuous monitoring of some process stream using activation analysis.

In an important but unusual application, ^{252}Cf sources have been used for human cancer radiotherapy (Mar 83). Implants of ^{252}Cf have been used successfully in the treatment of cancer of the cervix by such neutron radiotherapy.

9.5 PRODUCTION OF THE TRANSURANIUM ELEMENTS (Kel 83)

The premier facility for the preparation of transuranium elements has been the HFIR/TRU facility at Oak Ridge, Tennessee, U.S.A. The High Flux Isotope Reactor (HFIR), a light water moderated reactor, has used 93% enriched ^{235}U as its primary fuel to produce a neutron flux of 2–5 × 10^{15} n/cm^2 s. Because of its long operation span (since 1966), the HFIR uranium fuel rods contain (1989) appreciable amounts of plutonium, americium, and curium, making it a very efficient source of transuranium elements. In addition, special target assemblies can be used.

In this connection, one should note that the plutonium production reactors at Savannah River were diverted in the mid-1960s to the production of 3 kg of ^{244}Cm and large quantities of heavier nuclides through the irradiation of plutonium isotopes at a neutron flux exceeding 2 × 10^{15} n/cm^2 s. This enormous quantity of ^{244}Cm and the heavier nuclides also produced have served as special source material for further neutron irradiations at HFIR and as an independent source of heavy elements. Adjoining the HFIR is a Transuranium Processing Plant (TRU) in which the highly radioactive HFIR targets have been processed and the resulting transplutonium elements separated, purified and distributed to the research community. When the HFIR/TRU operates, it routinely produces significant quantities of the isotopes of curium through fermium (see Table 5.1).

REFERENCES

Ato 87	*Atomic Power in Space*, USDOE report DOE/NE/32117-H1, March 1987.
Ben 81	M. Benedict, T. Pigford, and H. Levi, *Nuclear Chemical Engineering* 2nd edn. (McGraw-Hill, New York, 1981).
Coc 84	T.B. Cochran, W.M. Arkin, and M.M. Hoenig, *Nuclear Weapons Databook*, Vol. I (Ballinger, Cambridge, 1984).
Cho 81	J. Choi and T.H. Pigford, Trans. Am. Nucl. Soc. **39**, 176 (1981).
Fol 86	G. Foldiak, Ed. *Industrial Application of Radioisotopes* (Elsevier, Amsterdam, 1986).
Gla 67	S. Glasstone, *Sourcebook on Atomic Energy*, 3rd edn. (Van Nostrand Reinhold, New York, 1967).
Jan 83	E.F. Janzow, Industrial Usage of Californium-252 in *Opportunities and Challenges in Research with Transplutonium Elements*, (National Academy Press, Washington, 1983).
Kel 83	O.L. Keller, *ibid*.

Kra 88   K.S. Krane, *Introductory Nuclear Physics* (Wiley, New York, 1988).

Ma 83   B.M. Ma, *Nuclear Reactor Materials and Applications* (Van Nostrand Reinhold, New York, 1983).

Mar 83   Y. Maruyama, Cf-252, New Radioisotope for Human Cancer Therapy in *Opportunities and Challenges in Research with Transplutonium Elements* (National Academy Press, Washington, 1983).

Mor 81   H. Morland, The Progressive, November 1979; *The Secret That Exploded* (Random House, New York, 1981).

Pat 69   J.H. Patterson et al., J. Geophys. Res. **74**, 6120 (1969).

Tur 73   A.L. Turkevich, Acc. Chem. Res. **6**, 81 (1973).

10

REFLECTIONS

The publication of this book follows closely the 50th anniversary of the discovery of nuclear fission and comes close to the 50th anniversary of the discovery of the first transuranium elements, neptunium and plutonium. One of the authors entered the field five years earlier than this. As a first year graduate student at Berkeley in 1934, he began to read the papers coming out of Italy and Germany describing the synthesis and identification of several elements thought to be transuranium elements. Perplexed and uneasy about this interpretation of those early experimental results, it was only natural that he should plunge into the field soon after these nuclides were shown to be fission products.

After the breakthrough into the real transuranium elements, in 1940, by McMillan and Abelson, the next higher element, plutonium was synthesized and identified at Berkeley within a matter of months. Quoting from an earlier book (Sea 58), "For many reasons this unusual element holds a unique position among the chemical elements. It is a synthetic element... the first synthetic element to be seen by man...It has unusual and very interesting chemical properties...It was discovered and methods for its production were developed during the last war, under circumstances that make a fascinating and intriguing story." The earlier book goes on to describe how "These methods for production were developed during the last war" at the University of Chicago's Metallurgical Laboaratory, including this author's role in the development of the chemical processes for plutonium's production and in the discovery there of the next two transuranium elements, americium and curium (95 and 96).

321

Following his return to Berkeley after World War II, the next six transuranium elements, berkelium through nobelium (97–102), were discovered during the following twelve years (1946–1958). Following a short stint as Chancellor of the Berkeley campus (1958–1961), he was called to Washington to serve as Chairman of the U.S. Atomic Energy Commission for ten years (1961–1971). During this time three more transuranium elements, lawrencium through hahnium (103–105) were discovered by Ghiorso and colleagues at Berkeley. As AEC Chairman his continuing interest in these elements encouraged unique advances in the transuranium field — for example, the dedication of plutonium production facilities at the Savannah River Plant and the use of the HFIR-TRU facilities at Oak Ridge for the production of an extraordinary quantity of transplutonium nuclides, and the dedication of a number of underground nuclear explosions in Nevada exclusively to the production and study of transuranium nuclides.

Although he resumed transuranium research upon his return again to Berkeley in 1971, only one additional transuranium element was discovered, element 106 in 1974. The initiative shifted to the GSI Laboratory in Germany where three more transuranium elements, 107, 108, and 109, were discovered in the 1980s. Investigators at the Dubna Laboratory in the U.S.S.R. have also been very active in the transuranium field and have made competing claims for the discovery of elements 104–109.

The greatest impact that the transuranium elements have had on chemistry was the recognition, in 1944, that their chemical properties demanded a change in the periodic table. It was recognized that the elements heavier than actinium should constitute an "actinide" series, analogous to the "lanthanide" elements. This concept had great predictive value, and its use to devise chemical identification procedures for many of the transuranium elements was the key to their discovery. The chemical properties of the transuranium elements through element 105 have been measured and found to be in conformance with this form of the periodic table. The validity of this concept makes possible the classification of the chemical properties of the undiscovered elements, through the noble gas, element 118.

The study of the transuranium elements has led to increased understanding of radioactive decay (especially spontaneous fission), nuclear structure, and nuclear reactions. Much of our understanding of alpha decay has come through studies of the decay of the transuranium elements. Spontaneous fission is almost exclusively a decay mode of these elements. It was the observation of a spontaneously fissioning isomer in ^{242}Am that led to the Strutinsky method of calculating the shell corrections to the liquid drop fission barrier. This development, in turn, led to the postulation of the existence of and discovery of the double-humped fission barrier with its accompanying implications for the stability of heavy nuclei and related effects upon the fission process. The general idea of deformation dependent shell corrections has had important implications in other areas of nuclear structure. The search for superheavy elements continues to be one of the severest tests of the predictive power of nuclear structure theories.

Turning to the future, the prediction of the form that a further extension of the periodic table would take seems relatively straightforward. Common-sense predictions, and calculations using modern supercomputers, suggest that the 8s electronic subshell should fill at elements 119 and 120, thus making these an alkali and alkaline earth metal, respectively. Next, a new inner transition series is expected to begin at element 121, a "superactinide" series (Sea 68). According to calculations, this series is expected to be more complicated than the actinide series because 6f, 5g, 8p, and 7d electrons will be added in mixed configurations (Table 3.3). But, for simplicity, it is common to retain the symmetry of the periodic table as much as possible (see previous comments). Therefore the superactinide series is said to end at element 153 with a "standard" 7d–8p series going from element 154 to element 168. Element 168 would be a noble liquid because its boiling point is predicted to be above room temperature. The extension to element 168 is, of course, far beyond the range of expected nuclear stability.

Prediction of the nuclear properties and the prospects for the synthesis and identification of elements beyond 109, which depend crucially on the efficacy of nuclear synthesis reactions, is less straightforward. It is interesting to recall the concluding paragraph of the earlier (1958) book (Sea 58):

> In summary, we can see that the synthesis and identification of future synthetic elements present a real challenge. By the time elements 104 and 105 are reached, we shall probably find that the longest-lived isotopes which can be made will exist barely long enough to permit traditional chemical identification. We can be quite sure that we shall be relying entirely on isotopes containing odd nucleons to make such chemical identifications, if these are possible at all for such high atomic numbers. It is likely that the present criteria and basic requirements for the discovery of a new element, namely complete chemical identification and separation from all previously known elements, will have to be changed at some point. Careful measurements of decay properties and determination of complete excitation functions and reaction mechanisms and the clever use of recoil techniques with the chemical identification of daughter isotopes should make it possible to make effective and satisfactory identification of isotopes with very short half-lives. The decay properties may have to be measured at the target area, on

recoil-product nuclei, during the bombardment. Reasonable, direct chemical (or equivalent) identification of the new element isotope can probably be made in some cases by using simple and fast methods involving migration of gaseous atoms or ions, volatility properties, reactions with surfaces, or gas-flow reactions. The identification of the first isotopes of all the new elements that will be discovered from now on probably will be accomplished through the use of such methods, and the production of isotopes of these new elements with sufficiently long half-lives to allow chemical identification by traditional methods, if possible at all, will follow later. Whichever method is used, acceptable evidence for the discovery of new transuranium elements should consist of reasonable establishment of the atomic number and *this demands more than the observation of predicted decay properties and yields*. Only the future can tell us if it will be necessary to settle for less than this criterion when the production of elements substantially farther up the atomic number scale is under investigation some years from now.

We would be more than satisfied if we could equal the prescience of these predictions. Thirty years ago the heaviest bombarding ion that had been used in the synthesis and identification of a new element (102, nobelium) was carbon. In the intervening years heavier ions have been used—nitrogen (for element 105) and oxygen (for element 106) and more recently, with closed shell target nuclei in the lead-bismuth region, chromium (for element 107) and iron (for elements 108 and 109). Fortunately, these experiments have established that the heaviest nuclei—from element 104 on—are much more stable (by many orders of magnitude) against decay by spontaneous fission than predicted by the simple liquid drop model. Also, in the intervening years, we have been blessed by the predictions of increased nuclear stability in the region of atomic number 114 and neutron number 184 and more recently, in the region of neutron number 162. The nemesis is still loss of the desired product (the new element) through competition with fission during the production process. This process becomes increasingly limiting as the atomic number of the desired product nucleus increases, leading to survival rates as low as one part in a billion or less.

Identification of the atomic number of putative new elements by establishing genetic relationships to known descendents by α-decay will probably continue to be used. Some special properties of the decay of superheavy elements may also be used if we are able to synthesize them.

Experimental studies of the chemistry of the heaviest elements will require the use of new techniques suited to the half-lives involved. For the first members of the transactinide series, we already know that the techniques of gas chromatography and automated rapid high-performance liquid chromatography serve us well. Using these relatively conventional methods it should be possible to determine for element 106 whether, and to what extent, its chemistry resembles that of tungsten. However for elements 107 and beyond, the response times of the techniques used in chemical studies may have to be milli- or micro- or nanoseconds. Such response times are well within the capabilities of modern laser-based techniques but special work will be needed to adopt these techniques to samples containing only a few atoms. Other possible techniques involving the use of volatility properties, reactions with liquids or surfaces, gas-flow reactions, the migration of gaseous atoms or ions, et cetera may come into play.

Such studies will require increased quantities of transactinide nuclei. We would like to see the development and construction of more powerful and versatile high flux reactors which will increase the supply of transplutonium nuclei, resulting in milligram amounts of ^{254}Es and nanogram amounts of ^{257}Fm. This increased availability of einsteinium and fermium will allow study of their macroscopic properties. Heavy ion accelerators and target technologies that would allow the use of heavy ion beams that are orders of magnitude more intense than those available today are desirable. Along with these developments, we look forward to the production of secondary radioactive beams of very neutron-rich nuclei in such accelerator complexes. Under such circumstances, it should be possible to synthesize superheavy nuclei, and to allow the identification of another half dozen chemical elements. We think these developments would be best done as part of an Institute or Laboratory for Transuranium Element Research, or several such Institutes or Laboratories.

While the developments outlined above will permit meaningful studies of the nuclear properties of the transuranium elements, we look forward to the most progress in the immediate future being made in studying the chemistry of the transactinide elements. This is a fascinating field that has been neglected until recently. There are many important chemical and physical questions to be explored.

The area of future practical applications is also worth considering.

There will be expanding applications of radionuclides made available in quantity through the use of nuclear reactors, fueled more by ^{235}U than by by-product ^{239}Pu. These include applications in industry, in agriculture, in the humanities, as an energy source in space and especially in medicine.

The present electricity-producing nuclear power reactors in the United States, and in most other countries, will continue to operate over their design lifetimes. We hope for a return in the United States and elsewhere to nuclear power as an option for needed future expansion in electric-generating capacity. (Some countries, such as France, never yet indulged in a hiatus in their dependence on this source of energy.) There are many reasons for such a return to this safe and reliable source of energy, such

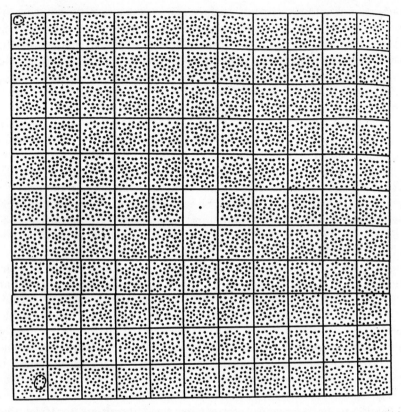

Figure 10.1 A realistic view of the world armaments situation. The chart shows the world's current firepower in terms of the firepower of World War II. The dot in the center square represents all the firepower of World War II (including the atomic bombs dropped on Hiroshima and Nagasaki): three megatons. The other dots represent the world's present nuclear weaponry. This comes to 18,000 megatons, which equals 6000 World War IIs. The US (and allies) and the SU (and allies) share this firepower approximately equally. The top lefthand circle enclosing nine megatons represents the weapons in just one Poseidon submarine. This is equal to the firepower of three World War IIs and is enough to destroy over 200 of the Soviet Union's largest cities. We have 31 such submarines and ten similar Polaris submarines. The bottom lefthand circle enclosing 24 megatons represents one new Trident submarine with the power of eight World War IIs: enough to destroy every major city in the Northern hemisphere. The Soviets have similar levels of destructive power. Just two squares on this chart (300 megatons) represent enough firepower to destroy all the large and medium size cities in the entire world (from Hof 85).

as avoidance of acid rain and air pollution, need to conserve fossil fuels as a source of chemicals, need to conserve oil as fuel for transportation, and increasingly, the need to avoid a worldwide catastrophe resulting from the "greenhouse" effect. However, we do not foresee any contribution from the operation of "breeder" reactors for decades to come.

In the long run, nuclear power reactors will surely be used in the exploration of the further reaches of space.

We conclude with an optimistic word about the "Sword of Damocles" that hangs over us all. Plutonium, unfortunately, serves as the efficient fuel for nuclear weapons. Produced as a by-product in nuclear power reactors, it can, in principle, be diverted for such weapons use. We believe that a more concerted support of the Nonproliferation Treaty (NPT) and greater support of the application of "safeguards" by the International Atomic Energy Agency (IAEA) can prevent such diversion. (We do not believe that the United States should forego nuclear power in the interest of nonproliferation—to the contrary, it is better to stay in the playing field to be better prepared to monitor and prevent proliferation.)

The greater danger is in the possibility of nuclear war between the superpowers. Such a catastrophe is almost unimaginable (see Figure 10.1), but a reality we must face. We believe that the process of nuclear arms control can prevent this war. But this process is necessary to achieve this goal. Arms control is different from disarmament. This does not seek to abolish nuclear weapons, which, realistically, is hardly possible at present and may never again be possible. Arms control seeks to defuse nuclear arms competition in various ways, such as eliminating the more provocative and threatening weapons or deployments, and by preventing dangerous, further stages in the arms race. Through this means, arms control aims to change the messages that nations send to each other and to bring about the conditions for political change and ultimately, perhaps a large degree of nuclear disarmament. The advent of "glasnost" and "perestroika" in the Soviet Union is a source of immense encouragement.

Above all, the purpose of arms control is to reduce and eliminate the chance of nuclear war. We believe that it will succeed, but only with the help of all of us.

REFERENCES

Hof 85   D.R. Hofstadter, *Metamagical Themas* (Basic, New York, 1985).
Sea 58   G.T. Seaborg, *The Transuranium Elements* (Yale University Press, New Haven, 1958) p. 1, pp. 293–294.
Sea 68   G.T. Seaborg, Annu. Rev. Nucl. Sci. **18**, 53 (1968).

APPENDIX I

Tables of Radioactive Decay Properties
of the Transuranium Nuclei*

*An updated version of the tables of I. Ahmad, *The Chemistry of the Actinide Elements*, 2nd Edn., J.J. Katz, L.M. Morss and G.T. Seaborg, Eds. (Chapman, London, 1986). All intensities are given on an absolute scale.

TABLE A1 Radioactive Decay Properties of Neptunium Isotopes

Mass Number	Half-life	Mode of Decay	Main Radiations (MeV)		Typical Method of Production
227 or 228	60 s	SF			^{22}Ne on ^{209}Bi
229	4.0 min	α	α	6.89	^{233}U (p, 5n)
230	4.6 min	α >3% EC <97%	α	6.66	^{233}U (p, 4n)
231	48.8 min	EC 98% α 2%	α γ	6.28 0.370	^{233}U (d, 4n)
232	14.7 min	EC	γ	0.327 (52%)	^{235}U (d, 6n)
233	36.2 min	EC <99% α ~10^{-3}%	α γ	5.53 0.312	^{233}U (d, 3n)
234	4.4 days	EC	γ	1.559 (18.4%)	^{233}U (d, 2n)
235	1.085 years	EC 99.9986% α 1.4×10^{-3}%	α	5.021 (0.00074%) 5.005 (0.00034%)	^{235}U (d, 4n)
236	1.55×10^5 years	EC 91% β^- 8.9% α 0.20%	γ	(0.160 (28%))	^{235}U (d, 3n) ^{235}U (d, 2n) ^{235}U (d, n)
236 m	22.5 h	EC 52% β^- 48%	γ	0.642 (0.9%)	^{235}U (d, n) ^{235}U (α, p2n)
237	2.14×10^6 years	α, SF ≤$(2.14 \times 10^{-10}$%)	α	4.799 (47%)	^{237}U daughter

238			4.771 (25%)	^{241}Am daughter
		γ	0.086 (12.3%)	
	2.117 days	β^-	1.29	^{237}Np (n, γ)
		γ	0.984 (27.8%)	
239	2.355 days	β^-	0.72	^{243}Am daughter
		γ	0.106 (27.2%)	^{239}U daughter
240	1.032 h	β^-	2.09	^{238}U (α, pn)
	7.4 min	γ	0.153, 0.271	
240 m	7.22 min	β^-	2.05	^{240}U daughter
		γ	0.555 (22%)	^{238}U (α, pn)
241	13.9 min	β^-	1.31	^{238}U (α, p)
		γ	0.175	^{244}Pu (n, p3n)
242 g	5.5 min	β^-	2.7	^{244}Pu (n,p)
		γ	0.786	^{244}Pu (n, p2n)
242 m	2.2 min	β^-	2.7	^{242}U daughter
		γ	0.736	
243	1.85 min	β^-	0.288	^{244}Pu(^{136}Xe, X)
244	2.29 min	β^-	0.217 (~39%)	^{244}Pu(^{136}Xe, X)
		γ	0.681 (~37%)	
		γ	0.163 (~34%)	

TABLE A2 Radioactive Decay Properties of Plutonium Isotopes

Mass Number	Half-life	Mode of Decay		Main Radiations (MeV)	Typical Method of Production
232	34 min	EC > 80%	α	6.60(62 rel)	^{233}U (α, 5n)
		α < 20%		6.54(38 rel)	
233	20.9 min	EC 99.88%	α	6.30	^{233}U (α, 4n)
		α 0.12%	γ	0.235	
234	8.8 h	EC 94%	α	6.200(4%)	^{235}U (α, 5n)
		α 6%		6.149(1.9%)	
235	25.3 min	EC 99.9973%	α	5.85	^{235}U (α, 4n)
		α 2.7 × 10^{-3}%	γ	0.049	^{233}U (α, 2n)
236	2.9 years	α	α	5.768(68.1%)	^{235}U (α, 3n)
		SF 8.1 × 10^{-8}%		5.721(31.7%)	^{236}Np daughter
237	45.17 days	EC 99.9958%	α	5.65(0.0007%)	^{235}U (α, 2n)
		α 4.2 × 10^{-3}%		5.33(0.0015%)	^{237}Np (d, 2n)
			γ	0.059(3.25%)	
238	87.7 years	α, SF	α	5.499(71.0%)	^{242}Cm daughter
		(1.9 × 10^{-7}%)		5.457(28.8%)	^{238}Np daughter
239	2.411 × 10^4 years	α, SF	α	5.157(73.1%)	^{239}Np daughter
		(4.4 × 10^{-10}%)		5.144(15.0%)	
			γ	0.129	

240	6.563×10^3 years	α, SF $(5.7 \times 10^{-6}\%)$	α	5.168(72.9%) 5.124(27.0%)	Multiple n capture from ^{238}U
241	14.4 years	β^- 99.99759% α 2.4×10^{-3}%	α β^- γ	4.897(0.002005%) 4.853(0.000292%) 0.021 0.149	Multiple n capture
242	3.73×10^5 years	α, SF (0.000550%)	α	4.900(76.5%) 4.856(23.5%)	Multiple n capture
243	4.956 h	β^-	β^- γ	0.581 0.084(23%)	Multiple n capture
244	8.00×10^7 years	α, SF (0.12%)	α	4.589(80.5%) 4.546(19.4%)	Multiple n capture
245	10.5 h	β^-	β^- γ	1.281 0.328(25%)	^{244}Pu (n, γ)
246	10.85 days	β^-	β^- γ	0.374 0.224(23%)	^{245}Pu (n, γ)

TABLE A3 Radioactive Decay Properties of Americium Isotopes

Mass Number	Half-life	Mode of Decay	Main Radiations (MeV)		Typical Method of Production
232	55 s	EC, α, EC-SF			^{230}Th (^{10}B, 8n)
234	2.3 min	EC, α, EC-SF	α	6.46	^{230}Th (^{10}B, 6n)
					^{10}B, ^{11}B on ^{233}U
237	1.22 h	EC 99.975%	α	6.042	^{237}Np (α, 4n)
		α 0.025%	γ	0.280 (47%)	^{237}Np (^{3}He, 3n)
238	1.63 h	EC >99%	α	5.94	^{237}Np (α, 3n)
		α 1.0×10^{-4}%	γ	0.0963 (28%)	
239	11.9 h	EC 99.990%	α	5.776 (0.00837%)	^{237}Np (α, 2n)
		α 0.010%		5.734 (0.001375%)	^{239}Pu (d, 2n)
			γ	0.278 (15%)	
240	2.117 days	EC >99%	α	5.378 (0.0001649)	^{237}Np (α, n)
		α 1.9×10^{-4}%		5.337 (2.28×10^{-5}%)	^{239}Pu (d, n)
			γ	0.988 (73%)	
241	432.7 years	α, SF(3.8×10^{-10}%)	α	5.486 (85.2%)	^{241}Pu daughter
				5.443 (12.8%)	Multiple n capture
			γ	0.059 (35.9%)	
242	16.01 h	B^- 82.7%	$β^-$	0.663	^{241}Am (n, γ)
		EC 17.3%	ν	0.042 (0.039%)	

242 m	141 years	IT 99.55% α 0.45%	α	5.207 (0.4%) 5.141 (0.026%) 0.0492 (0.19%)	^{241}Am (n, γ)
243	7.37 × 10³ years	α, SF(3.7 × 10⁻⁹%)	γ α	5.277 (88%) 5.233 (11%) 0.075 (68%)	Multiple n capture
244	10.1 h	β⁻	γ β⁻	0.387 0.746 (67%)	^{243}Am (n, γ)
244 m	26 min	β⁻ 99.964% EC 0.036	γ β⁻	1.50	^{243}Am (n, γ)
245	2.05 h	β⁻	β⁻ γ	0.895 0.253 (6.1%)	^{245}Pu daughter
246 m	25.0 min	β⁻	β⁻ γ	2.34 0.799 (24.7%)	^{246}Pu daughter
246	39 min	β⁻	γ	0.679 (53%) 0.205 (36%) 0.153 (25%)	^{244}Pu (α, d) ^{244}Pu (³He, p)
247	25.0 min	β⁻	γ	0.285 (23%) 0.227 (5.8%)	^{244}Pu (α, p)

TABLE A4 Radioactive Decay Properties of Curium Isotopes

Mass Number	Half-life	Mode of Decay		Main Radiations (MeV)	Typical Method of Production
238	2.4 h	EC >90%	α	6.52	^{239}Pu (α, 5n)
		α <10%			
239	~3 h	EC	γ	0.187	^{239}Pu (α, 4n)
240	27 days	α, SF(3.89 × 10^{-6}%)	α	6.291 (70.6%)	^{239}Pu (α, 3n)
				6.248 (28.8%)	
241	32.8 days	EC 99.0%	α	5.939 (0.689%)	^{239}Pu (α, 2n)
		α 1.0%		5.929 (0.181%)	
			γ	0.472 (71%)	
242	162.94 days	α, SF (6.3 × 10^{-6}%)	α	6.113 (74.0%)	^{239}Pu (α, n)
				6.070 (25.0%)	^{242}Am daughter
243	28.5 years	α 99.76%	α	5.786 (73.5%)	^{242}Cm (n, γ)
		EC 0.24%		5.742 (10.6%)	
			γ	0.278 (14.0%)	
244	18.10 years	α, SF (0.0001347%)	α	5.805 (77.0%)	Multiple n capture
				5.763 (23.0%)	^{244}Am daughter
245	8.5 × 10^3 years	α	α	5.362 (93.2%)	Multiple n capture
		SF (6.1 × 10^{-7}%)		5.304 (5.0%)	
			γ	0.175 (9.5%)	
246	4.7 × 10^3 years	α 99.97386%	α	5.386 (79%)	Multiple n capture
		SF 0.02614%		5.343 (21%)	
247	1.60 × 10^7 years	α	α	5.266 (13.8%)	Multiple n capture
				4.869 (71%)	
			γ	0.404 (72%)	
248	3.40×10^5 years	α 91.74%	α	5.078 (75.1%)	Multiple n capture
		SF 8.26%		5.034 (16.5%)	
249	1.0692 h	β^-	β^-	0.893	^{248}Cm (n, γ)
			γ	0.634 (1.5%)	
250	<1.13 × 10^4 years	SF			Multiple n capture
251	16.8 min	β^-	β^-	1.42	^{250}Cm (n, γ)
			γ	0.543 (10.9%)	

TABLE A5 Radioactive Decay Properties of Berkelium Isotopes

Mass Number	Half-life	Mode of Decay	Main Radiations (MeV)		Typical Method of Production
240	5 min	EC, EC-SF			^{232}Th (^{14}N, 6n)
242	7 min	EC			^{235}U (^{11}B, 4n)
					^{232}Th (^{15}N, 5n)
243	4.5 h	EC 99.85%	α	6.758 (0.0231%)	^{243}Am (α, 4n)
		α 0.15%		6.574 (0.0384)	
			γ	0.755 (10.0%)	
244	4.35 h	EC 99.994%	α	6.667 (0.003%)	^{243}Am (α, 3n)
		α 6×10^{-3}%		6.625 (0.003%)	
			γ	0.218	
245	4.94 days	EC 99.88%	α	6.349 (0.0181%)	^{243}Am (α, 2n)
		α 0.12%		6.145 (0.0248%)	
			γ	0.253 (29.1%)	
246	1.80 days	EC	γ	0.799 (61%)	^{243}Am (α, n)
			γ	5.712 (17%)	
247	1380 years	α	α	5.532 (45%)	^{247}Cf daughter
				0.084 (40%)	
			γ		^{244}Cm (α, p)
248[a]	23.7 h	β$^-$ 70%	β$^-$	0.55	^{248}Cm (d, 2n)
		EC 30%	γ	0.551 (5.0%)	
248[a]	>9 years	Decay not observed			^{246}Cm (α, pn)
249	320 days	β$^-$ 99.99855%	α	5.417 (0.00108%)	Multiple n capture
		α 1.45 × 10^{-3}%		5.390 (0.00023%)	
		SF 4.7 × 10^{-8}%	β$^-$	0.126	
			γ	0.327 (1.7 × 10^{-5}%)	
250	3.217 h	β$^-$	β$^-$	1.78	^{254}Es daughter
			γ	0.989 (45%)	^{249}Bk (n, γ)
251	56 min	β$^-$	β$^-$	~1.1	^{255}Es daughter
			γ	0.178	

[a]Not known whether ground state nuclide or isomer.

TABLE A6 Radioactive Decay Properties of Californium Isotopes

Mass Number	Half-life	Mode of Decay		Main Radiations (MeV)	Typical Method of Production
239	39 s	α	α	7.63	^{243}Fm daughter
240	1.06 min	α	α	7.59	^{233}U (^{12}C, 5n)
241	3.8 min	α	α	7.335	^{233}U (^{12}C, 4n)
242	3.5 min	α	α	7.385 (~80%)	^{233}U (^{12}C, 3n)
				7.351 (~20%)	^{235}U (^{12}C, 5n)
243	10.7 min	EC ~86%	α	7.06 (10%)	^{235}U (^{12}C, 4n)
		α ~14%		7.17 (4%)	
244	19.4 min	α	α	7.210 (75%)	^{244}Cm (α, 4n)
				7.168 (25%)	^{236}U (^{12}C, 4n)
245	43.6 min	EC ~70%	α	7.137	^{244}Cm (α, 3n)
		α 30%			^{238}U (^{12}C, 5n)
246	1.487 days	α ~99.99980%	α	6.750 (78%)	^{244}Cm (α, 2n)
		SF (0.00020%)		6.709 (21.8%)	^{246}Cm (α, 4n)
247	3.11 h	EC 99.965%	α	6.301	^{246}Cm (α, 3n)
		α 0.035	γ	0.294 (0.98%)	^{244}Cm (α, n)
248	334 days	α 99.9971%	α	6.262 (83.0%)	^{246}Cm (α, 2n)
		SF (0.0029%)		6.220 (17.0%)	

249	351 years	α	6.194 (2.17%)	²⁴⁹Bk daughter
			5.812 (84.4%)	
		SF	5.2 × 10⁻⁷%	
		γ	0.388 (66%)	
250	13.08 years	α 99.923%	6.031 (84.5%)	Multiple n capture
		SF 0.077%	5.989 (15.1%)	
251	898 years	α	5.851 (27%)	Multiple n capture
			5.677 (35%)	
		γ	0.177 (17.7%)	
252	2.645 years	α 96.908%	6.118 (81.6%)	Multiple n capture
		SF 3.092%	6.076 (15.2%)	
253	17.81 days	β⁻ 99.69%	5.979 (0.294%)	Multiple n capture
		α 0.31%	5.921 (0.016%)	
254	60.5 days	SF 99.69%	5.834 (0.256%)	Multiple n capture
		α 0.31%	5.792 (0.053%)	
255	1.4 h	β⁻		²⁵⁴Cf (n, γ)
256	12.3 min	SF		²⁵⁴Cf (t, p)

TABLE A7 Radioactive Decay Properties of Einsteinum Isotopes[a]

Mass Number	Half-life	Mode of Decay	Main Radiations (MeV)		Typical Method of Production
243	21 s	EC <70%			^{233}U (^{15}N, 5n)
		α >30%	α	7.89	
244	37 s	EC 96%	α	7.57	^{233}U (^{15}N, 4n)
		α ~4%			^{237}Np (^{12}C, 5n)
245	1.33 min	EC 60%	α	7.73	^{237}Np (^{12}C, 4n)
		α 40%			
246	7.7 min	EC 90%	α	7.35	^{241}Am (^{12}C, α 3n)
		α 10%			
247	4.7 min	EC ~93%	α	7.32	^{241}Am (^{12}C, α 2n)
		α ~7%			^{238}U (^{14}N, 5n)
248	27 min	EC 99.7%	α	6.87	^{249}Cf (d, 3n)
		α ~0.3%	γ	0.551	
249	1.70 h	EC 99.4%	α	6.770	^{249}Cf (d, 2n)
		α 0.6%	γ	0.380 (40%)	
250[a]	8.6 h	EC	γ	0.829 (74%)	^{249}Cf (d, n)
250[a]	2.22 h	EC	γ	0.989 (13%)	^{249}Bk (α, 3n)
251	1.38 days	EC 99.5%	α	6.492 (0.40%)	^{249}Bk (α, 2n)
		α 0.5%		6.462 (0.046%)	

[a]Not known whether ground state nuclide or isomer.

Mass	Half-life	Decay mode	Particle energies MeV (abundance)	Production method
252	1.291 years	α 76% EC 24%	α 6.632 (61%) 6.562 (10.3%) γ 0.785 (18%)	^{249}Bk (α, n)
253	20.4 days	α, SF (8.7 × 10^{-6}%)	α 6.633 (89.8%) 6.592 (6.6%)	Multiple n capture
254 g	275.7 days	α	α 6.427 (93%) 6.357 (2.6%) γ 0.065 (2.0%)	Multiple n capture, ^{253}Cf daughter
254 m	1.638 days	β$^-$ 99.59% α 0.33% EC 0.078%	α 6.382 (0.248%) 6.357 (0.027%) β$^-$ 1.171	^{253}Es (n, γ)
255	40 days	β$^-$ 92.0% α 8.0% SF 4 × 10^{-3}%	α 6.300 (7%) 6.260 (0.8%)	Multiple n capture
256[a]	25 min	β$^-$		^{255}Es (n, γ)
256[a]	~7.6 h	β$^-$		^{254}Es (t, p)

[a] Not known whether ground state nuclide or isomer.

TABLE A8 Radioactive Decay Properties of Fermium Isotopes

Mass Number	Half-life	Mode of Decay	Main Radiations (MeV)	Typical Method of Production
242	0.8 ms	SF		^{204}Pb (^{40}Ar, 2n)
243	0.18 s	α	α 8.546	^{206}Pb (^{40}Ar, 3n)
244	3.7 ms	SF		^{206}Pb (^{40}Ar, 2n)
				^{233}U (^{16}O, 5n)
245	4 s	α	α 8.15	^{233}U (^{16}O, 4n)
246	1.1 s	α 92%	α 8.24	^{235}U (^{16}O, 5n)
		SF 8%		^{239}Pu (^{12}C, 5n)
247[a]	35 s	α ≥ 50%	α 8.01 (43%)	^{239}Pu (^{12}C, 4n)
		EC ≤ 50%	8.06 (43%)	
247[a]	9 s	α	α 8.18	^{239}Pu (^{12}C, 4n)
248	36 s	α 99.9%	α 7.87 (80%)	^{240}Pu (^{12}C, 4n)
		SF 0.1%	7.83 (20%)	
249	2.6 min	α	α 7.53	^{238}U (^{16}O, 5n)
				^{249}Cf (α, 4n)
250	30 min	α	α 7.43	^{249}Cf (α, 3n)
				^{238}U (^{16}O, 4n)
250 m	1.8 s	IT		^{249}Cf (α, 3n)
251	5.3 h	EC 98.2%	α 6.832 (1.57%)	^{249}Cf (α, 2n)
		α 1.8%	6.783 (0.086%)	
252	1.058 days	α	α 7.040 (85%)	^{249}Cf (α, n)
			6.999 (15%)	
253	3.0 days	EC 88%	α 6.943 (5.1%)	^{252}Cf (α, 3n)
		α 12%	6.676 (2.8%)	
			γ 0.272 (2.6%)	
254	3.240 h	α 99.9408%	α 7.189 (85%)	^{254m}Es daughter
		SF 0.0592%	7.147 (14%)	
255	20.07 h	α	α 7.022 (93.4%)	^{255}Es daughter
		SF 2 × 10^{-5}%	6.963 (5.0%)	^{256}Md daughter
256	2.63 h	SF 91.9%	α 6.915	^{256}Es daughter
		α 8.1%		
257	100.5 days	α 99.79%	α 6.696 (3.4%)	Multiple n capture
		SF 0.21%	6.519 (93.5%)	
			γ 0.241 (10.3%)	
258	0.38 ms	SF		^{257}Fm (d, p)
259	1.5 s	SF		^{257}Fm (t, p)

[a] Not known whether ground state nuclide or isomer.

TABLE A9 Radioactive Decay Properties of Mendelevium Isotopes

Mass Number	Half-life	Mode of Decay	Main Radiations (MeV)	Typical Method of Production
247	3 s	α	α 8.43	^{209}Bi (^{40}Ar, 2n)
248	7 s	EC 80%	α 8.36 (~5%)	^{241}m (^{12}C, 5n)
		α 20%	8.32 (~15%)	^{239}Pu (^{14}N, 5n)
249	24 s	EC ≤ 80%	α 8.03	^{241}Am (^{12}C, 4n)
		α ≥ 20%		
250	52 s	EC 94%	α 7.82 (~2%)	^{243}Am (^{12}C, 5n)
		α 6%	7.75 (~4%)	^{240}Pu (^{15}N, 5n)
251	4.0 min	EC ≥ 94%	α 7.55	^{243}Am (^{12}C, 4n)
		α ≤ 6%		^{240}Pu (^{15}N, 4n)
252	2.3 min	EC		^{243}Am (^{13}C, 4n)
				^{238}U (^{19}F, 5n)
254[a]	10 min	EC		^{253}Es (α, 3n)
255	27 min	EC 92%	α 7.326	^{253}Es (α, 2n)
		α 8%	γ 0.453 (4.3%)	^{254}Es (α, 3n)
256	1.27 h	EC 90.7%	α 7.21 (5.9%)	^{253}Es (α, n)
		α 9.3%	7.140 (1.5%)	
257	5.2 h	EC 90%	α 7.064	^{254}Es (α, n)
		α 10%		
258	55 days	α	α 6.79 (28%)	^{255}Es (α, n)
			6.176 (72%)	
258 m	60 min	EC		^{255}Es (α, n)
259	1.6 h	SF		^{259}No daughter
260	32 days	SF		^{254}Es (^{22}Ne, X)

[a]Not known whether ground state nuclide or isomer.

TABLE A10 Radioactive Decay Properties of Nobelium Isotopes

Mass Number	Half-life	Mode of Decay	Main Radiations (MeV)	Typical Method of Production
250	0.25 ms	SF		^{233}U (^{22}Ne, 5n)
251	0.6 s	α	α 8.66 (18%) 8.59 (82%)	^{244}Cm (^{12}C, 5n)
252	2.3 s	α 73% SF 27%	α 8.415 (~55%) 8.372 (~18%)	^{244}Cm (^{12}C, 4n) ^{239}Pu (^{18}O, 5n)
253	1.7 min	α	α 8.02	^{246}Cm (^{12}C, 5n) ^{242}Pu (^{16}O, 5n)
254	55 s	α 90% EC 10% SF (0.17%)	α 8.10	^{246}Cm (^{12}C, 4n) ^{242}Pu (^{16}O, 4n)
254 m	0.28 s	IT		^{246}Cm (^{12}C, 4n) ^{249}Cf (^{12}C, α3n)
255	3.1 min	α 61.4% EC 38.6%	α 8.12 (27.9%) 8.08 (7.3%)	^{248}Cm (^{12}C, 5n) ^{249}Cf (^{12}C, α2n)
256	2.91 s	α ~ 99.5% SF 0.53%	α 8.45	^{248}Cm (^{12}C, 4n)
257	25 s	α	α 8.30 8.22	^{248}Cm (^{12}C, 3n)
258	1.2 ms	SF		^{248}Cm (^{13}C, 3n)
259	58 min	α ~ 78% EC ~ 22%	α 7.53 (18%) 7.50 (30%)	^{248}Cm (^{18}O, α3n)
260	106 ms?	SF		^{254}Es (^{18}O, X)
261	1.1 h?	SF, α	α 7.38	^{254}Es (^{22}Ne, X)
262	~5 ms	SF		^{254}Es (^{22}Ne, X)

TABLE A11 Radioactive Decay Properties of Lawrencium Isotopes

Mass Number	Half-life	Mode of Decay	Main Radiations (MeV)	Typical Method of Production
253	1.3 s	α	α 8.80	^{257}Ha daughter
254	13 s	α	α 8.460	^{258}Ha daughter
255	22 s	α	α 8.43 (40%) 8.37 (60%)	^{243}Am (^{16}O, 4n) ^{249}Cf (^{11}B, 5n)
256	26 s	α (>80%) EC (<20%)	α 8.52 (19%) 8.43 (37%)	^{243}Am (^{18}O, 5n) ^{249}Cf (^{11}B, 4n)
257	0.6 s	α (>85%) EC (<15%)	α 8.86 (82%) 8.80 (18%)	^{249}Cf (^{11}B, 3n) ^{249}Cf (^{14}N, α2n)
258	4.3 s	α	α 8.61 (25%) 8.59 (46%)	^{248}Cm (^{15}N, 5n) ^{249}Cf (^{15}N, α2n)
259	5.4 s	α	α 8.45	^{248}Cm (^{15}N, 4n)
260	3.0 min	α	α 8.03	^{248}Cm (^{15}N, 3n)
261	39 min	SF		^{254}Es (^{22}Ne, X)
262	216 min	EC (>50%)		^{254}Es (^{22}Ne, X)

TABLE A12 Radioactive Decay Properties of Rutherfordium Isotopes

Mass Number	Half-life	Mode of Decay	Main Radiations (MeV)	Typical Method of Production
253[a]	1.8 s	SF (~50% est.)		^{206}Pb (^{50}Ti, 3n)
254[a]	0.5 ms	SF		^{206}Pb (^{50}Ti, 2n)
255	1.4 s	SF (~50% est.)	α 8.715 (70%)	^{207}Pb (^{50}Ti, n)
256[a]	6.7 ms	SF (97.8%), α (2.2%)		^{208}Pb(^{50}Ti, 2n)
257	4.5 s	α 70%	α 9.013 (17%)	^{249}Cf (^{12}C, 4n)
		EC 16%	8.95 (13%)	
		SF ~14%		
258[a]	13 ms	SF		^{249}Cf (^{12}C, 3n)
				^{246}Cm (^{16}O, 4n)
259	3.0 s	α 91%	α 8.86 (60%)	^{249}Cf (^{13}C, 3n)
		SF 9%	8.77 (40%)	^{248}Cm (^{16}O, 5n)
260[a]	21 ms	SF		^{248}Cm (^{16}O, 4n)
261	1.08 min	α	α 8.28	^{248}Cm (^{18}O, 5n)
262[a]	47 ms	SF		^{248}Cm (^{18}O, 4n)

[a]The identity of this nuclide is not well established.

TABLE A13 Radioactive Decay Properties of Hahnium-109 Isotopes

Mass Number	Half-life	Mode of Decay	Main Radiations (MeV)		Typical Method of Production
Ha					
255*	1.5 s	SF (~20% est.)			^{207}Pb (^{51}V, 3n)
					^{206}Pb (^{51}V, 2n)
256	1.2 s	SF			^{260}107 daughter
257	1.4 s	α			^{209}Bi (^{50}Ti, 2n)
		SF	α	8.97 (33%)	
258	4.0 s	α 67%	α	9.17	^{262}107 daughter
		EC 33%		9.30	^{209}Bi (^{50}Ti, n)
260	1.52 s	α 90.4%	α	9.08 (25%)	^{249}Cf (^{15}N, 4n)
		SF 9.6%		9.05 (48%)	^{243}Am (^{22}Ne, 5n)
261	1.8 s	α ~75%	α	8.93	^{243}Am (^{22}Ne, 4n)
		SF ~25%			^{249}Bk (^{16}O, 4n)
262	32 s	SF 49%	α	8.53 (~7.6%)	^{249}Bk (^{18}O, 5n)
		α ~51%		8.45 (~38%)	
106					
259^a	480 ms	α	α	9.62	^{208}Pb (^{54}Cr, 3n)
260	3.6 ms	α (50%), SF (50%)	α	9.77	^{207}Pb (^{54}Cr, 2n)
261	0.26 s	α	α	9.56	^{208}Pb (^{54}Cr, n)
263	0.9 s	α	α	9.25 (3%)	^{249}Cf (^{18}O, 4n)
				9.06 (27%)	
107					
261	11.8 ms	α	α	10.10, 10.40	^{209}Bi (^{54}Cr, 2n)
		SF 15%			^{208}Pb (^{55}Mn, 2n)
262	102 ms	α	α	9.70, 10.06	^{209}Bi (^{54}Cr, n)
262 m	8.0 ms	α	α	10.37	^{70}Bi (^{54}Cr, n)
108					
264	76 μs	α	α	11.0	^{207}Pb (^{58}Fe, n)
265	1.8 ms	α	α	10.36	^{208}Pb (^{58}Fe, n)
109					
266	3.4 ms	α	α	11.0	^{209}Bi (^{58}Fe, n)

^{a}The identity of this nuclide is not well established.

NAME INDEX

van Aark, J., 186
Abelson, P. A., 9
Aguiar, C. E., 239
Ahmad, I., 154, 156, 157, 164, 166, 175, 329–346
Ahrland, S., 94, 97
Akap'ev, G. N., 53
Alaga, G., 170
Alder, K., 170
Aleixo, A. N., 239
Alexander, J. M., 237, 239
Allard, B., 297
Alonso, C. J., 54, 208, 282
Alonso, J. R., 54, 208, 232, 282
Arima, A., 245
Arkin, W. M., 311, 313, 315
Armbruster, P., 56–59, 135, 143, 208–211, 240, 242, 243, 275, 278–284, 286
Aronsson, P. O., 103
Asciutto, R. J., 245
Åström, B., 46
Atterling, H., 46
Audi, G., 123
Auerman, L. N., 85
Austern, N., 244
Ayik, S., 260

Back, B. B., 233
Baran, A., 144, 145, 274, 286
Barnes, C. A., 294
Barbosa, V. C., 239
Barnes, R. F., 178
Bass, R., 237, 243, 245, 246, 250
Baybarz, R. D., 107
Beckerman, M., 231
Behkami, A. N., 231
Belov, V. Z., 103, 115
Bemis, C. E., 52, 53, 122, 205, 208, 272
Benedict, M., 225, 305, 306, 309, 315
Bengtsson, R., 123
Bethe, H. A., 75
Bimbot, R., 211
Birkelund, J. R., 237, 250, 259, 277
Bjornholm, S., 132, 239, 243, 284
Bjornstad, J., 103
Blann, M., 231
Blatt, J. M., 168
Blocki, J., 237, 240
Bloomquist, C. A. A., 178, 255
Bock, R., 243
Bocquet, J. P., 190
Bohr, A., 154, 169, 171, 287
Bondietti, E. A., 297

NAME INDEX

Bondorf, J. P., 246
Brandt, R., 186
Breit, G., 245
Brewer, L., 81
Brink, D. M., 245
Brisset, R., 190
Britt, H. C., 131, 132, 136, 233
Broden, K., 103
Bromley, D. A., 250
Brosa, V., 186
Brown, D., 109
Browne, C. I., 35
Browne, E., 141, 147
Bruchertseifer, H., 57, 287
Brüchle, W., 81, 82, 103, 243, 255
Buklanov, G. V., 287
Burbridge, E. M., 292
Burbridge, G. R., 292
Burnett, D. S., 291, 292
Burnett, J. L., 107
Buning, K., 133, 144

Cameron, A. G. W., 231
Canto, L. F., 239
Carlsson, J. A., 208
Carnall, W. J., 82
Chan, F. S., 231
Chan, Y. D., 237, 239
Chasman, R. R., 154, 156, 164, 166, 175, 180, 182, 183, 187, 272
Chastelier, R. M., 249
Chelnokov, L. P., 53, 103, 115
Cherepanov, E. A., 234
Cherpigin, V. I., 287
Choi, J., 307, 308
Choppin, G. R., 41, 44
Choy, V. S., 287
Chuburkov, Y. J., 51, 81
Clayton, D. D., 292, 293, 294, 295
Cleveland, J. M., 297
Cochran, J. B., 311, 313, 315
Cohen, B. L., 154, 163, 244
Cohen, S., 233
Constantinescu, O., 57
Cook, J., 231
Cotton, F. A., 65
Cramer, B., 287
Crandall, J. L., 219, 223
Cunningham, B. B., 16, 21, 38, 201
Cwiok, S., 133, 144, 186, 272–274

Dahlinger, M., 138
Dahlman, R. C., 297
Danilov, N. A., 55, 56
David, F., 202
De, J. N., 237, 277
Delphin, W., 255
Demin, A. G., 53, 55–57, 206, 279, 283
Deruytter, A., 189
Desclaux, J. P., 75, 77, 82
Deshalit, A., 154
Diamond, H., 35, 38, 202
Dickmann, F., 246
Digregorio, D. E., 237, 239
Dittner, P. F., 52, 53, 208, 247
Domanov, V. P., 103, 115
Donangelo, R., 239
Donets, E. D., 49, 50
Dostrovsky, I., 232
Dougan, R. J., 186
van Driel, J., 246
Druin, V. A., 51, 53, 132, 146
Dufour, J. P., 211

Ebel, M. E., 245
Edelstein, N. M., 65
Edgington, D. N., 296
Eichler, B., 81
Ermakov, V. A., 49, 50
Erskine, J. R., 154, 156, 164, 166, 175
Eskola, K., 52, 53, 115, 208
Eskola, P., 52, 53, 115
Ewald, H., 203
Eyman, L. D., 297

Faber, M. E., 234
Fagan, P. J., 110
Faust, W., 203
Fefilov, B. V., 53
Feldmeier, H., 240
Feshbach, H., 154, 230
Fields, P. R., 35, 46, 178
Firestone, R. B., 141, 147
Fischer, R. D., 110
Fiset, E. O., 272, 273
Flerov, G. N., 46, 49, 51, 53, 55, 56, 60, 139
Fleury, A., 211
Flynn, K. F., 178
Foldiak, G., 316
Ford, J. L. C., 122
Forsling, W., 46

NAME INDEX 349

Fowler, M. M., 258
Fowler, W. A., 292–295
Fomichev, B., 132, 146
Fraenkel, Z., 232
Frank, L., 211
Franz, G., 255
Fred, M. S., 38
Freeman, A. J., 65, 82
Freiesleben, H., 250, 255, 258, 259
Fricke, B., 77, 81
Fried, S., 28
Fried, S. M., 35
Fried, M. S., 202
Friedlander, G., 151, 232
Friedman, A. M., 46, 154, 156, 164, 166, 175, 297
Friedman, H. G., 72
Friedmann, W. A., 250
Fursov, B. I., 133, 234

Gäggeler, H. W., 81, 203, 236, 243, 249, 258, 262, 263, 285
Gagne, M. R., 113
Gardes, D., 211
Garrett, J. D., 233
Gatti, R. C., 38
Gavrilov, K. A., 51, 53
Gavron, A., 232, 233
Ghiorso, A., 19, 22, 25, 31, 34, 40, 41, 44, 46, 49, 51–54, 62, 85, 115, 208, 211, 213, 233, 248, 282
Gibson, W. M., 213
Gil, S., 287
Gilat, J., 231, 233
Gilbert, A., 231
Gindler, J. E., 129
Glaser, R. E., 186
Glasstone, S., 313
Glebor, V. A., 81
Glendenin, L. E., 178
Glendenning, N. K., 244
Gonggrijp, S., 246
Goodman, L. S., 38, 202
Goodman, C. D., 52, 208
Gorbachev, V. M., 232
Graber, J., 295
Gregorich, K. E., 81, 116, 202, 249, 255
Groening, H. R., 231, 233
Gross, D. H. E., 246
Grossman, S., 186

Grover, J. R., 231, 233
Grozdez, B. A., 85
Gruber, J., 124
vonGunten, H. R., 249
Guttner, K., 203

Habs, D., 147, 148
Haefner, B., 103
Hageman, D. C. J. M., 246
Hahn, R. L., 52, 53, 247
Haire, R. G., 201
Håkonson, T. E., 296, 297
Halpern, I., 189
Hansen, O., 233
Hanson, W. C., 296
Harmon, B. A., 237, 239
Harris, J., 52, 53, 115, 208
Harvey, B. G., 8, 41, 44, 204
Hauser, W., 230
Henderson, R. A., 81, 249
Henderson, D. J., 98
Henry, E. A., 102
Hensley, D. C., 52, 53, 208
Herrmann, G., 8, 102, 103, 204, 258, 260
Hesler, J. P., 82
Hessberger, F. P., 133, 208, 209, 283
Heunemann, D., 148
Higgins, G. H., 34
Hildenbrand, K. D., 237, 239, 258, 277
Hilscher, D., 237, 239, 287
Hirsch, A., 35
Hockstra, R., 123
Hodgson, P. E., 245
Hoenig, M. M., 311, 313, 315
Hoff, R. W., 8, 204
Hoffman, D. C., 8, 51, 54, 81, 144, 178, 185, 192, 201, 204, 249, 274, 291
Hofmann, S., 203, 208, 209, 283
Hofstadter, D. R., 326, 327
Holhn, M. V., 122
Holm, L. W., 46
Holub, E., 237, 239
Hoover, A. D., 237
Horwitz, E. P., 98, 178, 255
Howard, W. M., 133, 137
Hoyle, F., 292
Hubener, S., 107
Hubert, F., 211
Huizenga, J. R., 7, 35, 129, 176, 231, 232, 237, 244, 250, 254, 259, 277

Hulet, E. K., 54, 85, 115, 185, 186, 201, 203, 208, 255, 282
Hunt, L. D., 52, 53
Hussonis, M., 57, 103, 115
Hyde, E. K., 8, 51, 54, 204

Iljinov, A. S., 55, 56, 203, 206, 234, 279, 283
Ingold, G., 287
Istekov, K. K., 234
Ivanov, M. P., 55, 56, 206
Ivascu, M. S., 146

Jacak, B., 249
Jackson, J. D., 233
Jackson, D. F., 228, 244
Jahnke, U., 237, 239
Jahnke, V., 287
James, R. A., 19
Janssens, R. V. F., 246
Janzov, E. F., 318
Jin, K. U., 81
Jing-Ye, Z., 123
Jones, G. A., 154
Jones, J. H., 291, 292
Jost, D. T., 81
Juger, E., 81

Kaffrell, N., 103
Kalnins, J., 211
Kamenshaya, A. N., 85
Karnaukov, N., 132, 146
Kasztura, L., 81
Katz, J. J., 8, 65, 82, 99, 105, 108, 110, 113, 204, 292, 297
Kaufman, R., 250, 251
Keller, C., 65
Keller, K. A., 152
Keller, O. L., 8, 51, 52, 54, 108, 113, 201, 202, 204, 247, 275, 276, 319
Kennedy, J. W., 13, 15, 151
Kharitonov, U. P., 53, 57
Klapdor, H. V., 175
Kolesnikov, N. N., 56
Kolesov, I. V., 53
Kondoh, T., 175
Konecny, E., 148, 233
Koop, E. A., 186
Korkisch, J., 99, 102
Korotkin, Y. S., 55, 57, 103, 115
Krane, K. S., 154, 244, 312
Krappe, H. J., 142

Kratz, J. V., 81, 103, 116, 237, 239, 243, 250, 255, 258, 277, 284, 285
Kraus, K., 101
Kubono, S., 245
Kupriyanov, V. M., 133, 234
Kuznetsov, V. I., 51
Kwiatkowski, K., 184

LaChapelle, T. J., 11
Landrum, J. H., 85, 115
Lane, L. J., 297
Larsh, A. E., 49, 51
Larsson, S. E., 272, 273
Lasarev, Y. A., 53
Latimer, R. M., 49, 51
Lawrence, F. O., 291
Lazzarini, A., 287
Lbov, A. A., 232
Leander, G. A., 133, 275, 281
LeBeck, D. F., 233, 248
Leber, R. E., 213
Lederer, C. M., 166, 173
Ledoux, B., 233
Lee, D., 81, 249, 255, 258
Lefort, M., 204, 244, 250, 252
Lehmann, M., 8, 250, 252, 287
Leino, M., 211
Lemmertz, P. K., 211
Lesko, K. T., 237, 239, 287
Levi, H., 225, 305, 306, 309, 315
Liljenzin, J. O., 94, 97, 258
Liran, S., 126, 151
Liu, Y., 249
Llabador, Y., 211
Lobanov, Y. V., 51, 53
Lojewski, Z., 133, 144, 145, 274, 286
Lougheed, R. W., 54, 85, 115, 186, 203, 208, 255, 282, 285
Loveland, W., 231–233
Lu, C. C., 208
Lucas, R., 237, 239, 277
Lukasiak, A., 246, 272, 273
Luo, C., 249
Lutzenkirchen, K., 243
Lynn, J. E., 132

Ma, B. M., 315
Maatta, E. A., 110
Macias, E. S., 151
Magda, M. T., 249, 250
Magnusson, L. B., 11

Malik, F. B., 208
Maly, J., 233, 248
Manhourat, M. B., 211
Manning, W. M., 35
Manriquez, J. M., 110
Marion, J. B., 263
Markov, B. N., 56
Marks, T. J., 110, 113
Marmier, P., 154, 244
Maruyama, N., 213, 318
Marx, D., 203
Mathews, G. J., 275, 295
Mazilu, D., 146
McDowell, W. J., 97
McFarland, R. M., 249
McGaughey, P. L., 232
McGowan, F. K., 122
McKelvey, D. R., 75
McMillan, E. M., 9, 13
McWherter, J. L., 291
Mebel, M. V., 234
Mech, J. F., 35
Metag, V., 146, 147, 148, 287
Metzinger, J., 175
Meyer, R. A., 102
Mikheev, N. B., 85, 132, 146
Miller, J. M., 154
Milner, W. T., 122
Milsted, J., 46
Möller, P., 123, 133, 135, 137, 138, 144, 173, 186, 272, 273, 275, 281, 285
Molitoris, J. D., 203
Moloy, K. G., 110
Moody, K. J., 175, 186, 203, 249, 255, 286
Del Moral, R., 211
Moretto, L. G., 233
Morgan, L. O., 19
Morland, H., 313, 314
Morrissey, D. J., 232
Morss, L., 65, 82, 99, 105, 108, 110, 113, 292, 297
Mosel, V., 274
Mottelson, B. R., 154, 169, 171
Münzenberg, G., 56, 57, 59, 60, 63, 203, 208, 209, 239, 283, 286
Mughabghab, S. F., 224
Mukhin, K. N., 263
Müller, A., 186
Müller-Westhoff, V., 111
Munzel, H., 152
Mustafa, M. G., 281

Myers, W. D., 123, 124, 125, 138, 281

Nash, K. L., 297
Naumann, R. A., 122
Nefedor, V. S., 81
Nelson, D., 101
Nestor, C. W., 208
Ngô, C., 250, 252
Nickel, F., 203
Nitschke, J. M., 54, 85, 115, 203, 204, 206, 208, 213, 282
Nix, J. R., 123, 133, 135, 142, 144, 173, 186, 272, 273, 275, 281, 285
Nolan, S. P., 113
Norman, E. B., 237, 239
Norris, A. E., 258
Nugent, L. J., 107
Nurmia, M., 51, 52–54, 81, 115, 151, 203, 208, 213, 233, 248, 249, 282
Nörenberg, W., 260

Oda, H., 175
Oganessian, Y. T., 51, 53, 55–57, 206, 235, 279, 282, 283
Orlova, O. A., 287
Otto, R. J., 258

Papanstassiou, D. A., 295
Parker, W. C., 203
Paskevich, V. V., 133
Patterson, J. H., 317
Patyk, Z., 133, 143, 144, 186, 272, 274
Patzell, P., 186
Penionzkevich, V., 206
Perelygin, V. P., 51
Percy, C. M., 231
Perlman, I., 21, 292
Peter, J., 250, 252
Peterson, J. R., 201
Petrzhak, K. A., 139
Philips, L., 38
Pigford, T. H., 225, 305, 306, 307, 308, 309
Pitzer, K. S., 75, 113
Plasil, F., 231, 233
Pleve, A. A., 55, 132, 146, 206, 287
Ploszajczek, M, 234
Plotko, V. M., 51, 53, 55, 56
Poenaru, D. N., 146
Poitou, J., 237, 239, 277
Polikanov, S. M., 132, 146
Pop, A., 249, 250

Popeko, G. S., 203
Pouliot, J., 237, 239
Powell, R. E., 75
Pyle, G. L., 35
Pyyko, P., 75

Randrup, J., 133, 144, 237, 255
Rasmussen, J. O., 154
Rees, T. F., 297
Reffo, G., 231
Reisdorf, W., 208, 209, 231, 233, 237, 239, 277, 283, 285
Reuter, W., 122
Reynolds, L. T., 110
Riedel, C., 260
Rivet, M. F., 211
Roberts, J. H., 231
Robinson, R. L., 122
Rose, H. J., 154
Rosenkevich, N. A., 85
Rossner, H. H., 237, 239, 287
Rourke, F. M., 291
Rozmej, P., 186
Rumer, I. A., 85
Ryan, J. L., 107
Rydberg, J., 94, 97, 103

Sahm, C. C., 208, 209, 283
Salpeter, E. E., 75
Samhoun, K., 202
Samyatnin, Y. S., 232
Sandulescu, A., 249, 250
Satchler, G. R., 228, 237, 239, 244
Scamster, V., 287
Schadel, M., 81, 203, 249, 255, 257–259, 262
Scherer, V., 81
Schegolev, V. A., 49, 50
Schillebeeckx, P., 189
Schmidt, K. H., 132, 143, 208, 209, 274, 283
Schneider, J. H. R., 208, 209, 283
Schock, L. E., 113
Schorstein, W., 103
Schött, H. J., 203
Schramm, D. N., 292
Schröder, W. U., 237, 250, 254, 256
Schwinn, E., 287
Schürmann, B., 260
Seaborg, G. T., 8, 11, 13–16, 19, 21–23, 25, 35, 41, 44, 46, 54, 63, 64, 82, 93, 99, 105, 107, 108, 110, 113, 124, 144, 173, 201–204, 208,, 235, 249, 255, 258, 275, 276, 282, 292, 297, 321, 323

Seeger, P. A., 293, 294, 295
Seglie, E. A., 245
Segre, E., 15
Seidel, W., 203
Seyam, A. M., 110, 113
Shalaevskii, M. R., 51, 103, 115
Shchegolev, V. A., 103, 115
Sheldon, E., 154, 244
Shera, E. B., 122
Shilov, B. V., 51
Shirley, V. S., 141, 147, 166, 173
Shirokovsky, I. V., 55, 57, 283
Shoun, R. R., 97
Siemens, P. J., 246
Siemssen, R. H., 246
Sierk, A. J., 142, 233, 284
Sikkeland, T., 46, 49, 51, 233, 238, 248
Silva, R. J., 51, 52, 53, 115, 208
Simbel, M. H., 234
Siwek-Wilczynski, K., 246
Sjoblom, R. K., 178
Skarnemark, G., 103
Skokelev, N. K., 53, 132, 146
Slatis, H., 203
Smirenkin, G. N., 133, 234
Smith, H. L., 35
Smith, M. H., 296
Sobiczewski, A., 133, 143, 144, 186, 272, 274
Soff, G., 77
Soloviev, V. G., 154
Somerville, L. P., 144, 178, 213
Specht, H. J., 147, 148
Spence, R. W., 35
Sperber, D., 237, 277
Sprilet, J. C., 202
Spitsyn, V. I., 85
Stanton, H. E., 38, 202
Stapanian, M. I., 291
Steffen, R. M., 122
Steinberg, E. P., 180, 182, 183, 187
Stelson, P. H., 122
Stender, E., 103
Stern, D., 113
Stewart, D. C., 214
Stokstad, R. G., 231, 237, 239, 278
Street, K., 25
Streitwieser, A., 111
Strutinsky, V. M., 125, 272
Studier, M. H., 35
Subbotin, N., 132, 146
Sümmerer, K., 243
Swant, J. A., 124, 295

Swiatecki, W. J., 123, 124, 133, 135, 138, 142, 144, 186, 233, 237, 239, 240, 243, 281, 284, 285
Szabo, A., 75

Taagepera, R., 151
Takahashi, K., 175
Tamain, B., 250, 252
Tanaka, S., 122, 249
Tarrant, J. R., 52, 53
Ter-Akopian, G. M., 60, 132, 146, 203, 287
Terrell. J., 193
Thomas, T. D., 230
Thompson, S. G., 22, 25, 27, 34, 41, 44, 93
Thuma, B., 208, 209, 283
Timokin, S. N., 81
Toth, K. S., 247
Trautman, N., 102, 103
Treiner, J., 123, 138
Tretyakov, Y. P., 51, 53, 55–57, 203, 206, 279, 283
Tsaletka, R., 51
Tsang, C. F., 237
Tubbs, L. E., 237, 277
Tucker, T. C., 208
Turkevich, A. L., 317, 318

Unik, J. P, 184, 185, 250, 255
Utyonkov, V. K., 55, 57, 283

Valens, E. G., 41
Vandenbosch, R., 146, 176, 183, 232, 244, 287
Vaz, L. C., 237, 239
Vermeulen, D., 132
Viola, V. E., 124, 144, 173, 184, 185, 250, 275, 295, 295
Vogel, U., 81

Wagemans, C., 189
Wahl, A. C., 13, 14, 15, 184, 189
Walker, M., 184
Wallman, T. C., 38
Walton, J. R., 46
Wapstra, A. H., 123
Warnecke, I., 255
Wasserburg, G. J., 295
Watters, R. L., 296, 297
Weber, J., 148, 233
Weis, M., 255
Weisskopf, V. F., 168, 218
Welch, R. B., 249, 255
Werner, L. B., 16, 21
Westmeier, W., 186
Whicker, F. W., 296
White, H. E., 75
Wilcke, W. W., 237, 259
Wilczynski, J., 246, 205, 253
Wild, J. F., 85, 115, 186
Wilding, R. E., 296
Wilhelmy, J. B, 233
Wilkins, B. D., 180, 182, 183, 187
Wilkinson, G., 65, 110
Wirth, G., 237, 239, 243, 255, 277
Wolf, K. L., 250, 255
Wolfgang, R., 250, 251
Wollersheim, H., 237, 259
Wolschin, G., 260

Yamada, Y., 175
Yashita, S., 211, 213
Yeremin, A. V., 212
Young, F. C., 263

Zeldes, N., 126, 151
Zhuikov, B. L., 81
Zumbro, J. D., 122
Zvara, I., 51, 81, 103, 115
Zvarova, T. S., 51

SUBJECT INDEX

Actinide concept, 66
Actinide contraction, 80
Alpha particle decay:
  centrifugal potential, 152
  half-lives, predicted, 151
    Keller-Munzel, 152
    Taagepera-Nurmia, 151
  hindrance factors, 153, 167
  K-selection rules, 167
  Q_α, 149
  selection rules, 152
Americium:
  discovery, 19
  first weighing, 21
  naming, 21
  radioactive decay properties, 334
Arms control, need, 327
Availability of transuranium elements, 199

Berkelium:
  discovery, 21
  first weighing, 27
  naming, 27
  radioactive decay properties, 337
Beta decay:
  half-lives, predicted:
    gross theory, 175

    Viola-Seaborg, 173
  Q_β, 173
  selection rules, 175
Bismuth phosphate process, 93

Californium:
  discovery, 21, 24
  first weighing, 27
  naming, 26
  radioactive decay properties, 338
^{252}Cf:
  fuel rod scanners, 318
  general properties, 318
  neutron activation analysis, 318
  radiotherapy, 318
  reactor startup sources, 318
Charged particle induced reactions, 226
  energetics, 226
Chart of transuranium nuclides, 150
Chemical methods of isolating transuranium nuclei, 204
Chemistry of small number of atoms, 201
Colors of aqueous cations, 82
Complete fusion reactions:
  dynamical hindrance, 239
  formalism, 228, 235
  nuclear structure effects, 243
  subbarrier fusion, 237

SUBJECT INDEX

Complex ion formation, 90
  chelates, 92
  kinetics of formation, 91
  stability, 90
  stereochemistry, 91
Computer-controlled chemistry, 202
Cross sections:
  compound nucleus formation, 228
  computer calculations, 231
Curium:
  discovery, 19
  first weighing, 21
  naming, 21
  radioactive decay properties, 336

Deep inelastic transfer:
  general features, 250
  heavy element synthesis, 255
  history, 250
  time scale, 252
Demography of transuranium elements, 3
Direct reactions:
  general, 243
  heavy element synthesis, 249
  heavy ion, 245
    multinucleon transfer, 245
    single nucleon transfer, 245
  theory, 244, 246
  Wilczynski sum rule model, 247
Discovery of elements, criteria for, 8
Drum systems, 206

Einsteinium:
  discovery, 28
  first weighing, 38
  magnetic moment, 38
  naming, 35
  radioactive decay properties, 340
  ^{254}Es, 213
Electromagnetic transitions:
  examples:
    e-e nuclei, 170
    odd nuclei, 172
  K-selection rules, 170
  reduced transition probabilities, 169
  selection rules, 168
  single particle lifetimes, 168
Electron configurations:
  actinides, 79
  transactinides, 81
Element 106:
  discovery, 54

  radioactive decay properties, 346
  synthesis, 282
Element 107:
  discovery, 56
  radioactive decay properties, 346
  synthesis, 56
Element 108:
  discovery, 57
  radioactive decay properties, 346
  synthesis, 283
Element 109:
  discovery, 59
  radioactive decay properties, 346
  synthesis, 59
Element 110
  first attempts to synthesize, 60
  properties, 285
  synthesis, 286
Emission spectra, 82
Environmental concentrations, Pu, 297
Extraction chromatography, 97
Extra-extra push energy, 239
Extra push energy, 239

f-Electrons:
  non relativistic orbitals, 71
    crystal field splittings, 74
    radial probability distributions, 74
    shape, 71
  relativistic orbitals, 75
    binding energies, 77
    effects on chemistry, 78
    shapes, 76
Fast separations, 101
  general, 101
  HPLC, 103
  gas chromatography, 103
  thermochromatography, 103
Fermium:
  discovery, 28
  naming, 35
  radioactive decay properties, 342
Fission:
  schematic view, 126
Fission barrier height:
  calculated, 133
  double-humped, 131
  general features, 131
  measured, 132
  "new path", 134
  shell correction effect, 134
  table, 134

Fission charge distributions:
  general, 189
  shell effects, 190
  σ, 191
  Z_p, 191
Fission gamma rays, 193
Fission mass distribution:
  heavier actinide, 178
  lighter actinides, 177
  shell effects, 177
  Wilkins, Steinberg, Chasman model, 180
Fission neutrons:
  energies, 192
  multiplicity, 191
  $v(A)$, 193
Fission total kinetic energy release:
  distributions, bimodal, 186
  heavy actinides, 185
  systematics, 184
  Wilkens, Steinberg, Chasman model, 186
Fissionability parameters, 127
Fuel reprocessing, 304

Γ_f/Γ_n, 232
  effects of angular momentum, 233
  Sikkeland formulation, 233
Genetic identification of reaction products, 207

Hahnium:
  discovery, 53
    controversy surrounding, 53
  group V chemistry, 53, 115
  radioactive decay properties, 346
Half-lives of transuranium elements, 4, 150
Halides, 109
Health physics, 213
  annual limits of intake, 214
  general hazards, 213
  ^{252}Cf, 213
  ^{254}Es, 213
Heavy element production in a high flux reactor, 37
Heavy ion reactions, 234
  cold fusion, 235
  cross section formalism, 236
  energetics, 234
Heavy particle radioactivity, 153
Helium Jets, 202, 205
HFIR/TRU, 319
Hydrolysis, 89

Ion-exchange, 99
  actinide/lanthanide separation, 99
  anion exchange, 101
  cation exchange, 99
  use in discovery of Bk, Cf, 23, 26
  use in discovery of Es, Fm, 34
  use in discovery of Md, 40
  use in discovery of No, 48
Ion exchange resin bead technique, 201
Ionic radii, 81

Lawrencium:
  discovery, 50
  naming, 51
  radioactive decay properties, 344

Magnetic properties, 84
Maximum likelihood method, 211
Mendelevium
  discovery, 38
  naming, 44
  radioactive decay properties, 343
Metallic state, 104
  bonding, 105
  criticality data, 107
  crystal structures, 105
  preparation, 104
  properties, 106
  purification, 105
Mike explosion:
  use in Es, Fm discovery, 28
  neutron capture sequence in, 35

Names of transuranium elements, 2
Naming of transuranium elements, 2
Natural abundances of actinide elements, 291
  environmentally significant, 296
  solar system abundances, 291
Neptunium:
  discovery, 8
  first isolation of macroscopic amount, 11
  naming, 11
  radioactive decay properties, 330
Neutron capture:
  calculations, 225
  synthesis of heavy nuclei in a reactor, 219
  synthesis of heavy nuclei in nuclear explosions, 224
Nobelium:
  discovery, 46

SUBJECT INDEX

Nobelium (*Continued*)
  false discovery, 46
  naming, 49
  radioactive decay properties, 344
Nuclear masses:
  decay cycles, 123
  Liran–Zeldes mass formula, 126
  semiempirical mass formula, 124
Nuclear pairing, 162
  BCS theory, 162
Nuclear power:
  need to return to, 325
  world use, 303
Nuclear reaction mechanisms:
  general classification, 218
  compound nucleus, 219
  direct reactions, 243
  deep inelastic transfer, 250
Nuclear reactors:
  advanced passive light water, 303
  breeder reactors, 303
  description, 301
  fuel, 302
  fuel reprocessing, 304
  liquid metal reactor, 303
  modular high temperature gas-cooled, 303
Nuclear reactor waste:
  quantity, 304
  toxicity, 304
Nuclear shapes, 121
Nuclear ship propulsion, 309
Nuclear structure, 154
  examples, 164
  Nilsson states, 155
  pairing effects, 162
  rotational levels, 158
  vibrational levels, 161
Nuclear weapons (fission), 309
  basic principles, 309
  construction, 311
  fissionable material requirements, 310
  gun-barrel design, 311
  implosion, 311
  numbers of, 326
  relation to nuclear power, 310
  size, 312
  supercriticality, achievement of, 311
Nuclear weapons (thermonuclear):
  basic principles, 312
  construction, 313
  neutron bomb, 313
  schematic diagram, 314
  yields, 313

Optimum Q value, 245
Organometallic compounds, 110
  cyclooctatetraene compounds, 111
  cyclopentadienyl compounds, 110
  hydrocarbyls, 112
  MO theory, 112
Oxidation states, 84
  environmental, 297
  Pu, 86
  stability of 3+ state, 85
Oxides, 108

Periodic Table:
  circa 1930s, 67
  circa 1944, 68
  circa 1945, 69
  evolution of, 65
  extension, 323
  limit on size, 7
  modern, 70
Plutonium:
  discovery, 11
    element, 11
    ^{239}Pu, 14, 292
  electricity generated from, 304
  first weighing, 16
  naming, 14
  oxidation states, 86
  polymeric, 90
  presence in nature, 296
  radioactive decay properties, 332
  separation of, use of redox cycles, 17
Precipitates, 109
^{244}Pu, abundance in nature, 291
Purex process, 96

Radioactive decay properties, 150
  tables of, 329
Radionuclide power sources, 315
  cardiac pacemakers, 315
  SNAP, 315
Recoil method:
  use in Md discovery, 39
  use in No discovery, 48
Redox potential, 86
r-process, 292
  heavy element abundances, 295

SUBJECT INDEX

Rutherford backscattering, 316
  basic principles, 316
  lunar surface analysis, 317
Rutherfordium:
  discovery, 51
    controversy surrounding, 51
  radioactive decay properties, 345
  solution chemistry, 52, 115
  thermochromatography, 103

SASSY, 211
Shell corrections, 125
SHIP, 208
Smoke detectors, 316
Solvent extraction, 94
  amine extractants, 97
  organophosphorus extractants, 95
  Purex process, 96
  types of extracting agents, 95
Specific activities of transuranium elements, 200
Spontaneous fission:
  calculations, 144
  discovery, 139
  half-lives, shell effects, 143
  half-lives, table, 140
  half-lives, theory, 139
  hindrance factors, 144
Spontaneously fissioning isomers:
  discovery, 146
  half-life systematics, 146
  spectroscopy, 148
  table, 147
Strutinsky method, 125
Superheavy elements:
  chemistry, 275
  decay properties, 274
  definition of, 287
  half-lives, 272
  history, 268
  island, 270
  laboratory synthesis:
    complete fusion reaction, 276
      excitation energies, 279
      fission, 280
      fusion hindrance, 278
      semiempirical fusion probability, 278
      subbarrier fusion, 277
      washing out of shell effects, 281, 284
    deep inelastic transfer, 285
  presence in nature, 274
Symbols for transuranium elements, 2

Tape systems, 207
Target, accelerator, transuranium, 203
  cooling, 203
  handling, 203
Techniques:
  general, 204
Thickness gauges, 316
Transactinides:
  chemical properties, 113

VASSILISSA, 212
Velocity filters, SHIP, 208

Wilczynski plots, 251

X-ray methods of identification, 208

RETURN

d Hall 542-3753

TDR

JAN 4 1991